VOLUME SIX HUNDRED AND EIGHTY EIGHT

METHODS IN ENZYMOLOGY

Crystallography of Protein Dynamics

METHODS IN ENZYMOLOGY

Editors-in-Chief

ANNA MARIE PYLE

Departments of Molecular, Cellular and Developmental Biology and Department of Chemistry
Investigator, Howard Hughes Medical Institute
Yale University

DAVID W. CHRISTIANSON

Roy and Diana Vagelos Laboratories
Department of Chemistry
University of Pennsylvania
Philadelphia, PA

Founding Editors

SIDNEY P. COLOWICK and NATHAN O. KAPLAN

VOLUME SIX HUNDRED AND EIGHTY EIGHT

METHODS IN ENZYMOLOGY

Crystallography of Protein Dynamics

Edited by

NOZOMI ANDO

Associate Professor, Department of Chemistry and Chemical Biology, Cornell University, USA

ELSEVIER

ACADEMIC PRESS

An imprint of Elsevier

Academic Press is an imprint of Elsevier
50 Hampshire Street, 5th Floor, Cambridge, MA 02139, United States
525 B Street, Suite 1650, San Diego, CA 92101, United States
The Boulevard, Langford Lane, Kidlington, Oxford OX5 1GB, United Kingdom
125 London Wall, London, EC2Y 5AS, United Kingdom

First edition 2023

Notices

Knowledge and best practice in this field are constantly changing. As new research and experience broaden our understanding, changes in research methods, professional practices, or medical treatment may become necessary.

Practitioners and researchers must always rely on their own experience and knowledge in evaluating and using any information, methods, compounds, or experiments described herein. In using such information or methods they should be mindful of their own safety and the safety of others, including parties for whom they have a professional responsibility.

To the fullest extent of the law, neither the Publisher nor the authors, contributors, or editors, assume any liability for any injury and/or damage to persons or property as a matter of products liability, negligence or otherwise, or from any use or operation of any methods, products, instructions, or ideas contained in the material herein.

ISBN: 978-0-443-15926-8
ISSN: 0076-6879

For information on all Academic Press publications
visit our website at https://www.elsevier.com/books-and-journals

Publisher: Zoe Kruze
Editorial Project Manager: Federico Paulo S. Mendoza
Production Project Manager: James Selvam
Cover Designer: Greg Harris

Typeset by MPS Limited, India

Contents

Contributors

Nozomi Ando
Department of Chemistry and Chemical Biology; Department of Physics; Graduate Field of Biophysics, Cornell University, Ithaca, NY, United States

Neti Bhatt
Department of Physics, Cornell University, Ithaca, NY, United States

Johannes P. Blaschke
National Energy Research Scientific Computing Center, Lawrence Berkeley National Laboratory, Berkeley, CA, United States

David A. Case
Dept. of Chemistry & Chemical Biology, Rutgers University, Piscataway, NJ, United States

Nathalie Colloc'h
Imagerie et stratégies thérapeutiques pour les cancers et tissus cérébraux (ISTCT), CNRS Université de Caen Normandie, Centre Cyceron, Caen, France

Anne-Claire Dhaussy
Normandie Université, CRISMAT UMR, Caen, France

Tzanko Doukov
Stanford Synchrotron Radiation Lightsource, SLAC National Accelerator Laboratory, Menlo Park, CA, United States

Siyuan Du
Department of Biochemistry; Department of Chemistry, Stanford University, Stanford, CA, United States

James S. Fraser
Department of Bioengineering and Therapeutic Sciences; Quantitative Biosciences Institute, University of California, San Francisco, CA, United States

Eric Girard
Univ. Grenoble Alpes, CEA, CNRS, IBS, Grenoble, France

Daniel Herschlag
Department of Biochemistry; Department of Chemical Engineering; Stanford ChEM-H, Stanford University, Stanford, CA, United States

Jan Kern
Molecular Biophysics and Integrated Bioimaging Division, Lawrence Berkeley National Laboratory, Berkeley, CA, United States

Thomas J. Lane
CHARM Therapeutics, London, United Kingdom

Hiroki Makita
Molecular Biophysics and Integrated Bioimaging Division, Lawrence Berkeley National Laboratory, Berkeley, CA, United States

Steve P. Meisburger
Cornell High Energy Synchrotron Source, Cornell University, Ithaca, NY, United States

Derek Mendez
Molecular Biophysics and Integrated Bioimaging Division, Lawrence Berkeley National Laboratory, Berkeley; Stanford Synchrotron Radiation Lightsource, SLAC National Accelerator Laboratory, Menlo Park, CA, United States

Ariana Peck
Linac Coherent Light Source, SLAC National Accelerator Laboratory, Menlo Park, CA, United States

Xiaokun Pei
Department of Chemistry and Chemical Biology, Cornell University, Ithaca, NY, United States

Frédéric Poitevin
Linac Coherent Light Source, SLAC National Accelerator Laboratory, Menlo Park, CA, United States

Billy K. Poon
Molecular Biophysics and Integrated Bioimaging Division, Lawrence Berkeley National Laboratory, Berkeley, CA, United States

Nicholas K. Sauter
Molecular Biophysics and Integrated Bioimaging Division, Lawrence Berkeley National Laboratory, Berkeley, CA, United States

Michael C. Thompson
Department of Chemistry and Biochemistry, University of California, Merced, Merced, CA, United States

Michael E. Wall
Computer, Computational and Statistical Sciences Division, Los Alamos, NM, United States

Haoyue Wang
Graduate Field of Biophysics, Cornell University, Ithaca, NY, United States

Stephanie A. Wankowicz
Department of Bioengineering and Therapeutic Sciences, University of California, San Francisco, CA, United States

Felix Wittwer
National Energy Research Scientific Computing Center, Lawrence Berkeley National Laboratory, Berkeley, CA, United States

David C. Wych
Computer, Computational and Statistical Sciences Division; Center for Nonlinear Studies, Los Alamos National Laboratory, Los Alamos, NM, United States

Filip Yabukarski
Department of Biochemistry, Stanford University, Stanford; Bristol-Myers Squibb, San Diego, CA, United States

Junko Yano
Molecular Biophysics and Integrated Bioimaging Division, Lawrence Berkeley National Laboratory, Berkeley, CA, United States

Iris D. Young
Molecular Biophysics and Integrated Bioimaging Division, Lawrence Berkeley National Laboratory, Berkeley, CA, United States

Miao Zhang
Molecular Biophysics and Integrated Bioimaging Division, Lawrence Berkeley National Laboratory, Berkeley, CA, United States

Preface

The first structures of macromolecules determined by X-ray diffraction led to the birth of structural biology. Since then, crystallography has made numerous contributions to modern biology by revealing intricately detailed snapshots of macromolecules. Key advances in the history of crystallography have been documented in *Methods in Enzymology*. Diffraction methods for biological macromolecules were first covered in Volumes 114 and 115, edited by Wyckoff, Hirs, and Timasheff in 1985. These volumes, which began with an introductory chapter by Max Perutz, covered important fundamentals of macromolecular crystallography, such as properties of crystals and their growth, measurement of diffraction, phasing and refinement, and how to draw protein structures both by hand and computationally. In the decade that followed, technological innovations led to a revolution in macromolecular crystallography and the rapid expansion of the Protein Data Bank. Volumes 276 and 277, edited by Carter, Jr., and Sweet, were published in 1997 as a major update to methods in macromolecular crystallography, covering topics such as detector technology, algorithmic advances, validation statistics, and the macromolecular crystallographic information file.

This current volume is dedicated to a topic that is at the forefront of crystallography, and indeed, all of structural biology: *protein dynamics*. Although crystal structures are often thought of as static, it has long been known that molecules can move within the lattice and that motion is important for biological function. In Volume 115, a chapter by D.I. Stuart and D.C. Phillips described how dynamic information can be derived from diffraction data. Based on their seminal structures of hen lysozyme, this chapter described how molecular motion can be inferred from B-factor analysis. Volume 277 foresaw "molecular cinematography" as a long-term goal of structural biology. Importantly, this volume contained pioneering contributions on diffuse scattering, Laue diffraction, and triggering methods. Over the past decade, crystallographic investigations of protein dynamics have reached new heights, catalyzed by the introduction of the director detector and X-ray free electron laser (XFEL). These advances led to a revival of non-cryogenic crystallography at both XFELs and synchrotrons and enabled accurate measurements of diffuse scattering. Volume 688, which comes 26 years after the previous volumes on crystallography, dives into these topics.

The first portion of this volume was inspired by the 57th International School of Crystallography, which was held in Erice, Sicily, in June 2022. Codirected by Andrew Goodwin and I, this course was the first Erice school on the topic of diffuse scattering. While periodic order is essential for crystallography, the departure from order is sometimes essential for material function. Crystallographically, the signature of disorder is the existence of a diffuse component to the scattering function, usually orders of magnitude weaker than the conventional Bragg scattering. The 57th Erice school brought 22 lecturers from a broad range of disciplines (from biochemistry to condensed matter physics) to discuss the various origins of diffuse scattering and their interpretation. In many ways, this meeting was a historic moment for the field. For the field of protein diffuse scattering, this meeting was especially meaningful because method developers worked together to debut new software and tutorials, in the form of Jupyter notebooks. Many of these workshop materials are now available to the public in Volume 688.

A common theme of this volume is that experiments are performed at non-cryogenic temperatures, which is an essential starting point for the study of protein dynamics. The first set of chapters focus on the diffuse component of crystal scattering and the current state of the art in measurement and interpretation. Chapters 1 and 2 cover data collection of diffuse scattering at non-cryogenic temperatures and data reduction. Chapters 3–5 focus on performing and interpreting the results of molecular dynamics (MD) simulations of protein crystals, while Chapter 6 covers non-MD models of diffuse scattering. Many of these chapters are highly interactive. For example, Chapters 2–4 are accompanied by educational Jupyter notebooks that first debuted at the Erice school, and Chapter 6 introduces *eryx*, a suite of computational tools that allow readers to become intimately familiar with models of diffuse scattering. The diffuse scattering section ends with Chapter 7, which discusses the use of diffraction simulators (also first introduced at the Erice school) to help researchers understand and plan experimental outcomes.

The remaining chapters represent a selection of state-of-the-art methods to extract information about protein dynamics from X-ray diffraction (the nondiffuse component of crystal scattering) collected at non-cryogenic temperatures. Chapter 8 walks readers through multi-conformer refinement of multi-temperature X-ray diffraction, while Chapter 9 provides a thorough discussion of experimental methods for combining X-ray crystallography with temperature perturbations.

Chapter 10 focuses on XFEL-based studies of metalloenzymes using serial crystallography and X-ray spectroscopy. Finally, the chapters on temperature perturbation crystallography are complemented by Chapter 11, which describes the use of pressure perturbation as a way to access conformationally excited states.

Finally, I would like to acknowledge the many people who made this volume possible. First and foremost, I am delighted by the excellent slate of authors for taking the time to share their expertise. Many of these authors have some connection to either the topics or authors from the previous *Methods in Enzymology* volumes. I am excited to see the tradition of crystallography being passed down and pushing new frontiers with these authors. Secondly, I am grateful to the reviewers, all of whom agreed to be named. The authors of Chapters 1–6 (Xiaokun Pei, Neti Bhatt, Haoyue Wang, Steve Meisburger, David Wych, Mike Wall, Dave Case, Ariana Peck, TJ Lane, Frédéric Poitevin) were reviewers for other chapters in the diffuse scattering section. In addition, we were lucky to have an excellent set of external reviewers (Doeke Hesktra, Chae Un Kim, Art Lyubimov, Allen Orville, and Bob Sweet) for Chapters 7–11. I am especially grateful to the reviewers who went above and beyond by testing software, and I would like to convey a special thanks to Ariana, Bob, and Dave for reviewing multiple chapters. Finally, thanks to the team at *Methods in Enzymology* for the invitation to edit this volume and for being willing to exchange multiple rounds of proofs to ensure that computer code and equations are presented accurately.

Nozomi Ando

CHAPTER ONE

Introduction to diffuse scattering and data collection

Xiaokun Pei[a,1], Neti Bhatt[b,1], Haoyue Wang[c], Nozomi Ando[a,b,c,*], and Steve P. Meisburger[d,*]

[a]Department of Chemistry and Chemical Biology, Cornell University, Ithaca, NY, United States
[b]Department of Physics, Cornell University, Ithaca, NY, United States
[c]Graduate Field of Biophysics, Cornell University, Ithaca, NY, United States
[d]Cornell High Energy Synchrotron Source, Cornell University, Ithaca, NY, United States
*Corresponding authors. e-mail address: nozomi.ando@cornell.edu; spm82@cornell.edu

Contents

Abstract

A long-standing goal in X-ray crystallography has been to extract information about the collective motions of proteins from diffuse scattering: the weak, textured signal that is found in the background of diffraction images. In the past few years, the field of macromolecular diffuse scattering has seen dramatic progress, and many of the past challenges in measurement and interpretation are now considered tractable. However, the concept of diffuse scattering is still new to many researchers, and a general set of procedures needed to collect a high-quality dataset has never been described in detail. Here, we provide the first guidelines for performing diffuse scattering experiments, which can be performed at any macromolecular crystallography beamline that supports

[1] Equal contribution.

Methods in Enzymology, Volume 688
ISSN 0076-6879, https://doi.org/10.1016/bs.mie.2023.07.007

room-temperature studies with a direct detector. We begin with a brief introduction to the theory of diffuse scattering and then walk the reader through the decision-making processes involved in preparing for and conducting a successful diffuse scattering experiment. Finally, we define quality metrics and describe ways to assess data quality both at the beamline and at home. Data obtained in this way can be processed independently by crystallographic software and diffuse scattering software to produce both a crystal structure, which represents the average atomic coordinates, and a three-dimensional diffuse scattering map that can then be interpreted in terms of models for protein motions.

1. Introduction

An important goal of structural biology is to relate protein structure and function, but this goal cannot be achieved without an understanding of how proteins move. In particular, the collective structural fluctuations of a protein are of fundamental importance for activity, yet they are very challenging to measure directly. Thus far, correlated protein motions have primarily been studied by nuclear magnetic resonance (NMR) spectroscopy, such as by correlating timescales of exchange rates (Lisi & Loria, 2016; Palmer, 2015). Structural techniques including X-ray crystallography and cryo-electron microscopy have been used to infer such motions by fitting ensemble models to electron density (Fraser et al., 2009; Fraser et al., 2011), analyzing difference distance matrices between multiple structures (Schneider, 2000), or analyzing variability among particle images (Zhong, Bepler, Berger, & Davis, 2021). However, only a subset of motions can be inferred in each case, either because of incomplete sampling of snapshots or because correlated motions whose amplitudes are smaller than the imaging resolution cannot be unambiguously determined. On the other hand, in crystallography, any deviation from a perfect crystal leads to a continuous signal that appears in the background of diffraction patterns, known as diffuse scattering (Fig. 1A). Crystal disorders, whether static (displacive, substitutive, etc.) or dynamic (atomic vibrations and correlated motion), produce distinct diffuse scattering patterns (Welberry & Weber, 2016). Thus, by combining diffuse scattering analysis with crystallographic structure determination, it is possible to simultaneously obtain a high-resolution structure and understand how atomic motions within the structure are correlated (Xu, Meisburger, & Ando, 2021).

However, it was only very recently that methods to process and analyze macromolecular diffuse scattering became available (Chapter 2: Meisburger & Ando, in press). In this chapter, we provide the first guidelines for

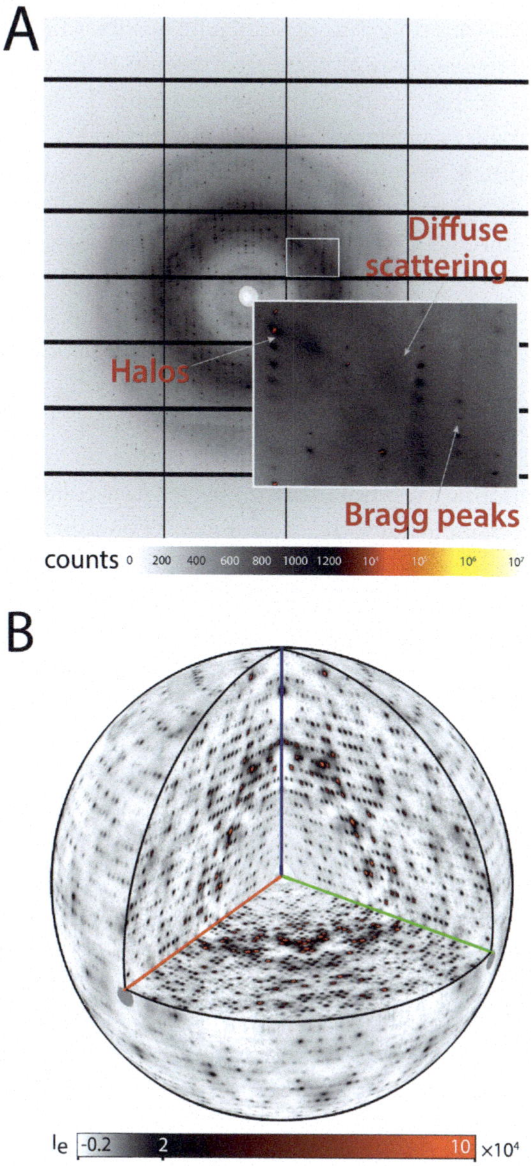

Fig. 1 Appearance of diffuse scattering in diffraction data from tetragonal lysozyme at room temperature. (A) To make the diffuse features clearly visible, a long still exposure was taken of a large crystal. The diffraction image shows Bragg peaks and diffuse scattering features including halos from lattice dynamics and the cloudy diffuse arising from correlated motions. (B) A three-dimensional diffuse scattering map reconstructed from a low-dose, high-redundancy diffraction dataset (Meisburger et al. 2023; reproduced with permission) showing both cloudy and halo features. Bragg peaks have been masked out, and the intensity is shown in absolute electron units (I_e) per asymmetric unit (ASU). A map of this quality is suitable for interpreting lattice dynamics and protein motion.

collecting diffuse scattering data of sufficient quality to generate a diffuse map for quantitative analysis of correlated motion (Fig. 1B). We begin by briefly introducing the theory of diffuse scattering. Following the theory section, we provide suggestions on how to conduct a successful diffuse scattering experiment at room temperature to best capture protein's conformational landscape at a physiologically relevant condition (Fraser et al., 2009, 2011). Although diffuse scattering can be measured at any macromolecular crystallography beamline in principle, special considerations are required in choosing and handling crystals, setting up the measurement, and minimizing excess background scattering and radiation damage. These considerations arise from the fact that diffuse scattering signals are extremely weak (orders of magnitude weaker than diffraction peaks on a per-pixel basis) and they are easily contaminated by the intense diffraction and background scattering. We note that the exact data collection setting will depend on the sample and the beamline. Therefore, we also discuss practical guidelines and potential alternatives for collecting data at various beamlines on different types of samples. Finally, we end by discussing preliminary steps in data processing, which can be followed by more rigorous methods (Chapter 2: Meisburger & Ando, in press) and interpreted by various models (Chapters 3–6: Case, in press; Peck et al., submitted; Wych & Wall, in press a; Wych & Wall, in press b).

2. Theory

A solid grasp of X-ray diffraction theory is important in diffuse scattering research. It is needed to understand the scattering features in detector images to enable better measurements, and for building appropriate models to interpret the data. The theory of macromolecular diffuse scattering has been reviewed recently (Meisburger, Thomas, Watkins, & Ando, 2017). Here, we provide a general overview and emphasize key results relevant to measurement and interpretation.

A typical macromolecular crystallography (MX) experiment consists of placing a protein crystal in an X-ray beam and measuring scattered X-rays using an area detector. The scattering results from the interaction between the X-rays and the electrons in the sample. Protein crystals are made up of repeating structural units, or unit cells. X-ray diffraction from these repeating unit cells causes constructive interference in certain directions, creating intense spots on the detector known as Bragg peaks. The positions of the Bragg peaks on the detector provide information on the size and shape of the unit cells, while the intensities of the Bragg peaks contain the information about the electron density within the unit cell. In conventional MX, these Bragg peak intensities are analyzed to reconstruct the arrangement of atoms in the unit cell: the protein's structure.

Bragg peaks are not the only features observed in X-ray diffraction images from protein crystals. The images also contain diffuse scattering, a continuous signal between and underneath the Bragg peaks. If all unit cells were identical (and the experimental setup perfect), Bragg peaks would be the only features observed. Diffuse scattering occurs whenever disorder is present, and protein crystals contain significant disorder. For instance, protein crystals have disordered solvent occupying space between protein molecules, defects in crystal packing, and intrinsic heterogeneity in terms of protein conformation, even at cryogenic temperatures. More interestingly, diffuse scattering has a distinctive pattern depending on how the disorder is correlated over distance: long-ranged disorders such as lattice distortions produce very different diffuse scattering from short-ranged disorders such as the thermal vibrations of atoms.

The scattering observed in diffraction images is not limited to diffuse scattering from the protein and solvent content of the crystal; it also includes experimental background. The largest source of background at MX beamlines is typically scattering from air and other materials in the beam path, such as sample mounting materials and excess solvent surrounding the crystal. Atoms can also fluoresce or scatter X-rays inelastically (Compton scattering). In accord with previous work (Meisburger et al., 2017; Meisburger, Case, & Ando, 2020), the diffuse scattering of interest is defined as the elastic component of the non-Bragg signal from all atoms in the crystal (i.e., the protein, ligands/ions, plus ordered/disordered solvent). Here we consider only the theory of elastic scattering for an infinitely large crystal, assuming that other contributions have been measured or otherwise corrected (see Section 4.3 and Chapter 2: Meisburger & Ando, in press).

The elastic scattering process that leads to Bragg diffraction and diffuse scattering is coherent. The scattered wave in any particular direction $\hat{\mathbf{s}}_1$ has an amplitude that is proportional to the structure factor, F, of the illuminated atoms. For X-rays, the structure factor is simply the Fourier transform of the electron density evaluated at a point in reciprocal space $\mathbf{q} = (2\pi/\lambda)(\hat{\mathbf{s}}_1 - \hat{\mathbf{s}}_0)$, where $\hat{\mathbf{s}}_0$ is the incident beam direction and λ is the X-ray wavelength. What is measured is not the X-ray scattering amplitude but rather its intensity, or the absolute square of the amplitude.

The intensity recorded by a detector, after correcting for background, geometric distortions, and other effects (see Chapter 2: Meisburger & Ando, in press) can be written as follows:

$$I_{\text{total}}(\mathbf{q}) = \langle |F_{\text{crystal}}(\mathbf{q})|^2 \rangle_{\text{time}}, \tag{1}$$

where F_{crystal} is the structure factor of the illuminated crystal volume, and the angle brackets denote the time average during the exposure. From the mathematics of Fourier transforms, the structure factor of a crystal will include a lattice of peaks, the reciprocal lattice, which is related to the lattice in real space formed by the repetitive arrangement of unit cells. These peaks correspond to the Bragg diffraction, and for convenience, they can be considered separately as follows:

$$I_{\text{total}} = I_{\text{Bragg}} + I_{\text{diffuse}}. \tag{2}$$

It can be shown using properties of Fourier transforms that I_{Bragg} depends only on the average structure factor of the unit cells (and thus the average electron density). Let F_{crystal} be written as a sum over structure factors of individual unit cells F_n at lattice positions $\mathbf{R}_n$ as follows:

$$F_{\text{crystal}}(\mathbf{q}) = \sum_n F_n(\mathbf{q}) e^{i\mathbf{q}\cdot\mathbf{R}_n}. \tag{3}$$

Then, the Bragg intensity is

$$I_{\text{Bragg}}(\mathbf{q}) = \langle |\overline{F_{\text{cell}}(\mathbf{q})}|^2 \rangle_{\text{time}} \left| \sum_n e^{i\mathbf{q}\cdot\mathbf{R}_n} \right|^2, \tag{4}$$

where $\overline{F_{\text{cell}}(\mathbf{q})}$ is the average unit cell structure factor. The first factor is a continuous function, the transform of the average unit cell, and this is multiplied by the second factor, which is a lattice of sharp peaks (delta functions in infinite crystal limit). Thus, the Bragg intensity is said to "sample" the unit cell transform at the reciprocal lattice points with integer Miller indices h, k, and l.

A similar expression for the diffuse intensity can be obtained by algebraic manipulation of the preceding equations:

$$I_{\text{diffuse}}(\mathbf{q}) = \left\langle \left| \sum_n e^{i\mathbf{q}\cdot\mathbf{R}_n} \left(F_n(\mathbf{q}) - \overline{F_{\text{cell}}(\mathbf{q})}\right) \right|^2 \right\rangle_{\text{time}}, \tag{5}$$

As this equation illustrates, the diffuse intensity depends on the difference between the instantaneous unit cell structure factors and the average, or equivalently, the instantaneous and average electron density. When these are identical (as in the case of a perfect crystal), no diffuse scattering is observed. Any source of disorder produces diffuse scattering.

In biological applications of diffuse scattering, disorder resulting from the motion of atoms in the protein is of particular interest. Such motions affect both the Bragg and diffuse intensities, and must be considered in models used for structure refinement against Bragg data. Typically, each atom in the model is assigned a Gaussian probability distribution describing its position in the unit cell. Neglecting disordered solvent for simplicity, the average unit cell structure factor is modeled as follows:

$$\overline{F_{\text{cell}}(\mathbf{q})} = \sum_j f_j(\mathbf{q})\, T_j(\mathbf{q})\, e^{i\mathbf{q}\cdot\bar{\mathbf{r}}_j}, \tag{6}$$

where the sum runs over all atoms in the unit cell, $\bar{\mathbf{r}}_j$ is the mean atomic position, f_j the atomic scattering factor, and T_j is the Debye–Waller factor that depends on disorder. The Debye–Waller factor is the Fourier transform of the Gaussian distribution of atomic displacements from the mean position, $\mathbf{u} \equiv \mathbf{r} - \bar{\mathbf{r}}$, as follows:

$$T_j(\mathbf{q}) = e^{-(1/2)\mathbf{q}\cdot\mathbf{U}_j\mathbf{q}}, \tag{7}$$

where $\mathbf{U}$ is the symmetric matrix,

$$\mathbf{U} = \langle \mathbf{u}\mathbf{u}^{\mathrm{T}} \rangle = \begin{pmatrix} \langle u_x u_x \rangle & \langle u_x u_y \rangle & \langle u_x u_z \rangle \\ \langle u_y u_x \rangle & \langle u_y u_y \rangle & \langle u_y u_z \rangle \\ \langle u_z u_x \rangle & \langle u_z u_y \rangle & \langle u_z u_z \rangle \end{pmatrix}. \tag{8}$$

The six unique elements of $\mathbf{U}$ are known as Atomic Displacement Parameters (ADPs). Often, ADPs are reduced to an isotropic B-factor:

$$B = (8\pi^2/3)\,\mathrm{trace}(\mathbf{U}). \tag{9}$$

Although ADPs or B-factors contain information about the displacement of an atom, they don't provide insight into the type of motion that is causing the displacement: many different types of motion potentially generate similar patterns of B-factor variation across the protein (Meisburger et al., 2017; Xu et al., 2021).

The diffuse scattering intensity generated by atoms with Gaussian-distributed displacements is as follows:

$$I_{\text{diffuse}}(\mathbf{q}) = \sum_{n}\sum_{n'} e^{i\mathbf{q}\cdot(\mathbf{R}_n-\mathbf{R}_{n'})}\sum_{j}\sum_{j'} f_j(\mathbf{q})f_{j'}(\mathbf{q})e^{i\mathbf{q}\cdot(\bar{\mathbf{r}}_j-\bar{\mathbf{r}}_{j'})} T_j(\mathbf{q})\,T_{j'}(\mathbf{q})\left[e^{(1/2)\mathbf{q}\cdot\mathbf{V}_{njn'j'}\mathbf{q}}-1\right], \tag{10}$$

where the sums run over pairs of atoms j and j' in unit cells n and n' respectively. Although the equation for the diffuse intensity is longer to write than its Bragg counterpart, most of the terms are identical. In fact, the only additional quantity is the matrix **V**. The **V** matrix elements are similar to ADPs, except that they describe the displacement *covariances* between pairs of atoms in the crystal, as follows:

$$\mathbf{V}_{njn'j'} = \left\langle \mathbf{u}_{nj}\mathbf{u}_{n'j'}^{\mathrm{T}}\right\rangle + \left\langle \mathbf{u}_{n'j'}\mathbf{u}_{nj}^{\mathrm{T}}\right\rangle. \tag{11}$$

If a pair of atoms move in an uncorrelated fashion with respect to each other, the **V** matrix elements for that pair will be zero. Consequently, in Eq. (10) the factor in square brackets becomes zero, and that pair of atoms will not contribute to diffuse scattering (an atom's motion is always correlated with itself, and this produces diffuse scattering, however it can be predicted from the ADPs and thus carries no new information). If, on the other hand, the motions are correlated, then diffuse scattering is produced. Thus, the unique information present in the diffuse intensity is the manner in which atomic motions are correlated, and the problem of analyzing diffuse scattering data amounts to determining the **V** matrix elements from measured data. Unfortunately, Eq. (10) cannot be inverted because the number of parameters exceeds the amount of data collected (nine parameters for every pair of atoms). For example, the number of unique atom pairs in a single molecule of lysozyme is approximately 2×10^6. If we then consider all atom pairs between unit cells, this number becomes orders of magnitude larger than the number of unique observations even for a finely sampled diffuse map, such as that reported for triclinic lysozyme ($\sim 50 \times 10^6$) (Meisburger et al., 2020). In general it is necessary to reduce the complexity of Eq. (10) by applying a model.

From Eq. (10), one can imagine how different kinds of disorder are likely to appear in diffraction images. If all the disorder is caused by protein motion such that displacements are not correlated between different unit cells, all the terms with $n \neq n'$ drop out, and the diffuse intensity becomes:

$$I_{\text{diffuse}}(\mathbf{q}) = N \sum_{j} \sum_{j'} f_j(\mathbf{q}) f_{j'}(\mathbf{q}) e^{i\mathbf{q}\cdot(\bar{\mathbf{r}}_j - \bar{\mathbf{r}}_{j'})} T_j(\mathbf{q}) T_{j'}(\mathbf{q}) \left[e^{(1/2)\,\mathbf{q}\cdot\mathbf{V}_{jj'}\mathbf{q}} - 1 \right], \quad (12)$$

where N is the number of unit cells. Unlike the equation for Bragg intensity (Eq. (4)), Eq. (12) has no dependence on the lattice vectors **R**. Thus, the diffuse pattern will not have any particular relationship with the reciprocal lattice. The texture of the diffuse pattern is determined by the phase factors $e^{i\mathbf{q}\cdot(\bar{\mathbf{r}}_j - \bar{\mathbf{r}}_{j'})}$, which are functions of the inter-atomic vector. Because atom pairs in this case are within a unit cell, there is a maximum distance cutoff that, in turn, sets how rapidly the phase factors can oscillate as a function of **q**. The result is a smoothly varying, or "cloudy" diffuse pattern on the scale of the reciprocal lattice.

On the other hand, if atomic displacements are correlated between unit cells, the lattice sums in Eq. (10) mean that diffuse scattering will be related to the reciprocal lattice. Certain models of lattice disorder, for instance, predict broad peaks (or "halos") that overlap with the Bragg peaks. In addition, the diffuse scattering features can be very sharp because the argument of the phase factor is no longer limited to short inter-atomic distances.

Diffuse scattering patterns from protein crystals contain both cloudy and halo features. From a modeling point of view, it is therefore necessary to account for both correlated motion within proteins, as well as lattice disorder, which produce the halo features. From an experimental point of view, the strategy is to measure the diffuse scattering at all values of **q** in reciprocal space (to obtain a complete, three-dimensional diffuse map) with sufficiently fine sampling to capture the halo features in addition to the cloudy pattern. Finally, we note that a high-quality structure determined from the Bragg data is a necessary starting point for diffuse scattering analysis. This is because most of the terms in Eq. (10) can be set using the Bragg data, and only the unknown covariance matrix elements need to be determined from the diffuse scattering.

3. Samples for diffuse scattering

In this section, we review best practices for room-temperature sample preparation and provide practical guidance for diffuse scattering based on our own experiences. We first describe considerations that go into sample requirements.

Low vs. high resolution diffraction. The conservation of total scattering (Eq. (2)) implies that crystals that diffract to low resolution will produce greater diffuse scattering. For example, crystals of *Thermosynechococcus vestitus* photosystem II that diffract to ~4.5 Å display strong diffuse scattering past the diffraction limit (Ayyer et al., 2016). However, crystals that diffract to low resolution often have problematic sources of disorder (see below). In contrast, crystals that have sufficient quality for conventional crystallography are generally useful for diffuse scattering studies because, as described above, most parameters needed to describe the diffuse scattering intensity (Eq. (10)) can be determined from the Bragg data. We therefore recommend at least starting with crystals that exhibit low mosaic disorder (see next section), and yield reproducibly high-resolution (~2.5 Å or less) structures at cryogenic temperature or room temperature.

Sources of disorder. The source of disorder that produces diffuse scattering should be considered. If diffuse scattering is observed in images from conventional structure determination, the crystal may be a promising candidate for further study. However, not all scattering is useful for understanding protein internal motions. "Messy" diffraction patterns that appear to have intense diffuse scattering often arise from high mosaicity and problematic forms of lattice disorder (Fig. 2D; detailed discussions in Section 4.1 and Xu et al., 2021), rather than from protein internal motion. These undesirable forms of disorder are commonly found in crystals that diffract to low resolution. Although total scattering analysis can be a valuable tool to characterize pathological crystals and in some cases to solve their structures (Lovelace & Borgstahl, 2020), pathological causes of diffuse scattering will only complicate the analysis of protein internal motion, and thus, such crystals are best excluded at the sample preparation stage. Likewise, other issues that are considered problematic in conventional crystallographic data processing, such as multiple lattices, twinning, and pseudo translation, should be avoided because Bragg diffraction and diffuse scattering will be difficult to separate from each other and the diffuse scattering from different crystalline domains will be averaged together.

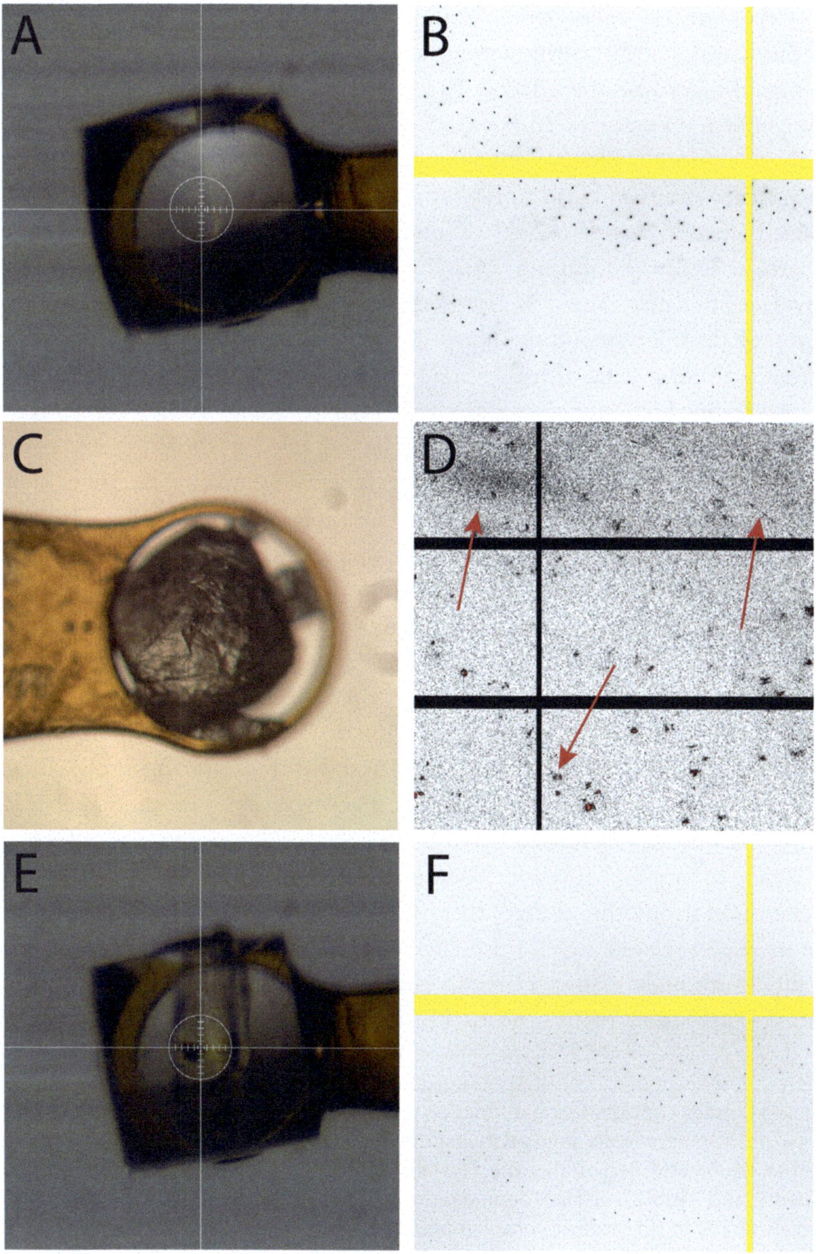

Fig. 2 Visual indicators of sample quality for diffuse scattering. (A) A large, high-quality crystal suitable for single-crystal diffuse scattering (tetragonal lysozyme). The circle in the center has a diameter of 100 μm. (B) Example X-ray diffraction from a crystal of similar quality to (A) showing sharp, well-separated Bragg peaks with no *(Continued)*

Cryogenic vs. room temperature. The development of cryo-cooling methods was critical to the widespread adoption of macromolecular crystallography because the effects of radiation damage are greatly reduced at cryogenic temperatures and frozen crystals can be shipped much more easily to remote facilities. However, cryo-cooling is also known to introduce lattice strain and lead to higher mosaicity or non-isomorphism (Halle, 2004; Juers & Matthews, 2001; Kriminski, Caylor, Nonato, Finkelstein, & Thorne, 2002; Thompson, Cascio, & Yeates, 2018; Vahedi-Faridi, Lovelace, Bellamy, Snell, & Borgstahl, 2003). Thus, to avoid undesirable forms of disorder (see above), we advocate non-cryogenic collection of diffuse scattering data, at least initially.

Achieving high signal-to-noise and high redundancy. In conventional crystallography, high redundancy may be achieved by utilizing crystals in high-symmetry space groups. For diffuse scattering analysis, a lower symmetry form is preferable if there are multiple options. This is because orientational averaging from multiple copies of the asymmetric unit will tend to diminish the diffuse signal from internal protein motions (Chapman et al., 2017). Additionally, the final form of processed diffuse scattering data is a three-dimensional diffuse map that samples *all* of reciprocal space within the resolution limits of the crystal of interest. Thus, it is advantageous to have many large (e.g., >100 μm in at least one dimension, see crystal sizes in Table 1), high-quality crystals for measuring such a large amount of data points (i.e., reciprocal space voxels). The increased scattering per unit dose afforded by large crystals makes a significant difference in signal to noise. Moreover, because the diffuse signal competes with background from other sources that may be difficult to eliminate or subtract, the increased relative signal from large crystals can improve the overall data quality. Although a single large crystal may be sufficient for a complete diffuse map, using many crystals has the benefit of extra redundancy for

Fig. 2—Cont'd contaminating diffraction or scattering. (C) Looping the crystal slowly in an environment with low relative humidity quickly dries out the surrounding mother liquor and dehydrates the crystal, resulting in precipitation of buffer components and visible cracking/degradation of the crystal. (D) Data collection from crystals similar to (C) results in problematic features indicated by red arrows, including split Bragg peaks, powder diffraction rings, and possibly contaminating scattering from polymeric components. (E) Crystal in (A) after radiation damage from a long φ-scan. (F) Diffraction image of the pre-exposed crystal taken with the same orientation, beam intensity, and exposure time as (B). Radiation damage causes significant decay in the intensity of Bragg peaks and likely alters diffuse scattering.

Table 1 Examples of crystals and their diffuse scattering dataset measurement conditions.

	Lysozyme (triclinic)	**Lysozyme (orthorhombic)**	**Lysozyme (tetragonal)**	**Sample A**	**Sample B**
Reference	**Meisburger et al. (2020)**	**Meisburger et al. (2023)**	**Meisburger et al. (2023)**	**Unpublished**	**Unpublished**
Sample					
Crystallization method	Microbatch under oil	Sitting drop vapor diffusion	Hanging drop vapor diffusion	Hanging drop vapor diffusion	Hanging drop vapor diffusion (5 μL drop size)
Crystal size (approx.)	Various: 100–500 μm	1000 × 400 × 200 μm^3	500 × 500 × 250 μm^3	500 × 100 × 50 μm^3	400 × 200 × 100 μm^3
Mount	Loop in PET capillary	Loop in PET capillary	Loop in PET capillary	Loop in PET capillary	Loop in PET capillary
Unit cell volume	26 nm^3	127 nm^3	243 nm^3	278 nm^3	2000 nm^3
Longest cell dimension	34.5 Å	73.9 Å	79.6 Å	85 Å	130 Å
Amino acids per unit cell	129	516	1032	1224	1400
Solvent content	33.8%	44.4%	41.9%	42%	60.8%
Mosaicity (apparent, sigma)	0.024°	0.03°	0.02°	0.043°	0.055°

(continued)

Table 1 Examples of crystals and their diffuse scattering dataset measurement conditions. (*cont'd*)

	Lysozyme (triclinic)	**Lysozyme (orthorhombic)**	**Lysozyme (tetragonal)**	**Sample A**	**Sample B**
Reference	**Meisburger et al. (2020)**	**Meisburger et al. (2023)**	**Meisburger et al. (2023)**	**Unpublished**	**Unpublished**
Beamline					
Location	CHESS F1	CHESS F1	CHESS F1	CHESS ID7B2	CHESS ID7B2
X-ray energy	12.693 keV	12.693 keV	12.693 keV	12.867 keV	12.867 keV
Bandwidth ($\Delta E/E$)	5×10^{-4}	5×10^{-4}	5×10^{-4}	10^{-4}	10^{-4}
Flux	2.5×10^{10}	2.5×10^{10}	2.5×10^{10}	1.7×10^{11}	1.7×10^{11}
Beam size	100 μm (round)	100 μm (round)	100 μm (round)	$70 \times 24\ \mu m^2$	$70 \times 24\ \mu m^2$
Beam profile	Top-hat	Top-hat	Top-hat	Gaussian	Gaussian
Detector model	Pilatus 3 6M	Pilatus 3 6M	Pilatus 3 6M	Eiger2 X 16M	Eiger2 X 16M
Sensor (thickness)	Si (1 mm)	Si (1 mm)	Si (1 mm)	Si (450 μm)	Si (450 μm)
Detector distance	230 mm	275 mm	250 mm	180 mm	180 mm

Data collection

Rotation per image	0.1°	0.1°	0.1°	0.1°	0.05°
Exposure per image	100 ms	100 ms	100 ms	12 ms	10 ms
Rotation rate	1°/s	1°/s	1°/s	8.3°/s	10°/s
Multi-motor scan	Sweep and translate	Sweep and translate	Sweep and translate	Vector: 500 μm/21 s	Vector: 400 μm/12 s
Rotation per sweep	50°	50°	50°	180°	60°
Number of sweeps (crystals)	11 (4)	8 (2)	8 (1)	1	1
Images per dataset	5500	4000	4000	1800	1200 (incomplete)

judging repeatability of the measurement (Su et al., 2021) and for reducing artifacts such as detector gaps and blind regions of reciprocal space. Merged datasets should be from isomorphous crystals of consistent quality. Because many crystallization trials may be needed for optimization, it is helpful if sample availability is not a limiting factor.

3.1 Crystallization and transport

It is important to establish a reproducible crystallization protocol when producing crystals for room-temperature experiments. Having good control over crystallization greatly facilitates optimization for producing large crystals, and it allows for samples to be prepared on demand prior to scheduled beamtime. It can be helpful to first map out the crystallization phase diagram and to understand the factors affecting crystal nucleation and growth rates, such as drop volumes and crystallization temperature. In cases where nucleation is difficult to control, seeding strategies can be used (Stura, 1999; Thaller et al., 1981).

The crystallization format should be considered carefully when producing crystals for data collection, especially if crystals will be transported to the synchrotron. It is important to note that crystallization format can influence growth in subtle ways. For instance, when vapor diffusion is used, the drop volume influences the rate at which the solution approaches supersaturation, the rate that protein is depleted from the drop once crystals nucleate and grow, and other factors. Thus, when switching between formats it may be necessary to re-optimize the growth conditions. For this reason, it is often wise to choose an appropriate format early in the project and stick with it. We have had good success with standard plastic 24-well hanging or sitting well trays, dialysis buttons, and 24- or 96-well sitting drop trays for micro-batch under paraffin oil. Strategies for optimizing growth conditions to produce large crystals have been reviewed in the neutron protein crystallography literature (Budayova-Spano, Koruza, & Fisher, 2020; Ng et al., 2015).

Shipping protein crystals at ambient temperature while maintaining diffraction quality is a challenge due to their sensitivity to humidity and temperature change. An ideal scenario would be to set up crystallization trays at the synchrotron. Another common solution is to transport lab-grown crystals in their crystallization trays. Ground transport (car or train) is preferred, because pressure changes during a flight make air travel problematic (Fischer, 2021). Vibrations should be minimized as much as possible. Securing trays in thick Styrofoam boxes can reduce such vibrations as well as prevent any abrupt temperature changes. If crystals are grown at 4 degrees, an ice slurry may be used to maintain this temperature during transport. Even with these precautions, it can be difficult to control temperature well enough to prevent condensation within hanging drop vapor-diffusion setups, so other crystallization formats such as sitting well and microbatch under oil may be better suited.

Another approach is to transport protein crystals pre-mounted for data collection. The simplest method is to place the crystals in glass capillaries sealed with epoxy, however the high background scattering from glass is not ideal for diffuse scattering measurements. We commonly use plastic capillaries with Kapton loops for sample mounting (MiTeGen MicroRT system). Other groups have reported successfully shipping pre-mounted crystals in this system, with the extra precaution of sealing the capillary to the magnetic base with vacuum grease. The Structural Molecular Biology program at Stanford Synchrotron Radiation Lightsource recommends adding 0.6–1% agarose to the humidity-stabilizing droplet in MicroRT tubing in order to keep the droplet in place during long distance shipping. In-situ crystallization plates have also been developed, which can be directly mounted on a goniometer. Diffuse scattering from crystals in such devices has not been reported, and it is likely that the increased background scattering from excess mother liquor around the crystal as well as X-ray window materials will need careful consideration. Regardless of whether the crystals are pre-mounted or not, control of temperature during shipping is of the utmost importance. A well-insulated thermal shipper has been developed by Crystal Positioning Systems that can maintain the box temperature at 20 °C (± 5 °C) for 7 days.

3.2 Sample mounting

There are many factors to consider when choosing which crystals to loop for a diffuse scattering study. Ideally, crystals should be free of any visible imperfections, such as twinning or cracks. These features—when visible—are not trivial at the ångström scale and require many unnecessary correction steps in data processing. Tiny imperfections can be difficult to discern by visual inspection and might only be discovered when manipulating crystals under the microscope. Ultimately, crystal quality cannot be judged by eye, and the most reliable method is to check the quality of the Bragg diffraction (i.e., conventional crystallographic data statistics) during data collection. Once high-quality, well-diffracting crystals are found, 10–20 nominally identical crystals are commonly needed for a successful study, especially given the need to establish optimal data collection parameters and the possibility of accidentally damaging samples during handling.

Looping crystals successfully takes practice. For experienced protein crystallographers, it may be second nature, but those new to the field should not underestimate the importance of learning excellent technique. It is important to avoid pressing, plunging, or pushing crystals as this may induce cracks or strain. A general method is to move the loop beneath the crystal then gently lift and glide out of the drop. The mechanical robustness of crystals can be very sample dependent, and different looping techniques may be required for different crystals.

After looping, the crystal must be maintained in an environment having a relative humidity as close as possible to its growth condition. With the MiTeGen MicroRT system, the mounted loop should be transferred immediately into a plastic capillary sleeve that has been pre-loaded with ~10 μL of well solution in the tip. The volume of this solution can be tuned, and the closer the liquid surface is to the crystal the better it preserves the humidity. When using capillary sleeves, it is important to check with the beamline scientist the maximum allowed length for the beamline geometry.

Some crystals are very sensitive to water loss and may degrade in the few seconds while being transferred to a capillary. An ideal solution is to loop crystals in a humid environment (Thorne, 2023) such as a cold room with a humidifier or a commercial sample-loading humid chamber. Alternatively, crystals can be transferred from mother liquor to oil to prevent dehydration, or looped after covering crystallization droplets with oil. Several types of oil have been successfully used for crystal mounting, and include polycarbon or modified polycarbon products (Bautis, 1975; Pflugrath, 2015) with a range of viscosity and vapor permeability. If the oil technique is used, several types should be screened to observe the effect on diffraction quality. Viscous oils such as type NVH immersion oil are useful for removing excess water (Warkentin & Thorne, 2009) and make an excellent vapor barrier, however more strain is introduced when looping crystals through it. For diffuse scattering experiments, it is also important to minimize the amount of oil on the surface of the crystal as it contributes background scattering. A common strategy is to quickly dip the looped crystals in oil and then wipe excess oil as much as possible by touching the loop on a glass surface under microscope. Finally, oil may not be a sufficient vapor barrier at room temperature. Oil may be combined with MicroRT tubing for a double layer of protection.

Finally, it is generally a good idea to visually inspect the crystal after looping and after data collection to assess whether additional imperfections appear (Fig. 2). These steps help to build a qualitative understanding of the relationship between crystal appearance, crystal sensitivity to certain environments, and diffraction quality.

4. Experimental design

Measuring diffuse scattering signals accurately requires careful attention to the experimental setup and data collection strategy. Until recently, accommodating non-ideal properties of X-ray detectors was a primary concern (Meisburger & Ando, 2017). In addition to having slow frame rates, the charge-coupled device (CCD) detectors used in earlier studies suffered from read-out noise and other artifacts. Nevertheless, 3D diffuse scattering maps of protein crystals were successfully measured using CCDs in pioneering studies (Meisburger et al., 2017; Wall, Clarage, & Phillips, 1997; Wall, Ealick, & Gruner, 1997). The introduction of hybrid photon counting (HPC) detectors in the last decade has made it feasible to collect diffuse scattering under similar conditions to modern Bragg data collection with continuous crystal rotation (Benschoten et al., 2016). With detectors no longer limiting, there has been renewed attention to other aspects of the experimental setup that affect data quality and methods to more accurately subtract the background (Meisburger et al., 2020; Xu et al., 2021). In addition, algorithms for scaling and merging have been developed that take full advantage of the single photon-counting sensitivity of HPCs (Meisburger et al., 2020), and diffuse scattering can now be collected using the modern approach of fine φ-slicing with a low dose per frame and high redundancy (Meisburger, Case, & Ando, 2023). In this section, we discuss in detail how various parameters of an MX beamline can be chosen or optimized for diffuse scattering, considerations for data collection strategy, and methods to evaluate data quality while at the synchrotron. For concrete examples, we refer to three previously published datasets from model protein crystals (triclinic, orthorhombic, and tetragonal lysozyme) and two of our unpublished datasets from more complex proteins (summarized in Table 1).

4.1 Beamline parameters

Current MX beamlines have not been designed with diffuse scattering in mind. However, many of the same considerations that improve Bragg data quality are also needed for diffuse scattering, such as a well-collimated and monochromatic beam, low-noise detector, and low background scattering. Thus, state-of the-art MX beamlines can be successfully used for diffuse scattering if they can accommodate room temperature data collection. When data of exceptional quality is needed, it can pay to choose a beamline with specific hardware and beam properties that will maximize data quality, and to collaborate with beamline staff to fully optimize the setup. Even if an "off the shelf" MX facility is used, certain beamline parameters must be chosen by the user during data collection, such as X-ray energy, beam size, and detector distance, and thus it is important to understand the likely effect of those choices on data quality.

As mentioned above, a high-quality X-ray detector is perhaps the most important beamline component for diffuse scattering measurements. HPCs such as EIGER 16M and PILATUS 6M are now widely available at MX facilities and provide the essential properties needed: single photon sensitivity, high dynamic range, fast frame rates, and single-pixel point spread function (Förster, Brandstetter, & Schulze-Briese, 2019). For facilities with very high instantaneous flux, such as X-ray free electron lasers (XFELs) and fourth-generation synchrotrons, charge-integrating hybrid pixel array detectors such as JUNGFRAU are now available with sufficiently low noise to be useful for diffuse scattering (Leonarski et al., 2018). Small pixels are useful in diffuse scattering because they allow for finely sampled maps, however they are of little advantage if the features in the diffraction pattern are intrinsically more than one pixel wide, for instance if the diffraction pattern is smeared by a mosaic crystal or a divergent beam.

In a diffuse scattering experiment, one of the primary challenges is to cleanly separate Bragg peaks from diffuse background. When the Bragg peaks are few and far between, this is relatively straightforward, and that is why early experiments focused on protein crystals with small unit cells. When crystals with larger unit cells are investigated, the demands on both crystal quality and measurement increase. The size and number of Bragg peaks appearing in a diffraction image is fundamentally limited by the crystal's mosaicity, a measure of angular broadening of the diffraction pattern due to slight misalignment of microscopic crystalline domains. Divergence of the X-ray beam produces a similar angular broadening, and often data processing software will report an "apparent" mosaicity, which is the combination of true mosaicity and other factors (Fig. 3) (Dauter, 1999; Snell, Bellamy, & Borgstahl, 2003). At room temperature, mosaicities can be extremely small (less than one hundredth of a degree) (Bellamy, Snell, Lovelace, Pokross, & Borgstahl, 2000; Snell et al., 1995). The divergence at

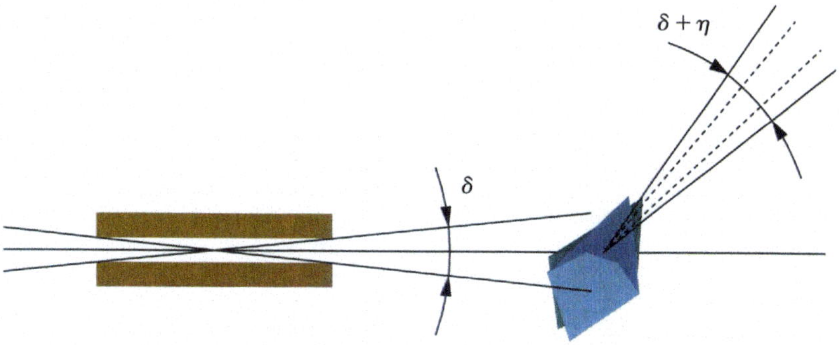

Fig. 3 Illustration showing how beam divergence (δ) and crystal mosaicity (η) contribute to the smearing of diffraction features. Reproduced with permission (Dauter, 1999).

MX beamlines with focusing optics may exceed the intrinsic mosaicity at room temperature, as was likely the case for the lysozyme datasets (Table 1). Finite energy bandwidth of the X-ray beam also smears the diffraction pattern, and for this reason high bandwidth beamlines (e.g., for Laue crystallography) are not suitable for diffuse scattering. For the best performance, crystal monochromators such as Si(111) are preferable to multilayer optics.

As diffuse scattering is carried out at non-cryogenic temperatures, the first obvious challenge is radiation damage, as the dose sensitivity for protein crystals at room temperature is approximately 1–2 orders of magnitude higher than at cryotemperatures (Southworth-Davies, Medina, Carmichael, & Garman, 2007; Thorne, 2023). To maximize the usable signal for a limited dose budget, it is important not to waste photons. Having control over beam size can help ensure that the entire crystal is irradiated evenly during the exposure. Ideally, the beam should not be larger than the crystal because the portion of the beam that does not interact with the crystal contributes extra background scattering (e.g., from the capillary and air in the beam path) without adding to the signal. Multi-motor scans can also be used to distribute the dose evenly when the crystal is larger than the beam. For instance, the tetragonal lysozyme dataset was collected from sweeps of 50 degrees spaced out over eight locations on a large crystal (Table 1, third column). The ability to perform vector or helical scans, where the goniometer translates and rotates at the same time, can be useful for distributing dose, particularly if the crystal is rod-shaped. These strategies are discussed further in Section 5.2.

The resolution of scattering that can be measured is determined by the X-ray energy, detector size, and detector distance. The considerations are similar to cryo-MX and have been reviewed elsewhere (Dauter, 1999). In conventional MX, it is common to choose the detector distance based on the resolution of diffraction so that the diffraction pattern fills the available area of the detector, maximizing the Bragg peak separation. For diffuse scattering, however, it may be preferable to place the detector close to the sample regardless of diffraction resolution. First, if the detector is too far away, diffuse scattering extending beyond the Bragg resolution would be missed. Second, for HPC detectors with negligible point spread function and small pixels, such as the EIGER, the Bragg peaks may be adequately separated even when the detector is close. High energy (short wavelength) can also be employed to collect high resolution data, with the advantage that the Ewald sphere is flatter, resulting in a smaller blind region of reciprocal space. However, the most common MX detectors use silicon sensors, and the quantum efficiency drops significantly at high energy. For example, the quantum efficiency of PILATUS with 450 μm thick silicon sensor drops to around 50% at 17 keV (Donath et al., 2013). In HPCs such as PILATUS and EIGER, the pixels are much thicker than they are wide, and thus a high energy photon striking the panel from an angle will be potentially absorbed by many adjacent pixels, broadening the pattern. HPCs with a high stopping power sensor, such as the EIGER CdTe, are now becoming available at MX beamlines and should enable better performance for diffuse scattering at atomic resolution.

Finally, the beamline should have low background scattering. Typically, air in the beam path between the final aperture and the beamstop is a major contributor to background at MX beamlines. This air path is often made somewhat large to accommodate hardware such as cameras and robotics. If possible, it should be reduced to a minimum. Of particular importance is the gap between the final aperture and the crystal, because air scattering in this region effectively makes a projection image of the sample mount on the detector, which is a non-trivial pattern that depends on the precise position of the sample during the scan and is difficult to subtract out. Air between the sample and beamstop contributes an isotropic scattering pattern that is most intense at low angles and is therefore less of a concern. However, it is often a simple matter to reduce the scattering by moving the beamstop closer to the crystal, at the expense of low-resolution Bragg data.

4.2 Data collection strategy

Once a crystal has been mounted at the beamline, the experimenter must choose the data collection parameters: the exposure time per image, starting angle and rotation speed of the goniometer (or φ increment per frame), total number of images to collect (or total angle of the φ scan), and perhaps other beamline settings such as beam size, κ angle, and attenuation. Assuming the goal of the experiment is to collect a detailed, three-dimensional diffuse map, the considerations are similar to those of conventional MX at room temperature. The dataset should be complete (covering as much of reciprocal space as possible without large gaps) with adequate signal-to-noise while carefully controlling the radiation damage to the crystal. With modern HPCs, parameter choices that improve Bragg data quality usually also improve diffuse scattering data quality. However, the relative importance of a particular parameter for Bragg data and diffuse maps may not be equal. For instance, the fine φ-slicing technique, where the φ angle increment per frame is chosen to be less than the mosaicity, will maximize the signal to noise ratio of the Bragg peaks by minimizing the amount of background scattering that contributes to the integrated peak intensity (Pflugrath, 1999). Fine φ-slicing is often advocated for Bragg data collection but is not essential. In contrast, for diffuse scattering the φ increment per frame determines the spatial resolution of the map (here, we use the term "spatial resolution" to refer to fineness of sampling in reciprocal space). Thus, the φ increment should be chosen small enough to achieve the desired map parameters (considering also the mosaicity, which fundamentally limits the spatial resolution). See Table 1 for examples and Chapter 2 for further discussion.

In MX, the term "strategy" usually refers to an optimal set of scan parameters to achieve high completeness given the crystal symmetry and orientation on the goniometer, and traditional strategies attempt to minimize the total number of diffraction images (Bourenkov & Popov, 2006; Dauter, 1999). At high flux beamlines with HPCs, minimizing the total number of frames is less of a concern, and thus modern strategies attempt to maximize data quality after merging. Such strategies typically involve high redundancy and low dose per frame (i.e., collecting 360 degrees of fine-sliced data) (Winter et al., 2019). When data are collected near room temperature, strategies to optimally distribute the dose gain extra importance, particularly if the crystals are small. For instance, the strategy used in serial oscillation crystallography is to collect small wedges (typically 1–10 degrees) from many crystals with random orientations and to merge the data to form a complete dataset (Hasegawa et al., 2017). For larger crystals at room temperature, the best strategy might be to use a large, uniform (top-hat) beam to uniformly irradiate the sample. Alternatively, if the diffraction is not uniform across the crystal, a small beam can be used with a multi-motor scan (such as a vector or helical scan) and frames with poor quality diffraction can be filtered out during processing (Flot et al., 2009; Polsinelli et al., 2017).

For measuring diffuse scattering from large crystals at room temperature, we advocate a strategy of high redundancy and low dose per frame with fine φ-slicing (Fig. 4). High redundancy (of a factor of ~10 or more) is especially important for diffuse scattering. It enables averaging over systematic errors such as detector pixel inconsistencies, allows outlier measurements to be identified with high confidence, and fills in the gap regions created by the arrangement of detector panels. In addition, high redundancy can be used in sophisticated scaling algorithms to further correct artifacts such as extra background scattering from solvent on the crystal surface or mounting materials (Meisburger & Ando, in, press; Meisburger et al., 2020). High redundancy can be achieved by collecting a large angular range (e.g., 360 degrees) from a single crystal with high symmetry (see tetragonal lysozyme in Table 1). Or, if the crystal is low symmetry, it can help to merge datasets from multiple crystals in different orientations, as this tends to cover reciprocal space more uniformly (see triclinic lysozyme in Table 1). Crystal shape is also an important consideration. Aligning the incident beam parallel to the longer crystal dimensions helps to improve signal to noise both for Bragg diffraction and diffuse scattering. For example, the diffuse scattering of plate-shaped isocyanide hydratase (ICH) crystals was measured after careful alignment so that the beam passed through the plane of the crystal as it rotated (Su et al., 2021), thus illuminating a larger number of unit cells in each exposure.

Finally, as room-temperature data collection is a dose-budgeting experiment, the dose limit needs to be determined before finalizing the data collection strategy. Often determining an optimal strategy is an iterative process of experimentation and analysis that may require more than one beamtime. The dose limit is determined by the highest flux and total exposure time that a crystal can tolerate while incurring minimal radiation damage. Once the dose tolerance of the sample is characterized (see Section 5.2), a data collection strategy can be designed to provide a complete dataset within the allowed dose. In general, one can either attenuate the beam and expose the crystal for longer and therefore increase

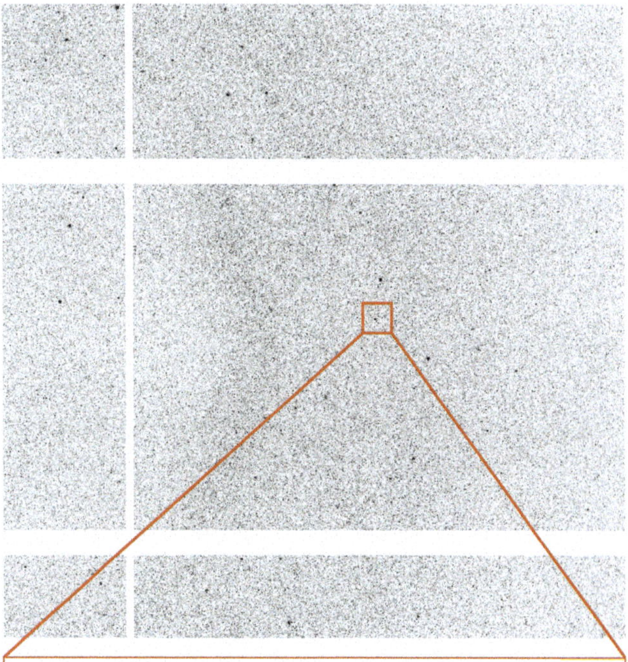

0	0	0	0	0	0	1	1	0	0	1	0	2	1	0	0
0	1	2	0	0	1	1	0	0	0	2	2	1	0	0	0
0	1	1	0	0	0	1	0	2	2	1	1	0	0	0	0
1	0	0	0	0	3	1	0	1	0	1	1	1	0	0	1
2	0	0	0	0	0	0	0	0	1	1	2	0	1	0	0
0	0	0	2	1	1	0	2	0	0	0	3	0	0	2	0
0	0	1	0	0	1	1	0	0	0	2	0	0	2	2	1
1	2	1	0	0	0	0	0	1	0	0	0	1	0	0	1
0	0	1	2	0	1	3	1	1	2	2	0	0	0	0	1
0	0	0	1	0	3	30	18	2	1	1	0	0	1	0	0
1	0	1	0	0	5	15	8	1	1	1	1	1	1	0	0
0	2	1	1	0	0	0	0	0	2	1	0	0	1	0	0
0	0	0	0	0	0	0	0	0	1	1	0	0	0	0	0
1	1	0	0	1	0	0	0	3	0	0	0	1	0	0	0
0	0	2	0	0	0	1	1	0	1	1	1	0	0	0	0
1	0	1	0	0	2	0	0	0	0	1	3	0	1	2	0

Fig. 4 Section of a typical diffraction image from a fine φ-sliced, low-dose diffuse scattering dataset (crystal B in Table 1). Bragg peaks and diffuse scattering are difficult

(Continued)

reciprocal space coverage in one single dataset, or decrease attenuation and expose the crystal for a shorter time and obtain higher I/sigma on each frame (see sample B in Table 1; beam not attenuated), possibly requiring multiple crystals for a complete dataset (see triclinic lysozyme dataset in Table 1). As discussed in Section 4.1, use of a larger beam size, a multi-motor scan (see sample A in Table 1), and a top-hat beam profile (rather than Gaussian), all help by distributing photons over a larger volume and therefore increase data collection efficiency.

4.3 Background measurement

The signal of interest in a diffuse scattering experiment is the elastic scattering from the atoms in the crystal (i.e., the proteins and the solvent occupying space between them). The ideal measurement conditions would consist of the protein crystal suspended in vacuum, so that no background scattering from the crystal mount or air is present in the diffraction images. Unfortunately, these conditions are difficult to achieve because of the need to maintain hydration, although we note that there is precedence for in-vacuum diffraction measurements using crystals wrapped in graphene (Warren et al., 2015; Wierman, Alden, Kim, McEuen, & Gruner, 2013). Fortunately, when large enough crystals are used, the background scattering from all sources at a typical macromolecular beamline is not overwhelming. In our experience, it can be made comparable to or less than the diffuse scattering signal from a large (>100 μm) crystal. Still, to obtain an accurate diffuse map, the background must be subtracted (Fig. 5).

The background has contributions from air in the beam path, sample mounting materials such as loops and capillaries, and excess water or oil on the crystal's surface. Air scatters isotropically (the intensity depends only on scattering angle, after accounting for polarization of the X-ray beam). However, as mentioned above, scattering from the air between the final aperture and the crystal makes a projection image of the sample mount on the detector. Consequently, the air scattering recorded by the detector may depend on the orientation of the sample during the scan (Meisburger et al., 2020). To measure the scan-varying background from air as well as the

Fig. 4—Cont'd to discern in the noisy image (top). Accumulated photon counts in each pixel for the zoomed-in region (bottom) show a compact Bragg peak that occupies several pixels and quickly decays to background. Most pixels in this image record zero or one photons. The diffuse signal is revealed only after merging to produce a three-dimensional map (not shown).

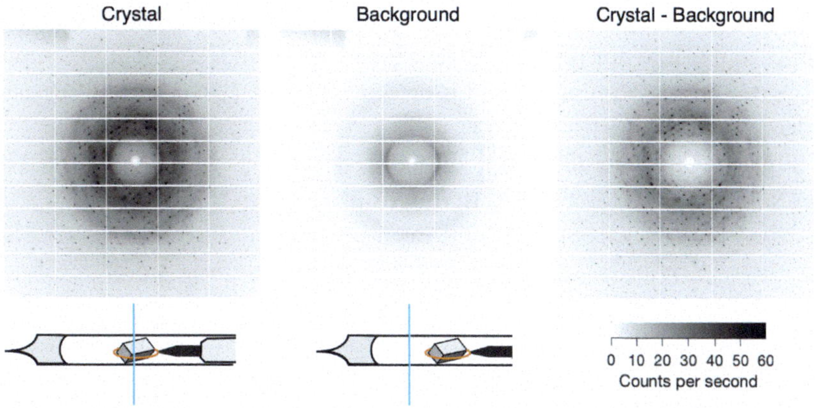

Fig. 5 Illustration of crystal and background data collection for triclinic lysozyme. Diffraction images are shown for crystal (left panel) and background measurements (middle panel) with equivalent exposures of 1 s and 1 degree of rotation. The paths of the X-ray beam (blue lines) relative to the crystal, loop, and capillary are indicated in the diagrams below each panel. The diffuse scattering intrinsic to protein crystals is obtained after background is subtracted (right panel). Shadows of the sample loop are visible in both left and central panels as a bright triangular area right of the beam center. Reproduced under CC BY 4.0 license (Meisburger et al., 2020).

capillary, we have found a reliable method is to translate the goniometer along the spindle axis by several hundred μm so the beam passes beyond the tip of the loop (but still through the capillary) (Fig. 5, middle panel). The measurement is done after each successful data collection, and it is performed over the same angular range. Because the background signal does not change rapidly with angle, the φ increment per image can be coarse (1 degree is adequate). After the background is measured, it is then subtracted from each frame during data processing (see Chapter 2).

After subtracting the air and capillary background scattering, contributions from extraneous water or oil around the crystal and the loop potentially remain. We have not attempted to measure the loop contribution (doing so would require removing the crystal and collecting an identical dataset from the same spot, a challenging task). Instead, we rely on scaling algorithms to estimate and subtract an excess isotropic signal. To use this method, two conditions must be met. First, the residual background scattering must be isotropic; for this reason, Kapton loops are preferred over nylon, whose scattering is strongly anisotropic. Second, the X-ray beam must pass through bare crystal during much of the scan, otherwise the scaling algorithm will not have sufficient information to determine the excess signal. We therefore strive to remove as much liquid from the crystal surface as possible during mounting, which has the added benefit of a more stable mount (if too much liquid is present, the crystal can slip during data collection).

Finally, even after all sources of non-crystal background are subtracted, there remains the Compton scattering from atoms in the crystal itself. It is arguably not background at all, since it originates from the sample. However, Compton scattering is incoherent and does not contribute structural information, and it is typically not included in disorder models that assume elastic scattering (introduced in Section 1). Fortunately, it is straightforward to predict Compton scattering knowing only the atomic inventory of the unit cell, as estimated for example from the protein sequence and solvent fraction. Before it can be subtracted, however, the data must be placed on an absolute intensity scale, the procedures for which we described previously (Meisburger et al., 2020). Having data on an absolute scale is particularly useful for fitting models of lattice disorder, because the overall scale factor can be used to determine the magnitude of the structural fluctuation (i.e., the ADP) in ångström units.

5. Data quality assessment

When collecting data at the synchrotron, how do you know if a diffraction dataset is sufficient to reconstruct a high-quality diffuse map? This question is best answered by performing the reconstruction. Unfortunately, at present, the reconstruction process cannot be done in real-time at the beamline. However, we have found that visual examination of the diffraction images and a careful assessment of Bragg data quality can give a good indication of whether a dataset is worthy of further analysis. In this section, we first identify general features we expect to see in diffraction images from high quality crystals, as well as common issues that can be identified visually. Second, we describe statistics reported during Bragg data processing of relevance to diffuse data quality, specifically to radiation damage. Finally, we introduce statistical measures used to evaluate diffuse map quality and benchmark model-data comparisons.

5.1 Examination of diffraction images

Although it is always good practice to examine diffraction images when collecting MX data, it has become less of a necessity in recent years with the prevalence of automated Bragg data processing pipelines (auto-processing). For diffuse scattering, auto-processing results are extremely useful: a complete, high quality Bragg dataset is a precondition for biophysical applications of diffuse scattering. However, they do not substitute for visual examination of diffraction images. Consider that a very small fraction of the detector pixels contributes to the integrated Bragg intensities, and integration programs have become effective at rejecting artifacts that don't fit the dominant lattice model (e.g., from twinning or multiple lattices). For diffuse scattering, we do not have the luxury of rejecting most detector pixels. Moreover, if multiple lattices or twins are present, the diffuse signals will overlap, and it will be difficult if not impossible to separate them.

Problematic diffraction can often be identified by eye. Diffraction from perfect crystals should have sharp and round Bragg peaks. If the Bragg peaks are very elongated, large, or smeared, it usually means that the crystal has high mosaicity. If closely adjacent Bragg peaks are observed, this usually indicates the existence of multiple lattices (Fig. 2D), which might mean that the crystal grew as a twin or the crystal cracked during looping or data collection. On the other hand, intense diffraction peaks that are far from each other usually suggests contamination of salt or ice crystals. If multiple lattices are present, this will often be flagged by Bragg data processing programs as a low fraction of indexed reflections. However, if the crystal is cracked so that the multiple-lattice peaks are very close together, the integration program may lump them into a single mosaic peak, and the issue would be missed without looking at the images.

Finally, we note that when using a strategy of high redundancy, fine φ-slicing, and low dose, the individual images will be extremely noisy (Fig. 4), and this complicates the examination process described above. One useful approach is to bin fine-sliced images together (for instance, every 1 degree) to increase signal-to-noise. Binning can be performed on-the-fly by commonly used image viewers, such as ADXV (http://www.scripps.edu/tainer/arvai/adxv.html) and ALBULA (https://www.dectris.com/detectors/albula-software/). Another approach is to collect longer exposures from test samples to assess diffraction quality before collecting full datasets from fresh crystals.

5.2 Quantifying and controlling radiation damage

Radiation damage processes begin as soon as the sample is exposed to X-rays, and the effects of this damage can be seen in the diffraction data. Radiation damage is known to cause reduction or radiolysis of a protein's chemical structure, which includes reducing of metal centers (Yano et al., 2005), breaking of disulfide bonds, decarboxylation of aspartate and glutamate, and dihydroxylation of tyrosine (Burmeister, 2000). In addition, when damage is pervasive, the crystalline order can be disrupted, causing a loss of Bragg peak intensity, increase in the Wilson B-factor, increase in mosaicity, and altered unit cell dimensions (Blake & Philips, 1962; Ravelli & McSweeney, 2000). Local and global damage may occur at very different rates, especially at cryogenic temperatures where the glassy solvent state slows the loss of crystalline order. Near room temperature, global damage progresses much more rapidly than at cryogenic temperatures (approximately 1–2 orders of magnitude faster) (Southworth-Davies et al., 2007). Diffuse scattering performed at room temperature must therefore contend with the possible effects of increased disorder caused by X-ray exposure. As Bragg peak intensities fade, a concomitant increase in the diffuse scattering is expected (the total scattering from a crystal is conserved). It is not yet understood how radiation damage would specifically impact diffuse scattering. However, it is unlikely to be a simple increase in background scattering that can be corrected in data processing. The dissociation of chemical bonds modifies the mechanical constraints of internal protein motion, and the disturbance of protein packing might introduce additional diffuse scattering from static disorder; both might influence the interpretation of observed diffuse scattering. Thus, for diffuse scattering it is especially important to monitor radiation damage effects and to choose a maximum dose conservatively.

Metrics have been proposed to quantify radiation damage in Bragg diffraction data. Both local and global damage potentially complicate diffuse scattering experiments at room temperature, where they are expected to occur at similar rates (Gotthard et al., 2019). Global damage results in the fading of high-resolution Bragg reflections, and this effect can be quantified using the dose at which intensity drops by half (Henderson, 1990) or the change in average B-factor between damaged and undamaged sample (Kmetko, Husseini, Naides, Kalinin, & Thorne, 2006). Evidence for site-specific damage can often be observed in the electron density maps, especially the appearance of difference electron density ($F_o - F_c$) in the vicinity of chemical species known to be radiation sensitive. Recently, sensitive metrics have been developed to automatically detect and quantify such features (Mora et al., 2020; Shelley & Garman, 2022).

For a given sample and measurement conditions, the amount of radiation damage assessed using metrics described above have been found to be roughly proportional to the applied dose (Kmetko et al., 2006). At cryogenic conditions, the rate of global damage progression for a given dose is nearly universal (Leiros, Timmins, Ravelli, & McSweeney, 2006; Owen, Rudiño-Piñera, & Garman, 2006; Sliz, Harrison, & Rosenbaum, 2003). In contrast, at room temperature the dose tolerance can be sample-dependent (Garman, 2010) and can also depend on the dose rate (Rajendran, Dworkowski, Wang, & Schulze-Briese, 2011; Southworth-Davies et al., 2007; Warkentin et al., 2012). For instance, it has been shown that very high flux data collection can outrun certain damage processes at room temperature, leading to an apparently higher dose tolerance under those conditions (Owen et al., 2012; Owen et al., 2014). The benefits of outrunning damage at a synchrotron are relatively modest (a factor of 1–4). At an XFEL, however, when diffraction data are collected in femtoseconds before the atoms have a chance to move, orders of magnitude higher dose may be used. For diffuse scattering experiments, the benefits of high flux data collection may be worth considering when only small crystals are available, since these experiments will be particularly signal-starved. However, we note that a boost in total signal does not solve the problem of high background scattering when small crystals are used. For this reason, it has been most common to collect diffuse scattering from large crystals using modest dose rates.

When starting data collection from a new sample and/or a new beamline, we recommend a systematic test of dose tolerance. A straightforward experiment is to collect narrow wedges (~10 degrees) repeatedly from a single spot on the crystal until the high-resolution peaks begin to disappear. After processing and scaling the data, the B-factor corrections from scaling can be plotted as a function of exposure time. A maximum tolerable B-factor increase can then be chosen, which sets a maximum exposure time for a dataset. For instance, when collecting diffuse scattering data from lysozyme polymorphs (Table 1), a maximum B-factor increase was set at $\sim 2\,\text{Å}^2$, leading to a maximum exposure time of 50 s. The dose limit estimated in this way represents a worst-case scenario, because the same spot on the crystal was irradiated continuously. The next step is to devise a strategy for obtaining a complete dataset by distributing the dose over the full crystal volume or from multiple crystals if necessary.

Because mitigation of radiation damage is so important for room-temperature data collection, here we review in greater detail strategies for dose distribution first introduced in Section 4.2. When thinking about dose, is important to consider that the X-ray intensity is not always evenly distributed through the cross section of the beam, and many beamlines employ focusing optics so that the profile is approximately Gaussian in one or two dimensions. The benefits of a top-hat (uniform) (Sutter et al., 2016) profile for more evenly distributing dose has gained attention in recent years. Furthermore, when a crystal is rotated during data collection, the dose at the rotation center is higher than at locations further away (Fig. 6A). Such non-uniform irradiation is not ideal for data quality because the signal becomes a mixture of signals from damaged and undamaged structures. Several strategies have been developed to alleviate these issues when the crystal is larger than the beam. For instance, the center of the crystal may be offset from the goniometer rotation center, such that the center is not over-irradiated (Fig. 6D) (Zeldin, Brockhauser, Bremridge, Holton, & Garman, 2013). Alternatively, the crystal can be translated while rotating (known as a 'helical scan') to distribute the dose over the entire crystal (Fig. 6C) (Flot et al., 2009). Similarly, mini-κ goniometers (Brockhauser, Ravelli, & McCarthy, 2013; Waltersperger et al., 2015) can be used to improve dose distribution compared with standard φ-scans. The program *RADDOSE-3D* (Bury, Brooks-Bartlett, Walsh, & Garman, 2018; Zeldin, Gerstel, & Garman, 2013) can be used to calculate dose distributions for different beam profiles, crystal contents, and data collection methods.

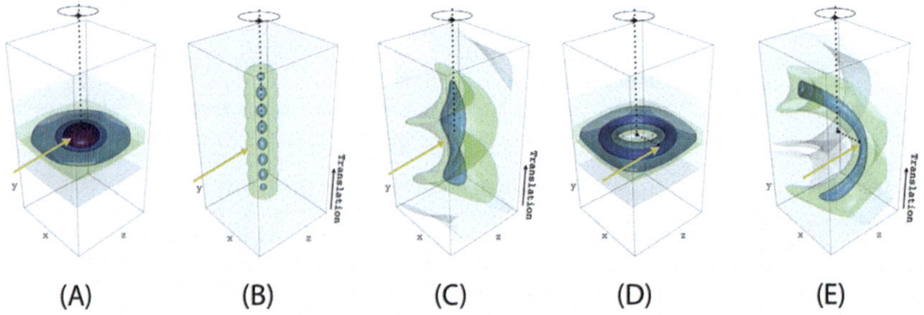

Fig. 6 Simulated dose distribution for various scan strategies. The accumulated X-ray dose on a cuboid crystal (boxes outlined in thin black lines) was simulated using RADDOSE-3D for 5 different scan types: (A) a typical single-point φ scan; (B) multi-point φ scan; (C) helical scan; (D) off-centering scan; and (E) a helical scan with offset centering. The model crystal was $100 \times 200 \times 100\ \mu m^3$ and the X-ray beam had a flux of 5×10^{11} photons/s, energy of 12.4 keV, and a Gaussian intensity profile of $20 \times 20\ \mu m^2$ (full width at half maximum). Contours correspond to dose of 0.0001 MGy (gray), 5 MGy (green), 10 MGy (light blue), 20 MGy (dark blue) and 30 MGy (red). Detailed parameters used for each simulation can be found at the original publication. Reproduced with permission (Bury et al., 2018).

After collecting a complete dataset, the global radiation damage should be assessed using the B-factor correction applied in scaling, as described above. If the dose distribution strategy was successful, the B-factor increase should be below the maximum tolerable value. If not, frames from overly damaged sample may be excluded from the dataset. However, excluding frames may result in an incomplete dataset, and in this case more diffraction data should be collected. A potential strategy to avoid incomplete data is to attenuate the beam (or collect shorter exposures) so that very little damage occurs during a 360-degree rotation, and to repeat the same scan multiple times. The later datasets can then be excluded once the B-factor increase threshold is reached without impacting completeness. If possible, the electron density should be examined for signs of local radiation damage especially at sites of known sensitivity (particularly disulfide bonds and metals). It may be necessary to revise the B-factor cutoff if too much damage is seen. Finally, since radiation damage issues are sometimes subtle and may go unnoticed during data collection, as a rule of thumb we recommend collecting more datasets than necessary to obtain a complete map.

5.3 Diffuse map quality assessment

The process of reconstructing a three-dimensional reciprocal space map from diffraction images follows similar steps as Bragg data reduction, including integration, scaling, and merging. These steps are discussed in greater detail in Chapter 2. Here, we review methods that have been employed to assess the quality of maps.

In general, two factors are important in data quality: precision and accuracy. The precision of a diffuse scattering dataset is related to how noisy it is. Because radiation damage limits the exposure times, the dominant source of noise in diffuse scattering is often the unavoidable Poisson statistics from photon-counting; the uncertainty in a measurement with N photons is $\sqrt{N}$. Precision improves when more data are averaged together according to the same rule of $\sqrt{M}$, where M is the number of measurements. Thus, to improve the precision by a factor of two, four equivalent datasets must be averaged together. Precision can also improve if background scattering is reduced, because the error in a measurement with background is $\sqrt{N_{\text{crystal}} + N_{\text{background}}}$. If the background can be reduced so it is significantly less than the scattering from the crystal, its effect on precision becomes negligible. On the other hand, even a small amount of background can ruin accuracy if it is not subtracted correctly. Although accuracy is harder to determine than precision, often reproducibility is a good proxy. When data can be reproduced under very different conditions (a different crystal, or even a different beamline), it is likely that the data are accurate. Thus, for both Bragg diffraction and diffuse scattering, statistics related to precision and reproducibility are used to assess data quality.

Because photon-counting statistics often dominate the noise in diffuse maps, data reduction algorithms propagate the Poisson error estimate through data reduction steps, giving an "error bar" for each data point (see Chapter 2 for details). The Poisson error represents a lower bound because other types of systematic error may also contribute. For instance, if a detector were poorly calibrated, the spread in equivalent observations from different detector panels might be larger than expected given Poisson statistics. To estimate the uncertainty including certain systematic errors, a common strategy is to split the dataset into equivalent halves that are merged separately and to compute the correlation between them. If the split is random, the statistic is called $CC_{1/2}$ (Karplus & Diederichs, 2012). For diffraction data, the split can also be performed according to symmetry operations and may be called CC_{Laue} or $CC_{Friedel}$ depending on how operations are grouped, or the data can be split between different crystals in the case of CC_{rep} (Su et al., 2021). These correlation coefficients measure the internal consistency of the dataset. Because signal-to-noise depends strongly on resolution, CCs are often calculated in bins of constant resolution. R-factors have also been used to assess internal consistency (Wall, Ealick, et al., 1997). In general, the switch to R-factors makes sense when correlation coefficients approach $\sim$1 because R-factors are more sensitive to discrepancies. In the context of Bragg diffraction, where the signal-to-noise ratio is very high, R-factors are the preferred statistic (except when judging resolution cutoffs). In contrast, most diffuse scattering datasets reported to date have relatively high noise, and so correlation coefficients are an appropriate statistic to use.

Although statistics used to assess Bragg data and diffuse data are superficially similar (such as $CC_{1/2}$), there is an important distinction. For Bragg data, the $CC_{1/2}$ is a universal measure of data consistency because only Bragg peak intensities (i.e., integrated *hkl* intensities) are compared. When $CC_{1/2}$ is applied to diffuse scattering, it is important to also specify exactly how the data were treated. For instance, if a very coarse map is generated by binning data down to one point per integer Miller index, the noise level will be very low, and $CC_{1/2}$ may be very high. In contrast, if an extremely fine map is generated from the same dataset, each point may be very noisy and $CC_{1/2}$ will be low. $CC_{1/2}$ will also depend on the kinds of diffuse scattering present. For example, if halo features are very intense, the $CC_{1/2}$ may be very high even if the cloudy diffuse pattern is noisy. Thus, for diffuse scattering, $CC_{1/2}$ is best used to compare datasets from nominally identical crystals that are processed identically (i.e., with the same mask and integration grid). There can be no general recommendation such as *the diffuse map must have* $CC_{1/2} > 0.95$ without also specifying the data processing details. Given these caveats, $CC_{1/2}$ is still a useful statistic. For instance, it has been used to judge whether a scaling algorithm made a meaningful improvement in data quality (see for example Fig. 7).

Finally, we note that $CC_{1/2}$ also plays an important role in quantifying model-data agreement. From $CC_{1/2}$, it is possible to estimate the best possible correlation between a model and the data if the model were perfect. This statistic, known as CC^* (Karplus & Diederichs, 2012), was originally developed for Bragg data, but it applies equally well to diffuse data (Fig. 8). It can be derived by considering how the noise in each half-dataset contributes to the correlation coefficient between them, compared with the noise level of the merged dataset vs. the model (which is noise-free). CC^* is a simple transformation of $CC_{1/2}$, as follows:

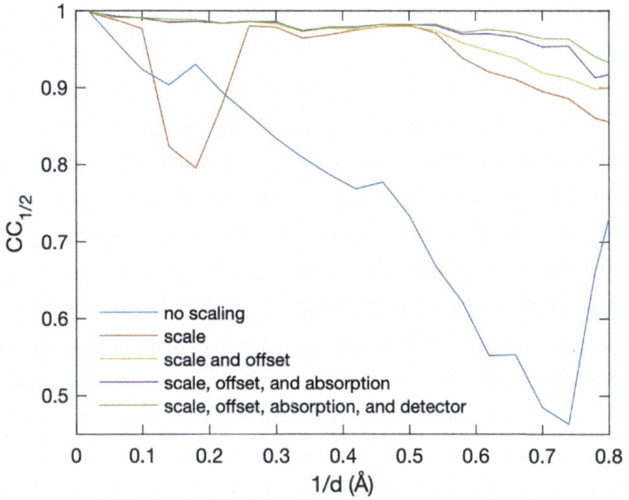

Fig. 7 Using $CC_{1/2}$ to measure data quality improvements during data processing. A scaling model with four correction types (scale, offset, absorption, and detector) was refined in mdx-lib (see Chapter 2) to merge a map of diffuse scattering from triclinic lysozyme with one sample per integer Miller index. The $CC_{1/2}$ statistic computed for each resolution shell (*d*) shows that the data become more self-consistent as each correction is added. Reproduced under CC BY 4.0 license (Meisburger et al., 2020).

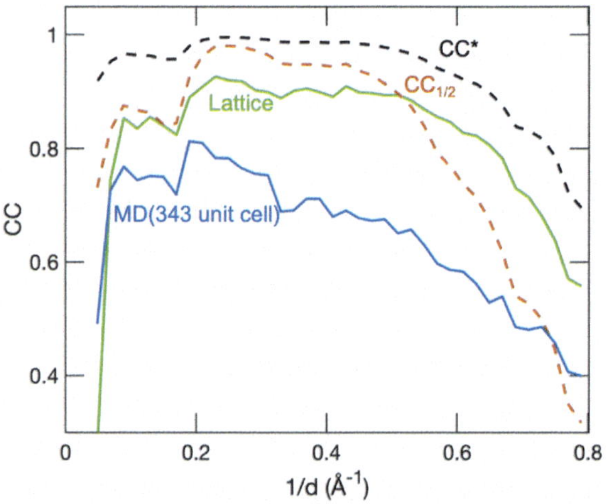

Fig. 8 Correlation coefficients (CC) for model-data comparison. CCs are computed between the experimental diffuse map from triclinic lysozyme and two simulations: a lattice dynamics model (solid green curve) and molecular dynamics (MD) simulation with 343 unit cells (solid blue curve). The CC* statistic (black dashed line) estimates the maximum possible model-data correlation and is calculated from $CC_{1/2}$ (dashed red line) using Eq. (13). Note that the lattice model CC exceeds $CC_{1/2}$ at high resolution (*d*) but always falls below CC*. *Adapted under CC BY 4.0 license (Meisburger et al., 2020).*

$$CC^* = \sqrt{\frac{2CC_{1/2}}{1 + CC_{1/2}}} \tag{13}$$

As the macromolecular diffuse scattering field matures, we expect further work will be done to address shortcomings of these indicators and to standardize the reporting of statistics in publications.

6. Conclusions

Recent years have seen major advances in the field of macromolecular diffuse scattering. Building on decades of seminal studies (Caspar, Clarage, Salunke, & Clarage, 1988; Phillips, Fillers, & Cohen, 1980; Polikanov & Moore, 2015; Wall, Clarage, et al., 1997; Wall, Ealick, et al., 1997), the widespread adoption of direct detectors at crystallography beamlines catalyzed the development of new experimental and computational methods. In 2020, it was shown with high confidence that diffuse scattering from protein crystals can be explained in terms of lattice dynamics and protein dynamics (Meisburger et al., 2020), achieving a goal that was first envisioned by Don Caspar and co-workers over 30 years ago (Caspar et al., 1988).

With the collective knowledge the field has gained over the past few years, it is now possible to provide general guidelines for collecting diffuse scattering data. In general, a crystal that is suitable for conventional crystallography meets the minimum requirement for a diffuse scattering study. However, as is generally true for all scattering experiments, the requirement for sample quality is more stringent for diffuse scattering than for conventional crystallography. This is because the signal of interest is spread throughout reciprocal space rather than being concentrated in diffraction spots. Additionally, to study protein motions without the complicating factors introduced by cryo-cooling crystals, it is recommended to start diffuse scattering experiments at non-cryogenic temperatures. Ultimately, the goals of diffuse scattering data collection are (1) to achieve the cleanest signal throughout reciprocal space without introducing unnecessary artifacts, (2) to collect data in a way that will resolve the most intense diffuse scattering features (e.g., halos), and (3) to collect data in a way that allows for sophisticated scaling and merging procedures (and generate data quality statistics in the process). Best practices, such as careful sample handling and beamline selection, low-dose data collection for mitigating radiation damage, precise background collection, stringent quality check on diffraction images, should be always taken in order to meet these goals.

There are still many unanswered questions about macromolecular diffuse scattering. For example, a systematic study of the effect of radiation damage has not yet been done. Additionally, it is not yet known whether diffuse scattering from a crystal that is considered "poor quality" can be interpreted in a robust manner. Such crystals may be highly mosaic (producing smeary diffraction) or feature highly disordered regions (leading to diffuse scattering that extends past the resolution limit of the diffraction pattern). Finally, measuring diffuse scattering from a serial crystallography experiment (such as that performed at an XFEL) has its own challenges, which need to be analyzed in detail. In addition to answering these questions, there is still room for improvement on the computational front, both in data processing and modeling. Although the field of macromolecular diffuse scattering has not yet matured, we hope that this chapter provides readers with the knowledge needed to design experiments that will produce datasets that can stand the test of time, such that they can be reanalyzed by newer software as they become available.

Acknowledgments

The authors are grateful to Dr. Lindsey Backman and Prof. Catherine Drennan (MIT) for sharing the image shown in Fig. 2D. Data shown in Figs. 1A, 2B, 2F and 4 were collected at beamline ID7B2 at the Center for High-Energy X-ray Sciences (CHEXS), which is supported by the at National Science Foundation (BIO, ENG and MPS Directorates) under award DMR-1829070, and the Macromolecular Diffraction at CHESS (MacCHESS) facility, which is supported by award P30GM124166 from the National Institute of General Medical Sciences, National Institutes of Health, and by New York State's Empire State Development Corporation (NYSTAR). This work was partially supported by the Cornell Center for Material Research with funding from the NSF MRSEC program (DMR-1719875) and primarily supported by National Institutes of Health grant GM124847 (to N.A.).

References

Ayyer, K., Yefanov, O. M., Oberthür, D., Roy-Chowdhury, S., Galli, L., Mariani, V., & Chapman, H. N. (2016). Macromolecular diffractive imaging using imperfect crystals. *Nature, 530*, 202–206.

Bautis, A. N. (1975). *Immersion oil formulations for use in microscopy and similar fields. U.S. Patent No. 3,929,667*. Washington, DC: U.S. Patent and Trademark Office.

Bellamy, H. D., Snell, E. H., Lovelace, J., Pokross, M., & Borgstahl, G. E. O. (2000). The high-mosaicity illusion: Revealing the true physical characteristics of macromolecular crystals. *Acta Crystallographica. Section D, Biological Crystallography, 56*, 986–995.

Benschoten, A. H. V., Liu, L., Gonzalez, A., Brewster, A. S., Sauter, N. K., Fraser, J. S., & Wall, M. E. (2016). Measuring and modeling diffuse scattering in protein X-ray crystallography. *Proceedings of the National Academy of Sciences of the United States of America, 113*, 4069–4074.

Blake, C. C. F., & Philips, D. C. (1962). Effects of X-irradiation on single crystals of myoglobin. In S. Eklund (Ed.). *Proceedings of the Symposium on the Biological Effects of Ionizing Radiation at the Molecular Level* (pp. 183–191). Vienna: International Atomic Energy Agency.

Bourenkov, G. P., & Popov, A. N. (2006). A quantitative approach to data-collection strategies. *Acta Crystallographica. Section D, Biological Crystallography, 62*, 58–64.

Brockhauser, S., Ravelli, R. B. G., & McCarthy, A. A. (2013). The use of a mini-κ goniometer head in macromolecular crystallography diffraction experiments. *Acta Crystallographica. Section D, Biological Crystallography, 69*, 1241–1251.

Budayova-Spano, M., Koruza, K., & Fisher, Z. (2020). *Large crystal growth for neutron protein crystallography. Methods in Enzymology, 634*, Elsevier, 21–46.

Burmeister, W. P. (2000). Structural changes in a cryo-cooled protein crystal owing to radiation damage. *Acta Crystallographica. Section D, Biological Crystallography, 56*, 328–341.

Bury, C. S., Brooks-Bartlett, J. C., Walsh, S. P., & Garman, E. F. (2018). Estimate your dose: RADDOSE-3D. *Protein Science, 27*, 217–228.

Case, D. A. (in press). MD simulations of macromolecular diffuse crystals: Implications for the analysis of Bragg and diffuse scattering. In *Methods in enzymology* (Vol. 688).

Caspar, D. L. D., Clarage, J., Salunke, D. M., & Clarage, M. (1988). Liquid-like movements in crystalline insulin. *Nature, 332*, 659–662.

Chapman, H. N., Yefanov, O. M., Ayyer, K., White, T. A., Barty, A., Morgan, A., & Pande, K. (2017). Continuous diffraction of molecules and disordered molecular crystals. *Journal of Applied Crystallography, 50*, 1084–1103.

Dauter, Z. (1999). Data-collection strategies. *Acta Crystallographica. Section D, Biological Crystallography, 55*, 1703–1717.

Donath, T., Brandstetter, S., Cibik, L., Commichau, S., Hofer, P., Krumrey, M., & Wernecke, J. (2013). Characterization of the PILATUS photon-counting pixel detector for X-ray energies from 1.75 keV to 60 keV. *Journal of Physics: Conference Series, 425*, 062001.

Fischer, M. (2021). Macromolecular room temperature crystallography. *Quarterly Reviews of Biophysics, 54*, e1.

Flot, D., Mairs, T., Giraud, T., Guijarro, M., Lesourd, M., Rey, V., & Mitchell, E. (2009). The ID23-2 structural biology microfocus beamline at the ESRF. *Journal of Synchrotron Radiation, 17*, 107–118.

Förster, A., Brandstetter, S., & Schulze-Briese, C. (2019). Transforming X-ray detection with hybrid photon counting detectors. *Philosophical Transactions of the Royal Society A: Mathematical, Physical and Engineering Sciences, 377*, 20180241.

Fraser, J. S., Clarkson, M. W., Degnan, S. C., Erion, R., Kern, D., & Alber, T. (2009). Hidden alternative structures of proline isomerase essential for catalysis. *Nature, 462*, 669–673.

Fraser, J. S., Van Den Bedem, H., Samelson, A. J., Lang, P. T., Holton, J. M., Echols, N., & Alber, T. (2011). Accessing protein conformational ensembles using room-temperature X-ray crystallography. *Proceedings of the National Academy of Sciences, 108*, 16247–16252.

Garman, E. F. (2010). Radiation damage in macromolecular crystallography: What is it and why should we care? *Acta Crystallographica. Section D, Biological Crystallography, 66*, 339–351.

Gotthard, G., Aumonier, S., De Sanctis, D., Leonard, G., Von Stetten, D., & Royant, A. (2019). Specific radiation damage is a lesser concern at room temperature. *IUCrJ, 6*, 665–680.

Halle, B. (2004). Biomolecular cryocrystallography: Structural changes during flash-cooling. *Proceedings of the National Academy of Sciences, 101*, 4793–4798.

Hasegawa, K., Yamashita, K., Murai, T., Nuemket, N., Hirata, K., Ueno, G., & Yamamoto, M. (2017). Development of a dose-limiting data collection strategy for serial synchrotron rotation crystallography. *Journal of Synchrotron Radiation, 24*, 29–41.

Henderson, R. (1990). Cryo-protection of protein crystals against radiation damage in electron and X-ray diffraction. *Proceedings of the Royal Society of London. Series B: Biological Sciences, 241*, 6–8.

Juers, D. H., & Matthews, B. W. (2001). Reversible lattice repacking illustrates the temperature dependence of macromolecular interactions. *Journal of Molecular Biology, 311*, 851–862.

Karplus, P. A., & Diederichs, K. (2012). Linking crystallographic model and data quality. *Science, 336*, 1030–1033.

Kmetko, J., Husseini, N. S., Naides, M., Kalinin, Y., & Thorne, R. E. (2006). Quantifying X-ray radiation damage in protein crystals at cryogenic temperatures. *Acta Crystallographica. Section D, Biological Crystallography, 62*, 1030–1038.

Kriminski, S., Caylor, C. L., Nonato, M. C., Finkelstein, K. D., & Thorne, R. E. (2002). Flash-cooling and annealing of protein crystals. *Acta Crystallographica. Section D, Biological Crystallography, 58*, 459–471.

Leiros, H.-K. S., Timmins, J., Ravelli, R. B. G., & McSweeney, S. M. (2006). Is radiation damage dependent on the dose rate used during macromolecular crystallography data collection? *Acta Crystallographica. Section D, Biological Crystallography, 62*, 125–132.

Leonarski, F., Redford, S., Mozzanica, A., Lopez-Cuenca, C., Panepucci, E., Nass, K., & Wang, M. (2018). Fast and accurate data collection for macromolecular crystallography using the JUNGFRAU detector. *Nature Methods, 15*, 799–804.

Lisi, G. P., & Loria, J. P. (2016). Solution NMR spectroscopy for the study of enzyme allostery. *Chemical Reviews, 116*, 6323–6369.

Lovelace, J. J., & Borgstahl, G. E. O. (2020). Characterizing pathological imperfections in macromolecular crystals: Lattice disorders and modulations. *Crystallography Reviews, 26*, 3–50.

Meisburger, S. P., & Ando, N. (2017). Correlated motions from crystallography beyond diffraction. *Accounts of Chemical Research, 50*, 580–583.

Meisburger, S.P., & Ando, N. (in press). Processing macromolecular diffuse scattering data. In *Methods in enzymology* (Vol. 688).

Meisburger, S. P., Case, D. A., & Ando, N. (2020). Diffuse X-ray scattering from correlated motions in a protein crystal. *Nature Communications, 11*, 1271.

Meisburger, S. P., Case, D. A., & Ando, N. (2023). Robust total X-ray scattering workflow to study correlated motion of proteins in crystals. *Nature Communications, 14*, 1228.

Meisburger, S. P., Thomas, W. C., Watkins, M. B., & Ando, N. (2017). X-ray scattering studies of protein structural dynamics. *Chemical Reviews, 117*, 7615–7672.

Mora, E., De La Coquelle, N., Bury, C. S., Rosenthal, M., Holton, J. M., Carmichael, I., & Weik, M. (2020). Radiation damage and dose limits in serial synchrotron crystallography at cryo- and room temperatures. *Proceedings of the National Academy of Sciences, 117*, 4142–4151.

Ng, J. D., Baird, J. K., Coates, L., Garcia-Ruiz, J. M., Hodge, T. A., & Huang, S. (2015). Large-volume protein crystal growth for neutron macromolecular crystallography. *Acta Crystallographica. Section F, Structural Biology Communications, 71*, 358–370.

Owen, R. L., Axford, D., Nettleship, J. E., Owens, R. J., Robinson, J. I., Morgan, A. W., & Evans, G. S. (2012). Outrunning free radicals in room-temperature macromolecular crystallography. *Acta Crystallographica. Section D, Biological Crystallography,* 810–818.

Owen, R. L., Paterson, N., Axford, D., Aishima, J., Schulze-Briese, C., Ren, J., & Evans, G. (2014). Exploiting fast detectors to enter a new dimension in room-temperature crystallography. *Acta Crystallographica. Section D, Biological Crystallography, 70*, 1248–1256.

Owen, R. L., Rudiño-Piñera, E., & Garman, E. F. (2006). Experimental determination of the radiation dose limit for cryocooled protein crystals. *Proceedings of the National Academy of Sciences, 103*, 4912–4917.

Palmer, A. G. (2015). Enzyme dynamics from NMR spectroscopy. *Accounts of Chemical Research, 48*, 457–465.

Peck, A., Lane, T. J., & Poitevin, F. (submitted). Modeling diffuse scattering with simple, physically interpretable models. In *Methods in enzymology* (Vol. 688).

Pflugrath, J. W. (1999). The finer things in X-ray diffraction data collection. *Acta Crystallographica. Section D, Biological Crystallography, 55*, 1718–1725.

Pflugrath, J. W. (2015). Practical macromolecular cryocrystallography. *Acta Crystallographica. Section F, Structural Biology Communications, 71*, 622–642.

Phillips, G. N., Fillers, J. P., & Cohen, C. (1980). Motions of tropomyosin. Crystal as metaphor. *Biophysical Journal, 32*, 485–502.

Polikanov, Y. S., & Moore, P. B. (2015). Acoustic vibrations contribute to the diffuse scatter produced by ribosome crystals. *Acta Crystallographica. Section D, Biological Crystallography, 71*, 2021–2031.

Polsinelli, I., Savko, M., Rouanet-Mehouas, C., Ciccone, L., Nencetti, S., Orlandini, E., & Shepard, W. (2017). Comparison of helical scan and standard rotation methods in single-crystal X-ray data collection strategies. *Journal of Synchrotron Radiation, 24*, 42–52.

Rajendran, C., Dworkowski, F. S. N., Wang, M., & Schulze-Briese, C. (2011). Radiation damage in room-temperature data acquisition with the PILATUS 6M pixel detector. *Journal of Synchrotron Radiation, 18*, 318–328.

Ravelli, R. B., & McSweeney, S. M. (2000). The 'fingerprint' that X-rays can leave on structures. *Structure, 8*, 315–328.

Schneider, T. R. (2000). Objective comparison of protein structures: Error-scaled difference distance matrices. *Acta Crystallographica. Section D, Biological Crystallography, 56*, 714–721.

Shelley, K. L., & Garman, E. F. (2022). Quantifying and comparing radiation damage in the Protein Data Bank. *Nature Communications, 13*, 1314.

Sliz, P., Harrison, S. C., & Rosenbaum, G. (2003). How does radiation damage in protein crystals depend on X-ray dose? *Structure, 11*, 13–19.

Snell, E. H., Bellamy, H. D., & Borgstahl, G. E. O. (2003). *Macromolecular crystal quality. 368*, 268–288.

Snell, E. H., Weisgerber, S., Helliwell, J. R., Weckert, E., Hölzer, K., & Schroer, K. (1995). Improvements in lysozyme protein crystal perfection through microgravity growth. *Acta Crystallographica. Section D, Biological Crystallography, 51*, 1099–1102.

Southworth-Davies, R. J., Medina, M. A., Carmichael, I., & Garman, E. F. (2007). Observation of decreased radiation damage at higher dose rates in room temperature protein crystallography. *Structure, 15*, 1531–1541.

Stura, E. (1999). Chapter 14. Seeding. In T. M. Bergfors (Ed.). *Protein crystallization: Techniques, strategies, and tips: A laboratory manual* (pp. 141–153). La Jolla, Calif: International University Line.

Su, Z., Dasgupta, M., Poitevin, F., Mathews, I. I., Van Den Bedem, H., Wall, M. E., & Wilson, M. A. (2021). Reproducibility of protein X-ray diffuse scattering and potential utility for modeling atomic displacement parameters. *Structural Dynamics, 8*, 044701.

Sutter, J. P., Alcock, S. G., Kashyap, Y., Nistea, I., Wang, H., & Sawhney, K. (2016). Creating flat-top X-ray beams by applying surface profiles of alternating curvature to deformable piezo bimorph mirrors. *Journal of Synchrotron Radiation, 23*, 1333–1347.

Thaller, C., Weaver, L. H., Eichele, G., Wilson, E., Karlsson, R., & Jansonius, J. N. (1981). Repeated seeding technique for growing large single crystals of proteins. *Journal of Molecular Biology, 147*, 465–469.

Thompson, M. C., Cascio, D., & Yeates, T. O. (2018). Microfocus diffraction from different regions of a protein crystal: Structural variations and unit-cell polymorphism. *Acta Crystallographica. Section D, Structural Biology, 74*, 411–421.

Thorne, R. E. (2023). Determining biomolecular structures near room temperature using X-ray crystallography: Concepts, methods and future optimization. *Acta Crystallographica. Section D, Structural Biology, 79*, 78–94.

Vahedi-Faridi, A., Lovelace, J., Bellamy, H. D., Snell, E. H., & Borgstahl, G. E. O. (2003). Physical and structural studies on the cryocooling of insulin crystals. *Acta Crystallographica. Section D, Biological Crystallography, 59*, 2169–2182.

Wall, M. E., Clarage, J. B., & Phillips, G. N. (1997). Motions of calmodulin characterized using both Bragg and diffuse X-ray scattering. *Structure, 5*(12), 1599–1612.

Wall, M. E., Ealick, S. E., & Gruner, S. M. (1997). Three-dimensional diffuse X-ray scattering from crystals of Staphylococcal nuclease. *Proceedings of the National Academy of Sciences, 94*, 6180–6184.

Waltersperger, S., Olieric, V., Pradervand, C., Glettig, W., Salathe, M., Fuchs, M. R., & Wang, M. (2015). PRIGo: A new multi-axis goniometer for macromolecular crystallography. *Journal of Synchrotron Radiation, 22*, 895–900.

Warkentin, M., Badeau, R., Hopkins, J. B., Mulichak, A. M., Keefe, L. J., & Thorne, R. E. (2012). Global radiation damage at 300 and 260 K with dose rates approaching 1 MGy s−1. *Acta Crystallographica. Section D, Biological Crystallography, 68*, 124–133.

Warkentin, M., & Thorne, R. E. (2009). Slow cooling of protein crystals. *Journal of Applied Crystallography, 42*, 944–952.

Warren, A. J., Crawshaw, A. D., Trincao, J., Aller, P., Alcock, S., Nistea, I., & Evans, G. (2015). *In vacuo* X-ray data collection from graphene-wrapped protein crystals. *Acta Crystallographica. Section D, Biological Crystallography, 71*, 2079–2088.

Welberry, T. R., & Weber, T. (2016). One hundred years of diffuse scattering. *Crystallography Reviews, 22*, 2–78.

Wierman, J. L., Alden, J. S., Kim, C. U., McEuen, P. L., & Gruner, S. M. (2013). Graphene as a protein crystal mounting material to reduce background scatter. *Journal of Applied Crystallography, 46*, 1501–1507.

Winter, G., Gildea, R. J., Paterson, N. G., Beale, J., Gerstel, M., Axford, D., & Hall, D. R. (2019). How best to use photons. *Acta Crystallographica. Section D, Structural Biology, 75*, 242–261.

Wych, D. C., & Wall, M. E. (in press a). Molecular-dynamics simulations of macromolecular diffraction, part I: Preparation of protein crystal simulations. In *Methods in enzymology* (Vol. 688).
Wych, D. C., & Wall, M. E. (in press b). Molecular-dynamics simulations of macromolecular diffraction, part II: Analysis of protein crystal simulations. In *Methods in enzymology* (Vol. 688).
Xu, D., Meisburger, S. P., & Ando, N. (2021). Correlated motions in structural biology. *Biochemistry, 60*, 2331–2340.
Yano, J., Kern, J., Irrgang, K.-D., Latimer, M. J., Bergmann, U., Glatzel, P., & Yachandra, V. K. (2005). X-ray damage to the Mn4Ca complex in single crystals of photosystem II: A case study for metalloprotein crystallography. *Proceedings of the National Academy of Sciences, 102*, 12047–12052.
Zeldin, O. B., Brockhauser, S., Bremridge, J., Holton, J. M., & Garman, E. F. (2013). Predicting the X-ray lifetime of protein crystals. *Proceedings of the National Academy of Sciences, 110*, 20551–20556.
Zeldin, O. B., Gerstel, M., & Garman, E. F. (2013). RADDOSE-3D: Time- and space-resolved modelling of dose in macromolecular crystallography. *Journal of Applied Crystallography, 46*, 1225–1230.
Zhong, E. D., Bepler, T., Berger, B., & Davis, J. H. (2021). CryoDRGN: Reconstruction of heterogeneous cryo-EM structures using neural networks. *Nature Methods, 18*, 176–185.

CHAPTER TWO

Processing macromolecular diffuse scattering data

Steve P. Meisburger[a,*] **and Nozomi Ando**[b,*]
[a]Cornell High Energy Synchrotron Source, Cornell University, Ithaca, NY, United States
[b]Department of Chemistry and Chemical Biology, Cornell University, Ithaca, NY, United States
*Corresponding authors. e-mail address: spm82@cornell.edu; nozomi.ando@cornell.edu

Contents

Abstract

Diffuse scattering is a powerful technique to study disorder and dynamics of macromolecules at atomic resolution. Although diffuse scattering is always present in diffraction images from macromolecular crystals, the signal is weak compared with Bragg peaks and background, making it a challenge to visualize and measure accurately. Recently, this challenge has been addressed using the reciprocal space mapping technique, which leverages ideal properties of modern X-ray detectors to reconstruct the complete three-dimensional volume of continuous diffraction from diffraction images of a crystal (or crystals) in many different orientations. This chapter will review recent progress in reciprocal space mapping with a particular focus on the

Methods in Enzymology, Volume 688
ISSN 0076-6879, https://doi.org/10.1016/bs.mie.2023.06.010

strategy implemented in the *mdx-lib* and *mdx2* software packages. The chapter concludes with an introductory data processing tutorial using Python packages *DIALS*, *NeXpy*, and *mdx2*.

1. Introduction

The fact that proteins and nucleic acids form crystals that diffract to (near) atomic resolution has had a profound impact on the biological sciences. X-ray diffraction data from macromolecular crystals is responsible for over 170,000 structures in the Protein Data Bank, the majority at 2.5 Å resolution or better. Although the X-ray scattering from individual molecules is weak, when they are aligned in a crystal, the diffraction interferes constructively in certain directions, causing sharp Bragg peaks to appear in diffraction images. The Bragg peak intensities depend on the repeating pattern of electron density in the crystal, and thus they are used for structure determination. However, not all X-rays scattered by a crystal go into Bragg peaks. In fact, for most protein crystals, the majority of the diffracted X-rays are found elsewhere, in a continuous diffuse pattern (Clarage & Phillips, 1997). Bragg diffraction and diffuse scattering are complementary: the former depends on regularity, the latter on deviations from the average. Intriguingly, the pattern of diffuse scattering encodes the statistical correlations of electron density fluctuations (Meisburger, Thomas, Watkins, & Ando, 2017). Such information is urgently needed in the modern era of structural biology, which seeks to explain function in terms of conformational dynamics, excited states, and ensembles (Ourmazd, Moffat, & Lattman, 2022). This need has motivated a renewed effort to develop methods to accurately measure diffuse scattering for studies of correlated motion (Xu, Meisburger, & Ando, 2021).

Recently, we reported detailed maps of diffuse scattering from lysozyme in triclinic, orthorhombic, and tetragonal space groups (Meisburger, Case, & Ando, 2020; Meisburger, Case, & Ando, 2023). These high quality crystals had sufficiently low mosaicity at room temperature to finely sample the diffuse scattering close to the Bragg peaks. The lysozyme datasets contain two features: a weak, slowly varying cloudy pattern, and intense halos of scattering around the Bragg peaks. Similar halo features have been observed in all diffuse maps reported to date with sufficient accuracy and precision to resolve them (De Klijn, Schreurs, & Kroon-Batenburg, 2019; Glover, Harris, Helliwell, & Moss, 1991; Moore, 2009; Peck, Poitevin, & Lane, 2018; Wall, Clarage, & Phillips, 1997), suggesting that halos are a universal feature of protein crystal diffraction. In lysozyme, the intensities and shapes of the halo features could be accounted for by modeling the vibrations of the crystal lattice. By measuring and accounting for the halos, the weaker residual scattering could be analyzed to fit models of internal motion (Meisburger et al., 2020).

The presence of halos in protein diffuse scattering has important implications for data processing. First, to resolve the halos accurately, the three-dimensional maps must be finely sampled. Lattice disorder models applicable to protein crystals predict halo shapes characterized by a sharp decay away from the Bragg peak on a scale much finer than the reciprocal lattice spacing. Second, the cloudy signal of interest is weak and partly obscured by the more intense halos, and therefore experimental artifacts that produce modulations in background scattering must be minimized experimentally or otherwise corrected in software. In this chapter, we first give a brief history of continuous diffraction measurements from protein crystals and review the diffraction theory relevant to data processing. We describe how data processing challenges are addressed computationally with a particular emphasis on the approach taken in *mdx-lib* and *mdx2*. Finally, we present an introductory tutorial where *mdx2* is used to reconstruct the three-dimensional map from a cubic insulin dataset collected at ambient temperature.

1.1 Brief history of macromolecular diffuse scattering

Macromolecular diffuse scattering has a long history (Meisburger et al., 2017). The modern approach of quantitative reciprocal space mapping (i.e., the process of not only mapping the detector coordinates to reciprocal space but also correcting intensities to accurately represent the crystal's elastic scattering) is rooted in a series of visionary studies by Donald Caspar and colleagues, beginning with their 1988 letter to Nature reporting the diffuse scattering from insulin (Caspar, Clarage, Salunke, & Clarage, 1988). The diffraction pattern from a cubic insulin crystal was recorded on X-ray film in a 40-h exposure (Bragg peaks were overexposed), digitized, and processed to correct for background, polarization, and the Compton and water scattering contributions (Fig. 1A). The processing revealed a diffuse pattern with two components: halos centered around the Bragg peaks, and cloudy "variational" scattering background. The halos were interpreted as displacement disorder of the lattice with correlations that decay exponentially with distance between unit cells. The variational scattering was explained using a similar function, but with correlated displacements

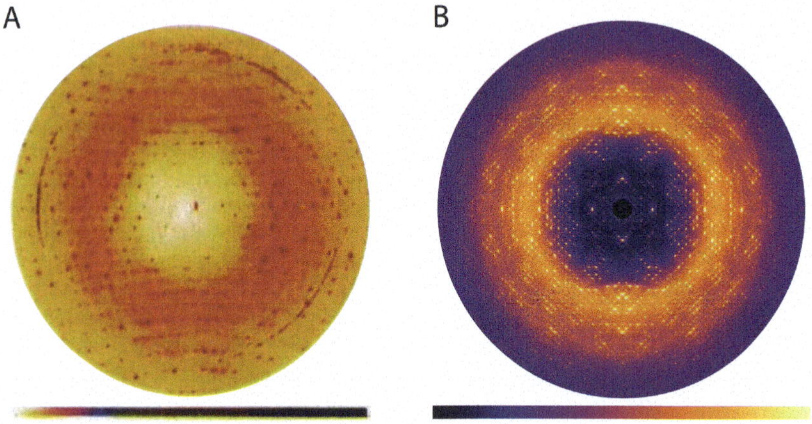

Fig. 1 Reciprocal space maps of cubic insulin. (A) A diffraction image recorded on X-ray film, digitized, and processed to remove background and polarization effects. A cloudy pattern and halos surrounding the Bragg peaks are superimposed on the largely isotropic diffuse scattering. Ring features are background scattering from the aluminum windows. (B) Central section (hk0) through the diffuse scattering map reconstructed using *mdx2* from fine-sliced rotation data collected using a photon-counting detector (see tutorial in Section 3). (A) Reproduced with permission from Caspar et al. (1988).

between atoms in the protein. Importantly, Caspar and colleagues demonstrated both that accurate measurements of diffuse scattering are feasible, and the technique's potential to reveal protein dynamics.

Caspar and his contemporaries analyzed two-dimensional diffraction patterns. Mathematically, such patterns sample a curved surface (the Ewald sphere) through a three-dimensional reciprocal space function aligned with the crystal. The next methodological breakthrough was to reconstruct the three-dimensional function from many diffraction patterns collected with the crystal at different orientations. The first such method was developed by Michael Wall in the lab of Sol Gruner, and it used a modified charge-coupled device (CCD) detector to record the diffuse signal without interference from over-exposed Bragg peaks. Computer memory limitations at the time made it infeasible to reconstruct the complete map at a sampling rate sufficient to resolve the halos (Wall, Clarage, et al., 1997), however coarse maps were generated for crystals of staphylococcal nuclease (Wall, Ealick, & Gruner, 1997), and later, the halos of a calmodulin crystal were resolved in a limited region (Wall, Clarage, et al., 1997). Three-dimensional mapping transformed the way experiment was compared with simulation and with data from other research groups; diffuse scattering became a quantitative and reproducible measurement (Wall, 1996; Wall, Ealick, et al., 1997), and statistics were developed to judge model quality analogous to R-factors in crystallography (Wall, Ealick, et al., 1997).

Modern approaches to reciprocal space mapping benefit from the ideal properties of hybrid photon-counting (HPC) detectors. Unlike CCDs and X-ray film from prior eras, the dynamic range of modern detectors is sufficient to record both Bragg peaks and diffuse scattering in the same image. Because of single-photon sensitivity, long exposures are not needed to overcome the intrinsic detector noise, and the diffuse signal can be accumulated over many observations. For instance, the reciprocal space maps of lysozyme mentioned in the previous section were reconstructed from fine-sliced datasets of 0.1 degrees per frame where most pixels recorded zero or one photon.

The introduction of intense X-ray free electron laser (XFEL) sources in the last decade has spurred interest in dynamic crystallography at non-cryogenic conditions (Shoemaker & Ando, 2018). Bright X-ray pulses from XFELs and fourth generation synchrotrons are sufficient to destroy the crystal in a single shot, requiring many crystals to obtain a complete dataset. At XFELs in particular, single-shot measurements of very small crystals are possible because the pulse duration is faster than the radiation damage process. XFELs have also proven useful for measuring very weak, non-Bragg scattering signals. For instance, interference fringes representing the crystal's shape transform have been visualized between the Bragg peaks of Photosystem I nanocrystals (Chapman et al., 2011). From larger crystals of Photosystem II, diffuse scattering beyond the limits of the Bragg diffraction resolution were observed in XFEL diffraction patterns and the diffuse scattering was used to improve the resolution of the electron density through iterative phasing techniques (Ayyer et al., 2016; Morgan et al., 2019).

1.2 Data processing software

Software is a key component of the reciprocal space mapping technique. A variety of software tools have been developed (Table 1). Although each is specialized for particular applications or mode of data collection, a common workflow emerges. First, the Bragg peak positions are analyzed using an indexing program such as *XDS* (Kabsch, 2010) or *DIALS* (Winter et al., 2022) to determine the orientation of the crystal during each X-ray exposure. Using this information, the pixels in each image are assigned reciprocal space coordinates. The data are further corrected for geometric

Table 1 Software for reciprocal space reconstruction.

Package	Location	Latest version	Contact	License
Lunus	github.com/lanl/lunus	0.2a	mewall@lanl.gov	open source (BSD-3)
EVAL15	www.crystal.chem.uu.nl/distr/eval	Oct. 26, 2021	l.m.j.kroon-batenburg@uu.nl	free for academic use
DIALS (dials.rs_mapper)	github.com/dials/dials	3.13.0	dials-support@lists.sourceforge.net	open source (BSD)
meerkat	github.com/aglie/meerkat	0.3.7	arkadiy.simonov@mat.ethz.ch	open source (MIT)
diffuse	github.com/apeck12/diffuse	Nov. 14, 2017	apeck@stanford.edu	open source
dspack	github.com/zhenwork/dspack	Nov. 20, 2022	zhensu@stanford.edu	open source
mdx-lib	github.com/ando-lab/mdx-lib	1.2.0	spm82@cornell.edu	open source (CC BY-NC-SA 4.0), requires MATLAB (commercial)
mdx2	github.com/ando-lab/mdx2	0.3.1	spm82@cornell.edu	open source (GPL-3.0)

effects such as solid angle and polarization. Finally, the corrected data are merged on a regular, three-dimensional grid, yielding a reciprocal space map.

Despite following similar workflows, current software tools differ in the types of experiments they can analyze, and the details of how experimental artifacts are treated. Thus, when choosing software it is important to consider the application. When reciprocal space mapping is used to diagnose or otherwise visualize pathological crystal packing, accuracy is less important than ease of use. One particularly convenient program is *dials.rs_mapper*, which generates 3D maps from indexed diffraction data (Winter et al., 2022). The output format for the 3D map is compatible with software developed for viewing electron density. The *EVAL* suite of programs (De Klijn et al., 2019; Schreurs, Xian, & Kroon-Batenburg, 2010) should be used when greater detail is needed, for instance to visualize sharp streaks or satellite reflections from modulated structures (Neviani, Lutz, Oosterheert, Gros, & Kroon-Batenburg, 2022). *EVAL* implements pixel splitting to produce densely sampled maps, and it includes an array of additional tools for analyzing non-standard diffraction data. Several software packages deal specifically with macromolecular diffuse scattering, and each has unique features. *Lunus* (Van Benschoten et al., 2016; Wall, 2009) has been optimized for high-performance computing architectures. *Diffuse* implements a novel scaling and background subtraction algorithm based on principal component analysis (Peck et al., 2018). *Dspack* has a streamlined command-line interface (Su et al., 2021). *Mdx-lib* is particularly suited to high-redundancy data, and it includes methods for background subtraction, scaling model refinement, and error propagation (Meisburger et al., 2020). Finally, diffuse scattering software designed for non-biological work is also available. In particular, *meerkat* (Simonov et al., 2020) has many similarities with the biological software tools described above.

Accessibility is also a consideration when choosing software. Most diffuse scattering projects are open source with permissive licenses (see Table 1). However, the level of effort required to use the software effectively on a new project is typically high. The user base for each diffuse scattering program is currently small, in most cases limited to the research group that developed it. Thus, there has not been a need to build general-purpose tools, streamlined interfaces, tutorials, or thorough documentation. For example, analyzing a new dataset using *mdx-lib* requires a high level of expertise in MATLAB to understand and modify example scripts. It is also important to note that many packages are in active development, and collaboration with the authors of the code may be required to implement necessary features, such as detector image formats.

As biological diffuse scattering transitions from a proof-of-concept experiment to a widely available technique, software tools are needed that are rigorous but user-friendly. Since experimental and algorithmic techniques are likely to evolve, the software should also be easy to extend and modify. Moreover, the software should be developed collaboratively and maintained by the community. The ideal software does not yet exist. Motivated by this need, we have developed a successor to *mdx-lib* called *mdx2*, which we describe for the first time in this chapter (Fig. 2). *Mdx2* re-implements the data processing workflow of *mdx-lib* in Python, but it does so without reinventing crystallographic computing (as *mdx-lib* did). Underneath, *mdx2* is lightweight, relying on *dxtbx* (Parkhurst et al., 2014) for data import and *cctbx* (Grosse-Kunstleve, Sauter, Moriarty, & Adams, 2002) for crystallographic computing. Since *DIALS* uses the same packages (Winter et al., 2022), *mdx2* and *DIALS* functions are interoperable. *Mdx2* includes a command-line interface (also inspired by *DIALS*), so familiarity with Python is not required. For more flexible workflows, *mdx2* can be imported as a Python package and run for instance in a Jupyter notebook. Finally, we have chosen to store all intermediate data files in *NeXus* format (Könnecke et al., 2015) to make them interoperable with other scientific software, especially the data visualization program *NeXpy* (The NeXpy Development Team, 2023).

The transition from *mdx-lib* to *mdx2* is not yet complete, and key features have not been implemented at the time of writing. Unlike *mdx-lib*, *mdx2* does not yet include a full scaling model, and it does not take advantage of multi-CPU architectures. However, *mdx2* is sufficiently capable to make it an excellent tool for data exploration and education, and it is used for the tutorial in Section 3.

2. Theory

The goal of reciprocal space mapping is to reconstruct the three-dimensional elastic X-ray scattering signal given many two-dimensional diffraction images of a crystal in different orientations (for example during a ϕ scan). From a geometric point of view, each pixel in each image must be related to a point in the three-dimensional reciprocal space of the crystal. However, geometry is not what makes reciprocal space mapping difficult. Instead, the difficulty lies in handling the non-ideal aspects of the data. Compared with the intense Bragg peaks, diffuse scattering is a weak signal

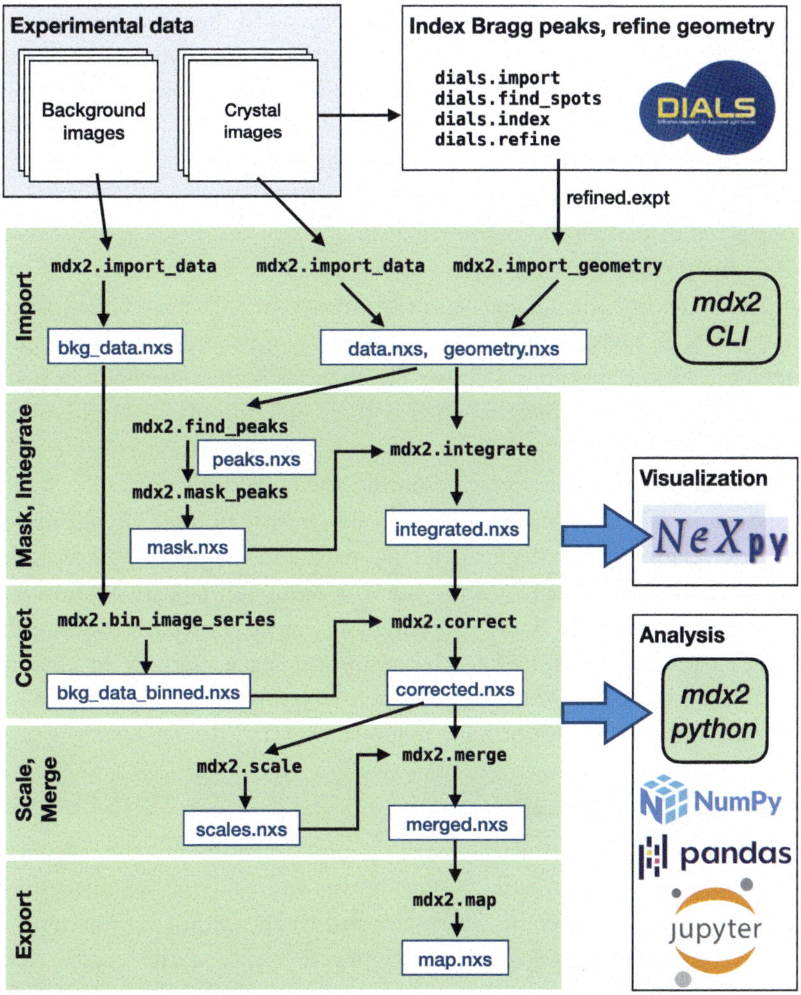

Fig. 2 Workflow for reciprocal space reconstruction using *mdx2*. Rotation data from crystals and corresponding background images are collected using procedures described in Chapter 1 (Pei et al., 2023). Bragg peaks in the crystal images are processed using *DIALS* in order to refine the diffraction geometry. Using the *mdx2* command-line interface (CLI), the *DIALS* geometry and diffraction data are imported in NeXus-formatted hdf5 files (.nxs extension) for convenient visualization using *NeXpy*. Subsequent *mdx2* CLI programs (green shading) integrate, correct, scale, and merge the data to produce a three-dimensional reciprocal space map (see tutorial in Section 3). At each stage of processing, analysis is performed in Python using *mdx2* together with standard scientific packages (see Section 3.7).

because it is spread throughout reciprocal space, and thus in any individual image, the signal may not be visible above the noise. Diffraction images also contain "background," either from materials other than the crystal, or from inelastic scattering of atoms within the crystal itself (Compton scattering). Finally, the illuminated crystal volume (and also the background scattering) potentially varies throughout the data collection, and this must be corrected as part of the reconstruction.

Recently, HPC detectors have enabled a data collection strategy of low dose per frame, fine slicing, and high redundancy [see Chapter 1 (Pei, Bhatt, Wang, Ando, & Meisburger, 2023)]. This strategy is particularly advantageous for reciprocal space mapping. Because the angle increment per frame is small, reciprocal space can be finely sampled without splitting pixels. Furthermore, since many redundant observations are made, the systematic errors can be averaged out or otherwise corrected during scaling and merging.

In Section 2.1, we first review the Ewald construction for visualizing the geometric relationship between detector space and reciprocal space, as this intuition is essential for appreciating the data reduction task. In Section 2.2, we discuss the physics of X-ray scattering and detection needed to relate the number of photons measured to the reciprocal space intensity in electron units. Finally, in Section 2.3, we describe the approach taken by *mdx-lib* to handle the unique features of high-redundancy HPC datasets.

2.1 Scattering geometry

In a typical macromolecular crystallography experiment, the crystal is placed in a monochromatic X-ray beam and diffraction images are recorded on a planar detector. First, we consider the geometric relationship between a detector pixel and the reciprocal space of the crystal. The incident X-rays are described by a *wavevector* $\mathbf{s}_0$ in the direction of the beam. Similarly, a scattered wavevector $\mathbf{s}_1$ points from the crystal toward a detector pixel. For the elastic scattering process considered here, the wavevectors have the same length $|\mathbf{s}_0| = |\mathbf{s}_1| = \lambda^{-1}$, where λ is the X-ray wavelength. The *scattering vector* $\mathbf{s}$ is defined as the difference between the incident and scattered wavevectors as follows:

$$\mathbf{s} \equiv \mathbf{s}_1 - \mathbf{s}_0. \tag{1}$$

Importantly, the intensity of interest is a function of the scattering vector, $I = I(\mathbf{s})$. Thus, Eq. (1) relates the geometry of the experiment ($\mathbf{s}_0$ and $\mathbf{s}_1$) to the reciprocal space of the sample ($\mathbf{s}$). This relationship can be visualized intuitively using the Ewald construction (Fig. 3A, B).

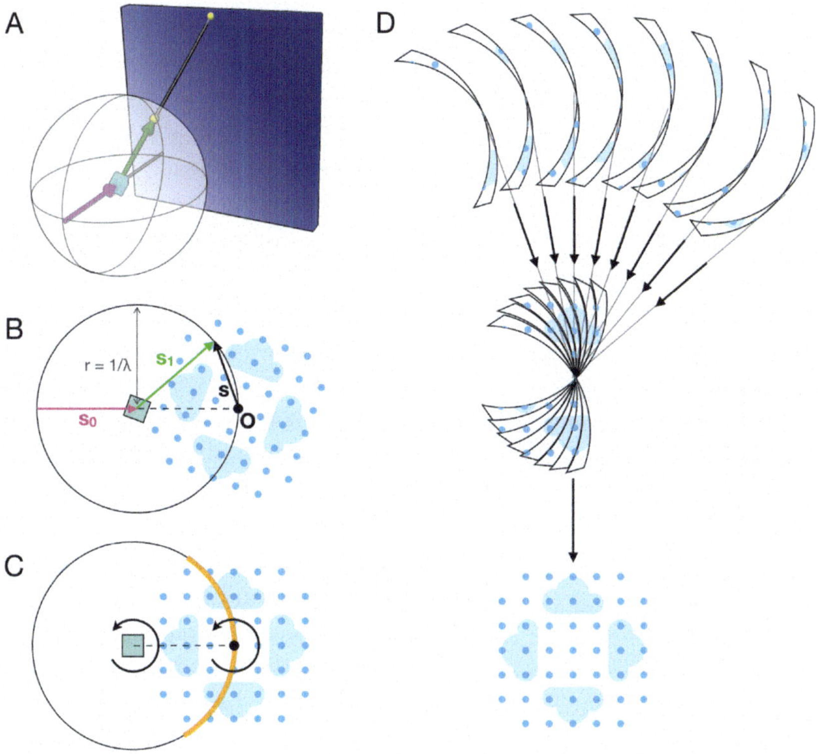

Fig. 3 Geometry of reciprocal space reconstruction. (A) An X-ray beam (magenta arrow) is incident on a crystal (cyan) and scattering is recorded on a two-dimensional detector (blue rectangle). Each detector pixel is projected onto the surface of the Ewald sphere, drawn centered on the crystal. The projection (yellow points) is illustrated for a particular scattering direction (green arrow). (B) The same geometry as (A) viewed in the plane formed by the incident and scattered wavevectors ($\mathbf{s}_0$ and $\mathbf{s}_1$). In the Ewald construction, the origin of reciprocal space (**O**) is placed as shown so that the scattering vector **s** lies on the surface of the sphere. The blue pattern represents a slice through the reciprocal space intensity $I(\mathbf{s})$ with lightly shaded clouds representing diffuse scattering and darker circles representing Bragg peaks at the reciprocal lattice points. (C) The pattern of scattering depends on the intersection of $I(\mathbf{s})$ with the Ewald sphere (orange region), which rotates as the crystal rotates. (D) Scattering data recorded at different orientations (slices are drawn with finite thickness to illustrate continuous rotation during exposure) are merged in three dimensions to recover the reciprocal space map (first set of arrows), and known symmetries may be used to further complete the reconstruction (bottom arrow).

If the crystal is rotated (for instance during a ϕ scan), the reciprocal space coordinate system rotates along with it in the same direction (Fig. 3C, arrows). The scattering vector for a particular pixel remains stationary in the laboratory frame, and therefore it samples a different part of reciprocal space. Taken together, the detector pixels map onto a curved surface in reciprocal space, the Ewald sphere (Fig. 3C, orange).

Conventionally, the reciprocal space coordinate axes are fixed to the reciprocal unit cell vectors ($\mathbf{a}^*$, $\mathbf{b}^*$, and $\mathbf{c}^*$), such that Bragg peaks have integer coordinates (Miller indices h, k, and l). Note that for certain space groups the reciprocal lattice vectors are not orthogonal. The change of basis between this *fractional* reciprocal space and the Cartesian, laboratory frame can be written compactly in matrix notation as follows (Busing & Levy, 1967; Waterman et al., 2016):

$$\mathbf{s} = \mathbf{RUBh}, \tag{2}$$

where $\mathbf{h} = (h,k,l)$ is a vector of Miller indices, $\mathbf{B}$ is a matrix whose rows are the reciprocal unit cell vectors, and $\mathbf{U}$ and $\mathbf{R}$ are pure rotation matrices. $\mathbf{U}$ describes the orientation of the crystal at a particular reference point in the scan (for instance at $\phi = 0$), and $\mathbf{R}$ describes the scan-varying rotation (for instance, the rotation about a spindle axis by the amount ϕ). The inverse of Eq. (2) relates the laboratory coordinates (i.e. a pixel location via Eq. (1)) to the fractional reciprocal space:

$$\mathbf{h} = \mathbf{B}^{-1}\mathbf{U}^{-1}\mathbf{R}^{-1}\mathbf{s}. \tag{3}$$

Thus, from a series of images collected with the crystal in different orientations, it is possible to reconstruct the three-dimensional reciprocal space (Fig. 3D).

Scattering from a crystal obeys certain symmetries according to its space group (Rupp, 2009), and thus the full reciprocal space can often be reconstructed without sampling all possible orientations. Neglecting resonant (anomalous) scattering for simplicity, the symmetry of reciprocal space is described by the crystal's Laue group, which is the point group of the crystal plus a center of inversion (the point group contains only the rotation operators of the space group; the center of inversion comes about because Fourier transforms of a real function are centro-symmetric). For instance, the space group $P\,2_1$ has the general positions (x, y, z) and $(-x, y + 1/2, -z)$, with a rotational part corresponding to the point group 2. The corresponding Laue group is $2/m$, with general positions (h, k, l), $(-h, -k, -l)$, $(h, -k, l)$, and $(-h, k, -l)$. Thus, only one fourth of reciprocal space is unique (the rest can be generated from the reciprocal space asymmetric unit by applying the Laue group operators). Furthermore, if the crystal is rotated by 360 degrees, each point in reciprocal space is sampled twice (e.g., once by each half of the detector). Neglecting the curvature of the Ewald sphere for simplicity, a 360 degree dataset from a $P\ 2_1$ crystal would be 8-fold redundant (equivalent points are measured 8 times during the scan). Such redundancy can be used to correct for systematic errors because generally the equivalent observations are made by different detector regions and at different crystal orientations (see Section 2.3).

2.2 Components of diffraction

As discussed above, a detector pixel records scattered X-rays corresponding to a particular location in the crystal's reciprocal space. The number of X-ray photons recorded depends on the three-dimensional intensity function of interest, but also on the physics of the scattering process, response of the detector, and background scattering from other sources.

Consider an X-ray image taken as the crystal rotates continuously by a small amount $\Delta\phi$ about an axis $\hat{\mathbf{m}}$ for time Δt during which counts are accumulated by a detector pixel with solid angle $\Delta\Omega$. When projected onto the Ewald sphere, the pixel sweeps out a small reciprocal space volume Δv^* during the rotation:

$$|\Delta v^*| = \Delta\phi\Delta\Omega\lambda^{-3}|\hat{\mathbf{m}}\cdot(\hat{\mathbf{s}}_0 \times \hat{\mathbf{s}}_1)|, \tag{4}$$

where $\hat{\mathbf{s}}_0$ and $\hat{\mathbf{s}}_1$ are unit vectors in the direction of the incident and scattered wavevectors respectively. The intensity recorded by the pixel can be written as an average over the swept volume, defined as follows:

$$\langle I \rangle_{\Delta v^*} \equiv \frac{1}{|\Delta v^*|}\int_{\Delta v^*} d^3\mathbf{s} I(\mathbf{s}). \tag{5}$$

If the intensity varies smoothly (i.e., constant over the exposure) the integral simplifies and the recorded intensity is effectively a sample of $I(\mathbf{s})$ at the mean value of $\mathbf{s}$, which we call $\overline{\mathbf{s}}$:

$$\langle I \rangle_{\Delta v^*} = I(\overline{\mathbf{s}}). \tag{6}$$

In contrast, consider what happens when a Bragg peak is integrated. The total scattering (Bragg plus diffuse) can be modeled as follows (Meisburger et al., 2017):

$$I(\mathbf{s}) = I_{\text{diffuse}}(\mathbf{s}) + I_{\text{Bragg}}(\mathbf{s})\sum_{\mathbf{h}} v_c^*\delta(\mathbf{s} - \mathbf{g}_{\mathbf{h}}), \tag{7}$$

where $\mathbf{g_h}$ are nodes of the reciprocal lattice and v_c^* is the reciprocal unit cell volume. Assuming the diffuse scattering varies smoothly and a Bragg peak with Miller index **h** is fully recorded, the average intensity is as follows:

$$\langle I \rangle_{\Delta v^*} = I_{\text{diffuse}}(\overline{\mathbf{s}}) + \frac{v_c^*}{|\Delta v^*|} I_{\text{Bragg}}(\mathbf{g_h}). \tag{8}$$

Note that the Bragg intensity is modified by a geometric prefactor (proportional to the Lorentz correction factor). Although Lorentz corrections are applied during Bragg peak integration, they should not be used when correcting diffuse scattering. The Lorentz factor is needed only when features of interest are smaller than the volume of reciprocal space integrated by the detector pixels. In general, it is important to remember that while Bragg and diffuse data processing are similar, there are subtle differences arising from the distinct signal properties. For the purposes of this chapter, we assume Bragg peak integration has already been performed (e.g. using conventional software like *XDS* or *DIALS*), and focus instead on processing the continuous scattering. Note that for Eq. (6) to be valid, care must be taken to integrate the reciprocal space map on a fine grid and to exclude pixels containing Bragg peaks from the analysis.

The number of photons recorded by the detector pixel can be calculated based on the geometric considerations, properties of the detector, and the physics of X-ray diffraction. Previous reviews may be consulted for a complete derivation (Meisburger et al., 2017; Meisburger et al., 2020); here we give the result. Let the beam illuminate a volume V_0 of the crystal with an incident flux (photons per second) J_0. The number of photons recorded n is as follows:

$$n = n_{\text{background}} + \Delta t \Delta\Omega EAPJ_0 r_e^2 (V_0/v_c)(I_{\text{diffuse}} + I_{\text{inelastic}}), \tag{9}$$

where $n_{\text{background}}$ are the photon counts due to background, E is the efficiency of the detector pixel, A is the attenuation factor due to materials between the crystal and the detector, P is the polarization factor, r_e is the classical electron radius, v_c is the unit cell volume, and I_{diffuse} and $I_{\text{inelastic}}$ are intensities per unit cell. Thus, to measure I_{diffuse}, all of the unknown quantities in Eq. (9) must be measured or calculated theoretically.

The factors $\Delta\Omega$, E, A, and P can be calculated from the diffraction geometry and known material properties. The solid angle of a pixel is:

$$\Delta\Omega = (a/d^2)\cos\omega, \tag{10}$$

where a is the pixel's area, d is its distance from the crystal, and ω is the angle between the detector surface normal and the incident X-ray. The pixel's absorption efficiency depends on the material and thickness δ as follows:

$$E = 1 - \exp(-\kappa\delta/\cos\omega), \tag{11}$$

where κ is the absorption coefficient (e.g. of Silicon at the X-ray wavelength used). The material between the sample and detector is typically air, whose transmission can be calculated knowing the absorption coefficient μ as follows:

$$A = \exp(-\mu d). \tag{12}$$

Finally, the polarization effect modifies the scattering efficiency as follows:

$$P = p(1 + (\hat{\mathbf{s}}_1 \cdot \hat{\mathbf{s}}_0)^2) + (1 - 2p)(1 - (\hat{\mathbf{s}}_1 \cdot \hat{\mathbf{n}})^2), \tag{13}$$

where $\hat{\mathbf{n}}$ is a unit vector normal to the plane of polarization. Synchrotron X-ray sources are typically polarized in the plane of the ring with a polarization factor $p \sim 1$.

With geometric factors calculated from first principles, the remaining unknown quantities in Eq. (9) are the incident flux, the illuminated volume, the background scattering, and the inelastic scattering from the crystal. These quantities are estimated in later stages of data processing, for instance by utilizing the redundant observations to derive relative scale factors for different diffraction images (see Section 2.3). In the mean time, an intermediate quantity I_{measured} is computed from the corrected photon counts as follows:

$$I_{\text{measured}} = \frac{n/\Delta t - r_{\text{background}}}{\Delta\Omega EAP}, \tag{14}$$

where $r_{\text{background}}$ is the background scattering rate, which has been measured separately, e.g. by translating the crystal out of the beam [see Chapter 1 (Pei et al., 2023)].

2.3 Estimating and correcting errors

Error estimation is an essential part of data processing, and it begins during integration. After masking out the Bragg peaks, pixels from adjacent images are grouped together according to which voxels they contribute to in the (unsymmetrized) map, and their counts are summed. The uncertainty due to Poisson statistics for an observation with n photon counts is $\sqrt{n}$. This uncertainty is propagated through all subsequent correction steps (Eq. (14)) so that each measured intensity value has a corresponding error estimate. The errors are used as weighting factors for scaling, as described below.

Typically, each voxel (or its symmetry equivalent) is observed multiple times in the dataset at different points in the scan. This redundancy can be used to estimate unknown scale factors in Eq. (9). The scaling algorithm minimizes the least-squares difference between $I_{\text{predicted}}$ and I_{measured} weighted by the uncertainty estimates (σ):

$$\chi^2 = \sum_i \frac{|I_{\text{measured}}(i) - I_{\text{predicted}}(i)|^2}{\sigma_i^2}, \tag{15}$$

where the sum runs over all observations i. $I_{\text{predicted}}$ comes from the scaling model: it is a function of the scaling parameters and of the merged intensities I_{merged}. Initially, both the scale factors and the merged intensities are unknown, and both must be refined simultaneously. In general, the optimization problem is nonlinear but it can be solved by an alternating least squares (ALS) algorithm (Hamilton, Rollett, & Sparks, 1965), which alternates between two steps: merging the data with scale factors fixed, and determining scale factors given the merged data.

Let the scaled (but not merged) intensities and uncertainties be I_{scaled} and σ_{scaled}. Assuming a linear relationship between I_{measured} and I_{scaled} (which is a general feature of scaling models for X-ray diffraction), Eq. (15) can be rewritten as follows:

$$\chi^2 = \sum_i \frac{|I_{\text{merged}}(\mathbf{h}_i) - I_{\text{scaled}}(i)|^2}{\sigma_{\text{scaled}}(i)^2}, \tag{16}$$

where $\mathbf{h}_i$ refers to the index of the observation mapped to the asymmetric unit of reciprocal space. Let the index j refer to all observations mapping to a particular voxel $\mathbf{h}$. The value of I_{merged} that minimizes Eq. (16) is as follows:

$$I_{\text{merged}}(\mathbf{h}) = \frac{\sum_j w_j I_{\text{scaled}}(j)}{\sum_j w_j}, \tag{17}$$

where $w = \sigma_{\text{scaled}}^{-2}$. The corresponding uncertainty in the merged intensity is estimated as follows:

$$\Delta I_{\text{merged}}(\mathbf{h}) = \left(\sum_j w_j\right)^{-1/2}. \tag{18}$$

In *mdx-lib*, the following scaling model is refined:

$$I_{\text{predicted}}(i) = a(x_i, y_i, \phi_i)\, d(p_i)\, [b(\phi_i)\, I_{\text{merged}}(\mathbf{h}_i) + c(s_i, \phi_i)]. \tag{19}$$

The model relates the merged intensities for a certain voxel in the scaled and merged map $I_{\text{merged}}(\mathbf{h}_i)$ to a predicted value in the unscaled, unmerged dataset $I_{\text{predicted}}(i)$. Each observation has three scale factors (a, d, and b) and one offset term c. The scale factor a accounts for self-absorption of the crystal, and its value depends on the position of the observation on the detector (x_i, y_i) and the rotation angle of the crystal ϕ_i. The scale factor d corrects for detector flat-field errors, and its value depends on which detector chip recorded the counts (index p_i). The scale factor b depends only on the rotation angle, and it corrects for changes in illuminated volume and incident beam intensity during the scan.

Programs for scaling Bragg data, such as Aimless (Evans & Murshudov, 2013) and XDS (Kabsch, 2010), include scale factors similar to those in Eq. (19). However, the offset term c is unique to diffuse scattering. Its purpose is to compensate for extra background scattering that is not included in the background measurement, for instance from liquid on the crystal surface or the loop materials. The scaling model assumes that the added background is isotropic (varying as a function of resolution, s_i). When the scaling model is refined, the value of c is restrained so that it is always positive (or zero).

The scaling algorithm solves for the unknowns in Eq. (19) by minimizing χ^2 (Eq. (15)), while also enforcing smoothness of the scale factors to avoid overfitting (Meisburger et al., 2020). The terms in the scaling model are physically-motivated, and thus after refinement the scale factors should be inspected to verify that they make physical sense (Fig. 4).

By refining the scaling model, the intensities of redundant observations are brought into agreement, however the absolute scale factor has not been determined (i.e., the product of photon flux and illuminated volume in Eq. (9)). Depending on the application, it may be necessary to place the map on an absolute scale; for instance, it is required to subtract a simulated inelastic term in Eq. (9). We previously developed a method based on comparing the simulated and experimental total scattering (Meisburger et al., 2020, 2023) which has been implemented in *mdx-lib* (for examples, see https://github.com/ando-lab/mdx-examples).

3. Data processing tutorial

To provide a concrete example of the steps involved in data processing outlined above, we present an introductory tutorial using the *mdx2* software.

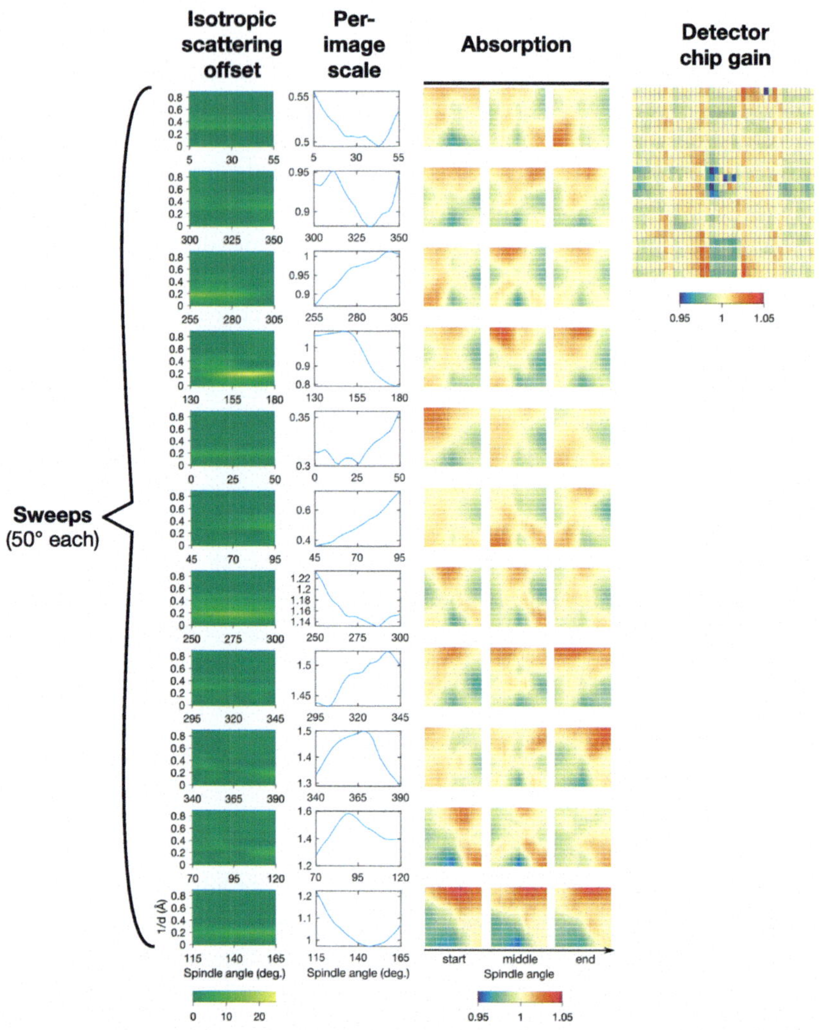

Fig. 4 Scaling model for a triclinic lysozyme dataset from refinement using *mdx-lib*. Panels are arranged with rows corresponding to the eleven sweeps of data distributed over four crystals (see original publication for details) and columns corresponding to the four terms in the scaling model in Eq. (19) (left to right): *c*, *b*, *a*, and *d*. Reproduced under CC-BY-4.0 license from Meisburger et al. (2020).

The tutorial can be completed by following instructions and typing commands in this chapter, or by downloading and stepping through the accompanying Jupyter notebooks (see Appendix A.1). A relatively small dataset was chosen for this tutorial so that it can be completed on a personal computer (specific requirements are given in Appendix A.1). A room-temperature X-ray diffraction dataset from bovine insulin in space group $I2_13$ (Fig. 5) was collected at the Cornell High Energy Synchrotron Source (CHESS) beamline F1 following protocols described previously (Meisburger et al., 2020, 2023) and summarized in Chapter 1 (Pei et al., 2023). The insulin dataset consists of a ϕ scan with 500 diffraction images (0.1 degrees each, 50 degrees total) and 50 background measurements (1 degree each over the same angular range). The dataset is available online (see Appendix A.2).

The first step is to install the software. The recommended method is to create a stand-alone Python environment using *miniconda3* (Appendix A.3) and then install *mdx2* from the GitHub repository using *pip* (Appendix A.4). In addition to *mdx2*, this data processing tutorial uses *DIALS* for pre-processing, *NeXpy* for visualization, and *Jupyter* for interactive notebooks. These programs can be installed and run separately, but for convenience we recommend installing them in the same Python environment as *mdx2* (Appendix A.5).

3.1 Indexing and geometry refinement

The first step in data processing is to determine the diffraction geometry and crystal orientation by analyzing the locations of strong Bragg peaks. These steps are performed using *DIALS*, a suite of programs for integrating and scaling Bragg data (Winter et al., 2022). For our purposes, we will only use indexing and geometry refinement tools. Normally these steps would be followed by integration and scaling, but we will skip them to save time. Detailed tutorials are available on the *DIALS* website (https://dials.github.io).

3.1.1 Metadata import

The image file headers contain metadata specifying the experimental geometry such as X-ray wavelength, detector distance, oscillation increment per image. The command `dials.import` reads the image headers and outputs a json-formatted text file `imported.expt` containing the relevant metadata. In the terminal, change to the working directory (containing the `images` folder) and run the following command to import the insulin dataset:

```
dials.import images/insulin_2_1
```

3.1.2 Beamstop mask

The beamstop mask must also be generated and added to `imported.expt`. A beamstop mask is not essential for Bragg data processing, but it is important to create one for later import into *mdx2*. The mask can be drawn using the

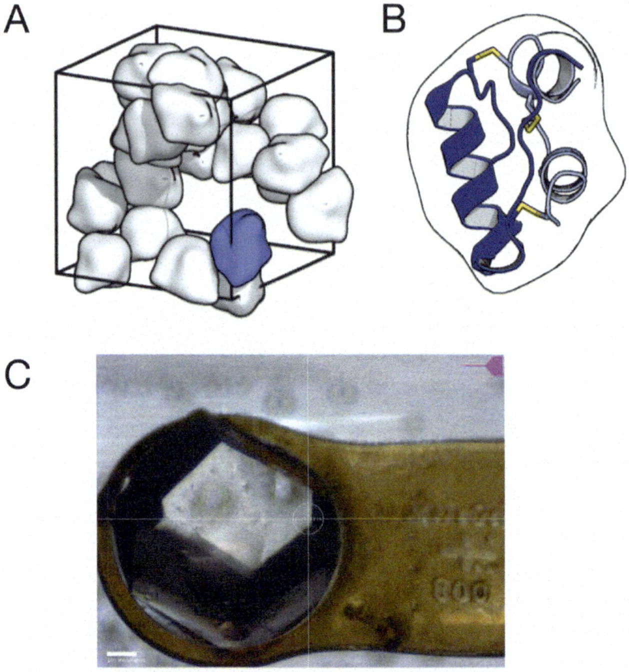

Fig. 5 Insulin crystal that produced the dataset used in the tutorial. (A) Insulin from *Bos taurus* was crystallized using a Zn-free condition in the space group $I2_13$. The unit cell (wireframe box) contains 24 copies of the asymmetric unit (smooth molecular surfaces). The packing arrangement has large channels occupied by disordered solvent. (B) The asymmetric unit, drawn in the same orientation as the blue blob in (A), contains a single insulin molecule consisting of two alpha-helical chains (blue and gray cartoons) cross-linked by three disulfide bonds (yellow sticks). (C) Image taken during data collection of the ~800 μm crystal mounted on a Kapton loop and held within a plastic capillary to maintain hydration. The dataset processed in this tutorial consists of a single 50-degree sweep (0.1 degrees and 0.1 s per frame). Because of the large crystal size and high symmetry, a single sweep is sufficient for a complete reciprocal space map.

graphical image viewer that comes with *DIALS*:

```
dials.image_viewer imported.expt
```

Alternatively, the following will generate an appropriate circular mask for the insulin dataset

```
dials.generate_mask imported.expt \
   untrusted.circle=1264,1242,50
dials.apply_mask imported.expt input.mask=pixels.mask \
   output.experiments=imported.expt
```

Note that the beamstop shadow mask is relatively simple in this case because the beamstop was suspended on an X-ray transparent Kapton film. It is more common to mount the beamstop on a rigid support whose shadow must be more carefully masked.

3.1.3 Spotfinding

Next, `dials.find_spots` will read all of the images and locate Bragg peaks. This requires reading in all of the diffraction images, and it can take a while depending on computational resources.

```
dials.find_spots imported.expt
```

3.1.4 Indexing

The indexing step assigns Miller indices to each diffraction spot. When processing an unknown crystal, it is usually necessary to re-index the data after space group determination. Here, the space group is known (the insulin crystal is $I2_13$, space group number 199), and we can avoid having to re-index later by passing this information to `dials.index`.

```
dials.index imported.expt strong.refl space_group=199
```

For diffuse scattering, it's important to verify that all (or nearly all) peaks are indexed. There might be split Bragg reflections, multiple lattices, or salt crystals contaminating the signal. Or, the crystal might have slipped during data collection.

Check that indexing was successful by examining the *dials.index* output. Find the table with "% indexed."

```
+------------+------------+--------------+------------+
|   Imageset |  # indexed |  # unindexed | % indexed  |
|------------+------------+--------------+------------|
|          0 |      50694 |          417 | 99.2%      |
+------------+------------+--------------+------------+
```

The fraction of indexed peaks should be close to 100%. A small percentage may suggest incorrect experimental geometry, multiple lattices, twinning, salt or ice diffraction, or other issues. The indexing results should be inspected by running `dials.image_viewer indexed.expt`. In our case, 99.2% of strong peaks were indexed (there were 417 unindexed peaks and 50,694 indexed peaks).

Find the table called "RMSDs by experiment."

```
RMSDs by experiment:
+-------+--------+----------+----------+-----------+
|   Exp |   Nref |   RMSD_X |   RMSD_Y |    RMSD_Z |
|    id |        |     (px) |     (px) |  (images) |
|-------+--------+----------+----------+-----------|
|     0 |   4999 |  0.21024 |   0.2213 |   0.20059 |
+-------+--------+----------+----------+-----------+
```

This table includes the root mean square displacement (RMSD) between the predicted and observed spot centroids in units of pixels (in x, y directions) and images (in z direction). If the RMSDs are less than ~1, this indicates that the indexing solution was successful and a single set of geometric parameters can describe the entire dataset accurately. However, if the crystal slips slightly during data collection or the lattice constants change, the spot predictions can be improved by fitting a scan-varying geometric model.

3.1.5 Geometry refinement

Next we use `dials.refine` to fit a scan-varying model of the crystal geometry. This can be important if the lattice constants change due to radiation damage, or if the crystal is not held rigidly in the loop and slips during data collection.

```
dials.refine indexed.expt indexed.refl
```

To see if the spot predictions improved, find the table "RMSDs by experiment" in the text output:

```
RMSDs by experiment:
+-------+--------+----------+----------+-----------+
|   Exp |   Nref |   RMSD_X |   RMSD_Y |    RMSD_Z |
|    id |        |     (px) |     (px) |  (images) |
|-------+--------+----------+----------+-----------|
|     0 |  45511 |  0.17561 |  0.18017 |   0.19491 |
+-------+--------+----------+----------+-----------+
```

Compared with the corresponding table from `dials.index`, the RMSDs have improved as we expected. However the improvement is modest. In our case, the indexing solution already had small RMSDs and refinement was not strictly necessary.

3.1.6 Diagnostic statistics and plots

The `dials.report` function generates an html file with interactive plots. The report should be examined closely to catch any issues that would potentially complicate analysis of the diffuse scattering.

```
dials.report refined.expt refined.refl
```

Open the file `dials.report.html` in an internet browser.

Expand the tab called "Analysis of scan varying model" (Fig. 6A, B). These plots should be checked to see whether the unit cell axes change significantly during data collection. Abrupt changes in the orientation parameters may indicate that the crystal slipped during data collection.

Expand the tab called "Analysis of reflection centroids." Ideally, the error in x and y positions should appear random across the detector face. In our dataset, the errors cluster in rectangular blocks corresponding to the detector modules. If sub-pixel accuracy is desired, the module alignments can be calibrated. Next, examine the plots of spot centroid errors (Fig. 6C, D). The spread of centroids around the mean should show a compact peak. If there are multiple peaks observed, then the crystal may be twinned or multiple lattices may be present (e.g. due to cracking).

3.1.7 Background image metadata

An additional dataset was acquired for background subtraction in *mdx2*. We'll use `dials.import` and `dials.apply_mask` to create a metadata file `background.expt` that will be used in Section 3.4.

```
dials.import images/insulin_2_bkg \
   output.experiments=background.expt
dials.apply_mask background.expt input.mask=pixels.mask \
   output.experiments=background.expt
```

3.2 Data exploration

In this section, we prepare data files for reciprocal space mapping with *mdx2* and demonstrate interactive tools for data exploration.

3.2.1 Importing diffraction data

First, we introduce the *mdx2* command line programs for importing diffraction images and geometric corrections from *DIALS* in Section 3.1. Most features of *mdx2* needed for the tutorial are available in command-

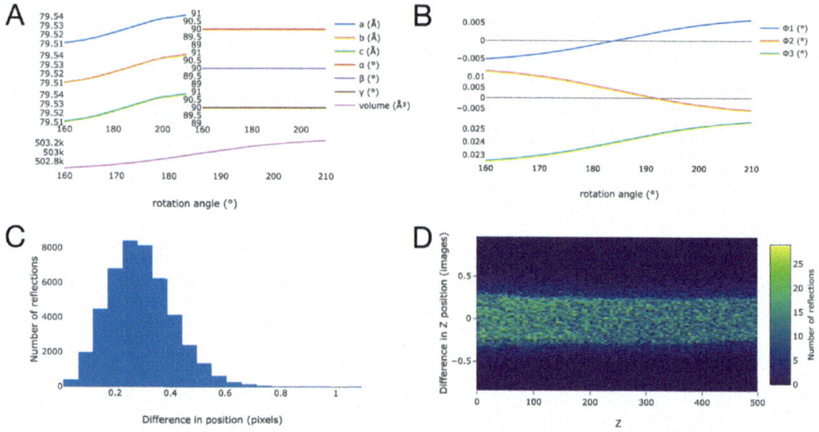

Fig. 6 Plots from the *DIALS* report that illustrate a successful geometry refinement. The scan-varying model allows for the unit cell constants (A) and setting angles (B) to vary smoothly throughout the scan. Refinement minimizes the difference between observed and predicted coordinates of the Bragg peaks on the detector surface (C) and image number *Z* in the scan (D). The parameter changes are minimal during the scan, and the errors are much less than one pixel or image. Furthermore, there are no sudden jumps or outliers in panel (D), which might indicate that the crystal had slipped or cracked during data collection.

line programs. Run the following program to verify that *mdx2* is installed and activated:

```
mdx2.version
```

The version number (0.3.1) should be printed.

The first step in the *mdx2* workflow (Fig. 2) is to import the diffraction images. Unlike *DIALS*, *mdx2* does not read diffraction data directly from image files. Instead, the data are imported (i.e. copied) into a three-dimensional array (image stack) and stored in a NeXus-formatted hdf5 file. There are two advantages to this: first, subsequent *mdx2* programs can be very simple because they don't need to handle diverse image formats, and second, the data can be compressed in a way that improves performance when it is subsequently read.

From within the directory containing the *DIALS* output files (`refined.expt`), run the following:

```
mdx2.import_data refined.expt --chunks 20 211 493
```

The program reads image files (the file path is stored in `refined.expt`) and saves the raw data and associated metadata in the file `data.nxs`. The `--chunks` parameter controls the size of the sub-arrays that will be compressed in the hdf5 file. Chunking allows for reading or writing small segments of the array without decompressing the entire dataset. Here, we chose a size of $20 \times 211 \times 493$. Roughly, 211×493 corresponds to a panel on the Pilatus 6M, and 20 specifies a stack of 20 frames (2 degrees of rotation). The particular chunk shape is not critical, but it can affect performance of subsequent operations.

Note that operations involving raw image data can be time consuming. Data import may take several minutes. For slow operations, *mdx2* produces print statements to track its progress. In this case, it prints the frame number as files are being read (0–499). Every 20 frames, the program writes chunks incrementally to the data file (`data.nxs` will increase in size until it reaches approximately 1.1 Gb).

The contents of any NeXus file can be viewed using the program `mdx2.tree`. For instance,

```
mdx2.tree data.nxs
```

prints the contents of the data file (Fig. 7A).

3.2.2 Importing diffraction geometry

The next step is to import the geometric information necessary to map each pixel to its location in reciprocal space. Run the following program:

```
mdx2.import_geometry refined.expt
```

The output file `geometry.nxs` contains crystal lattice information, symmetry operators, and pre-computed arrays of correction factors and Miller indices. To conserve disk space, the array values are not computed at every pixel, but instead on a coarser grid. By default, the corrections are sampled every 1 degree of rotation and every 10 pixels in the x and y directions. If needed, these parameters can be changed using the optional `--sample_spacing` argument. Help text for any `mdx2` program can be printed by adding `--help`. For instance, running `mdx2.import_geometry --help` will print the following:

A

```
data.nxs: root:NXroot
  @HDF5_Version = '1.12.2'
  @file_name = '/cdat/tem/spm82/erice-2022-data-reduction/data.nxs'
  @file_time = '2023-01-17T21:30:10.881376'
  @h5py_version = '3.7.0'
  @nexusformat_version = '0.7.8'
  entry:NXentry
    image_series:NXdata
      @axes = ['phi', 'iy', 'ix']
      @mdx2_class = 'ImageSeries'
      @mdx2_module = 'mdx2.data'
      @signal = 'data'
      data = int32(500x2527x2463)
      exposure_times = float64(500)
      ix = float64(2463)
      iy = float64(2527)
      phi = float64(500)
```

B

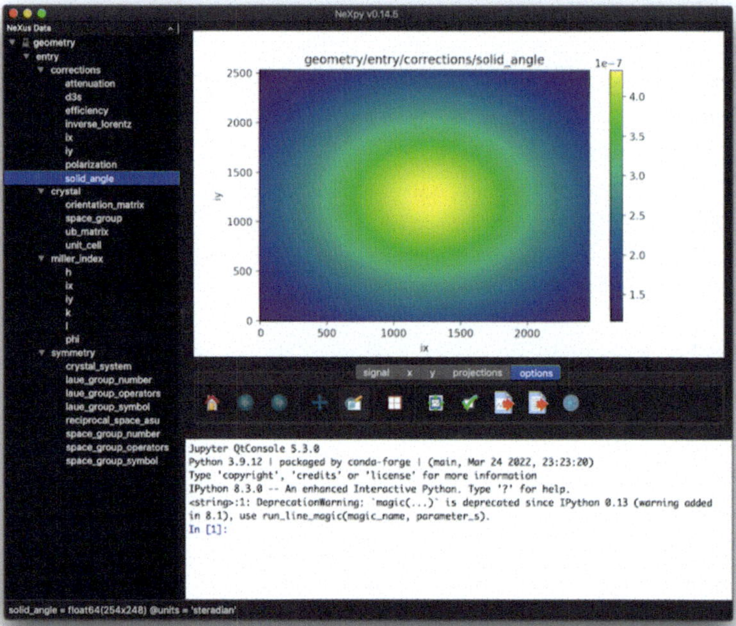

Fig. 7 NeXus files produced by *mdx2* and visualization using *NeXpy*. (A) The file `data.nxs` contains the raw diffraction data as a 3-dimensional array with associated metadata in an NXdata object. The text representation of the file was produced by the function `mdx2.tree`. (B) The more complex file `geometry.nxs` has many NeXus objects organized in a hierarchical fashion (tree view, left panel). Objects can be plotted by double-clicking (here, the `solid_angle` correction mapped over the detector surface). Axes are generated automatically from the metadata.

```
usage: mdx2.import_geometry [-h] [--sample_spacing PHI IY IX]
                            [--outfile OUTFILE]
                            expt

Import experimental geometry using the dxtbx machinery

positional arguments:
  expt                  dials experiments file, such as refined.expt

optional arguments:
  -h, --help            show this help message and exit
  --sample_spacing PHI IY IX
                        inverval between samples in degrees or pixels
                        (default: [1, 10, 10])
  --outfile OUTFILE     name of the output NeXus file (default: geometry.nxs)
```

Under "optional arguments" above, the help text explains that the default interval for sampling is every 1 degree of rotation and every 10 pixels on the detector.

The contents of `geometry.nxs` can be viewed using `mdx2.tree` as above. However, the output is limited to the metadata such as the array types and sizes. Much more information is available using *NeXpy*. For instance, to view the solid angle correction, follow these steps:

- In the terminal type `nexpy` to launch the program
- Click `File` > `open`, navigate to your working directory
- Open `geometry.nxs`
- Expand the tree in the left panel to locate `solid_angle`.
- Double-click it to plot.

The solid angle per pixel (steradian units) should be displayed as a colormap on the detector surface (Fig. 7B).

It can be educational to browse the data objects in `geometry.nxs` as many of the quantities introduced in Section 2 are pre-computed during the import step. Crystal geometry objects include the `orientation_matrix` (**U** in Eq. (2)), ub_`matrix` (product **UB** in Eq. (2)), and Miller index maps h, k, and l (computed using Eq. (3) from the *DIALS* scan-varying model). The 3×3 matrices in `laue_group_operators` represent general positions of the Laue group, the 3×4 matrices in `space_group_operators` represent general positions of the space group operators [the extra column is needed to represent translations

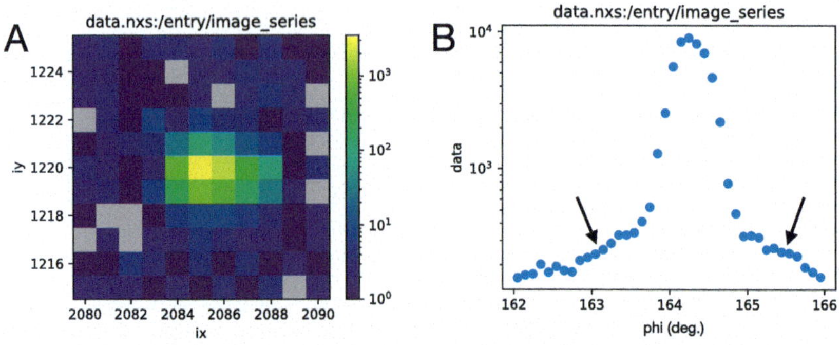

Fig. 8 Rocking curve showing Bragg diffraction and diffuse scattering. An intense peak was selected that passed slowly through the Ewald sphere (lying close the rotation axis in reciprocal space). (A) The Bragg peak in frame 42 where its' intensity is maximum. The intensity decays quickly to background over several pixels (note the logarithmic color scale; pixels with zero counts are colored gray). (B) The total number of counts within the region around the Bragg peak in (A) as a function of rotation angle. The central bell-shaped peak is the Bragg diffraction, and the gradual tails are diffuse halos (arrows). The plots were generated using built-in functions of the *nexusformat* package (see Section 3.2).

(Rowicka, Kudlicki, Zelinka, & Otwinowski, 2004)], and `reciprocal_space_asu` is a mathematical expression defining the asymmetric unit of reciprocal space. Correction factors introduced in Section 2.2 include `solid_angle` (Eq. (10)), `efficiency` (Eq. (11)), `attenuation` (Eq. (12)), `polarization` (Eq. (13)), d3s (Eq. (4)), and `inverse_lorentz` (the absolute value factor in Eq. (4)).

3.2.3 Calculating a rocking curve in Python

Finally, we demonstrate how saving diffraction data in NeXus files can enable rapid data exploration in Python. The following lines in Python are all that is needed to plot a rocking curve for an intense reflection (Fig. 8):

```
from nexusformat.nexus import nxload
image_series = nxload('data.nxs')['/entry/image_series']
limits=((20,60),(1215,1225),(2080,2090))
image_series.project(0,limits=limits).logplot()
```

Here, *project* is a method associated with the NXData object called `image_series` that calculates a projection of a high dimensional array to a subset of the axes. The pixels surrounding the Bragg peak are summed onto axis 0 of the dataset (frame number). The range of pixels considered is given by the `limits` argument: frames 20 through 60, pixels 1215–1225 in the y direction, and 2080–2090 in the x direction. A plot is produced similar to Fig. 8B. Further examples of working with NeXus files in Python can be found in the *nexusformat* documentation (http://nexpy.github.io/nexpy/pythonshell.html).

Mdx2 does not currently provide a means of locating particular Bragg peaks of interest based on their Miller indices. To find a particular Bragg spot, we recommend overlaying the spot predictions with Miller index labels on diffraction images using dials.image_viewer.

3.3 Bragg peak masking

Because different corrections are needed when integrating Bragg peaks and diffuse data (see Eq. (8)), it is important to ensure that the signals are separated before processing. Note that a Bragg peak's extent in ϕ can vary depending on the distance from the spindle axis in reciprocal space, and Bragg peaks may not be perfectly symmetric Gaussian peaks. Peak width depends on crystal mosaicity, and this can even vary during a scan. In addition, the diffraction geometry may not be modeled exactly right, leading to further errors in peak position. A masking function must account for all of these effects.

Designing a perfect Bragg peak mask is a tricky problem, and different researchers have approached it various ways. With *mdx2* we take a simple empirical approach. Here is the algorithm:

- List all of the pixels in the image stack above a certain count threshold ("hot pixels").
- Index all of the hot pixels, converting their position in image space to Miller index (h, k, l).
- Subtract the nearest integer from each Miller index leaving only the fractional part (Δh, Δk, Δl) which varies from -0.5 to 0.5. Because peaks have finite extent, the values of Δh, Δk, Δl will not all be zero, but will look like a cloud of points (Fig. 9A).
- Fit all of the points to a single anisotropic 3D Gaussian distribution (Fig. 9B).
- Create a binary mask for all pixels in the image stack that excludes pixels falling within the 3D Gaussian region according to a sigma threshold (Fig. 9C), and any hot pixels falling outside these regions.

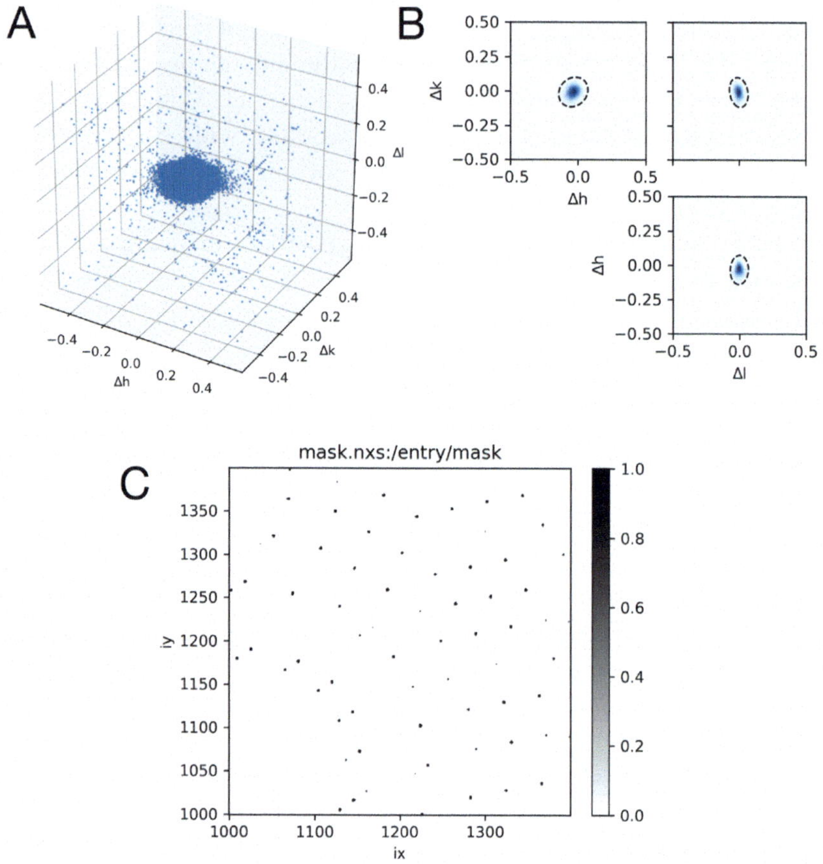

Fig. 9 Bragg peaking masking. (A) Strong pixels (those exceeding a count threshold) are mapped into reciprocal space and plotted as a function of distance from the nearest Bragg peak (axes are fractional Miller indices). (B) The points are fit to a three-dimensional Gaussian probability distribution, here shown as a dashed contour line at 3 standard deviations superimposed on orthogonal projections of the distribution in (A). The Gaussian fit accounts for potential anisotropy due to crystal properties and errors in the model of diffraction geometry. (C) A mask is generated for all pixels in each image according to the Gaussian fit.

To execute the masking algorithm in *mdx2*, first perform the strong pixel search:

```
mdx2.find_peaks geometry.nxs data.nxs --count_threshold 20
```

A threshold of 20 was chosen to be 10 times the background level of ~2 photons/pixel, which works well in this case. *Mdx2* does not currently have a method to automatically estimate the threshold. The images should be inspected manually to determine an appropriate value. When the search finishes, a Gaussian is automatically fit to the point cloud:

$$p(\mathbf{r}) \propto \exp[-(1/2)\,|\boldsymbol{\sigma}^{-1}(\mathbf{r}-\mathbf{r}_0)\,|^2\,],$$

where $\mathbf{r} = (\Delta h, \Delta k, \Delta l)$. The best-fit values according to `mdx2.find_peaks` are printed and saved to a file (`peaks.nxs`):

```
6602 peaks were rejected as outliers
GaussianPeak model: r0 = [-0.033588 -0.006928 -0.002245]
sigma =
[[-0.025445 -0.025525 -0.001434]
 [-0.029339  0.021216  0.004702]
 [ 0.004484 -0.006027  0.022625]]
```

Note that the peak center $\mathbf{r}_0$ is not exactly zero, which may indicate a slight error in the diffraction geometry. This error, as well as the extent of the point cloud ($\boldsymbol{\sigma}$), should be taken into account when designing a reciprocal space grid for integration. The sampling should not be finer than the geometric errors allow. Here, the Bragg peaks are well localized; the width of the Gaussian distribution is approximately 1/20 of the reciprocal unit cell dimension (~two times the diagonal elements of $\boldsymbol{\sigma}$, above). Thus, a sampling rate perhaps as fine as $20 \times 20 \times 20$ voxels per Bragg peak could be justified. Note that such fine maps consume a lot of computer memory, and thus for the purposes of this tutorial, a coarser map will be used. In practice, there is not a one-size-fits-all method to determine the optimum sampling as it depends on the signal-to-noise needed and the application (e.g., fitting a power law requires that there should be at least 3 points—ideally more—between the Bragg peak and the edge of the Brillouin zone).

Next, create the mask:

```
mdx2.mask_peaks geometry.nxs data.nxs peaks.nxs \
   --sigma_cutoff 3
```

The `sigma_cutoff` parameter controls the size of the masked region around the Bragg peak. A value of 3 means that pixels are masked within 3 standard deviations of the peak center according to the model fit above. The function also masks outliers flagged previously. The resulting masks are saved as an image stack in `mask.nxs` and should be inspected using *NeXpy* before proceeding (Fig. 9C).

Note that in *mdx2* version 0.3.1, `mdx2.mask_peaks` does not account for systematic absences. This is a particular issue for space group $I2_13$, because the reflection condition $h + k + l = 2n$ means that half of the Bragg peaks are absent. Although the mask generated above excludes twice as many spots as necessary, in this case relatively few pixels are excluded so the effect on the final map is minimal.

3.4 Background subtraction

Diffuse scattering experiments are different from Bragg data collection in that background must be measured carefully and subtracted (Pei et al., 2023). For insulin, background images were collected using 1 s exposures every 1 degree of rotation (compared with 0.1 s/0.1 degree for the crystal). The background dataset was previously imported using *DIALS* in Section 3.1 to produce the file `background.expt`. First, import the dataset using `mdx2.import_data`.

```
mdx2.import_data background.expt --outfile bkg_data.nxs \
    --chunks 10 211 493
```

Inspect the images in `bkg_data.nxs` using *NeXpy*. The background pattern includes scattering from air, diffuse rings from the capillary surrounding the crystal, and a shadow of the pin (Fig. 10A).

Because the background features vary gradually across the detector and with rotation angle, the noise can be reduced by smoothing. In *mdx2*, a downsampling (binning) method is used.

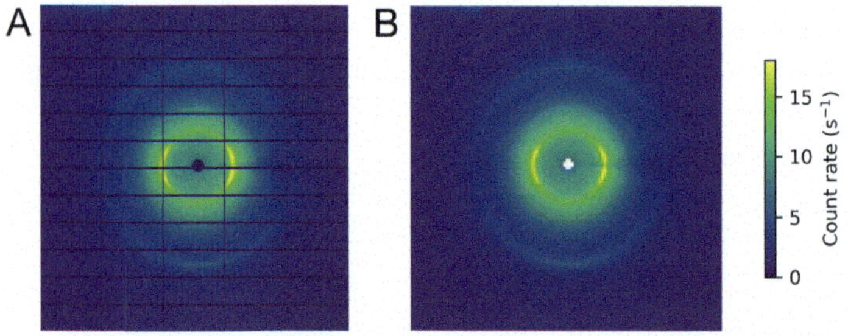

Fig. 10 Image processing for background subtraction. (A) An example background image (1 degree oscillation, 1 s exposure) from the insulin rotation dataset. (B) The same image after binning to improve signal-to-noise. Background corrections are computed for each voxel in the reciprocal space map by interpolation of the binned image stack.

```
mdx2.bin_image_series bkg_data.nxs 10 20 20 \
   --valid_range 0 200 --outfile bkg_data_binned.nxs
```

The function has three mandatory arguments after the input file name to specify the bin size (here, 10 degrees by 20 pixels by 20 pixels). In addition, the optional argument `valid_range` is used to mask any pixel with counts outside the given interval. Here, a maximum of 200 counts was chosen to be 10 times the nominal background level of ~20 counts per pixel (the background level is 10-fold higher in these images than in the crystal images due to the longer exposure time per frame). The threshold is used to reject broken pixels and stray diffraction if present (e.g., from tiny salt crystals), and is similar to `count_threshold` in `mdx2.find_peaks` described in Section 3.3.

The resulting binned image stack is saved as `bkg_data_binned.nxs` and can be viewed using *NeXpy* (Fig. 10B). The binned images will be used after integration when applying geometric corrections according to Eq. (14) (see Section 3.5).

3.5 Integration, scaling and merging

3.5.1 Integration

Integration refers to accumulating photon counts on a three-dimensional reciprocal space grid. For the tutorial, a grid was chosen with 4 subdivisions in each direction, meaning voxels will be centered at Miller indices $h = 1$, 1.25, 1.5, 1.75, etc. This is by no means the optimal choice: the limit depends on the mosaicity, experimental geometry, and most importantly on how the diffuse scattering map will be used. However, when experimenting with different choices for the subdivisions, be aware that finer maps require more memory and disk space.

```
mdx2.integrate geometry.nxs data.nxs --mask mask.nxs \
   --subdivide 4 4 4
```

For each voxel, the integrated photon counts and other data used in corrections are stored in a table in `integrated.nxs`. After integration, the pre-computed geometric and background corrections are applied (Eq. (14)):

```
mdx2.correct geometry.nxs integrated.nxs \
   --background bkg_data_binned.nxs
```

The corrected intensities are saved as a table in `corrected.nxs`. The following Python code can be used to load the table as a *pandas* dataframe and preview the first several lines:

```
from mdx2.utils import loadobj
tab = loadobj('corrected.nxs','hkl_table')
# convert to pandas dataframe
df = tab.to_frame().set_index(['h','k','l','op','n'])
# show the first several lines
df.head()
```

The columns contain the following information:

- `h`, `k`, `l`: Miller indices mapped to the *asymmetric unit*
- `op`: index of the symmetry operator that mapped the observation point to the asymmetric unit
- `s`: scattering vector magnitude (= 1/resolution)
- `intensity`, `intensity_error`: measured intensities and errors corrected using Eq. (14).
- `rs_volume`: volume of reciprocal space recorded as a fraction of the reciprocal unit cell. Here, a voxel that is fully recorded would have `rs_volume` $= 4^{-3} \sim 0.0156$.

3.5.2 Scaling

Finally, the corrected intensities can be scaled and merged. *Mdx2* implements the scaling algorithm from *mdx-lib* described in Section 2.3, however in *mdx2* version 0.3.1 only the per-image scale factor b is included in the model (Eq. (19)). Refine b as follows:

```
mdx2.scale corrected.nxs --smoothness 1
```

The model is refined using regularization to ensure that the correction factors vary gradually. The `smoothness` parameter sets the relative weight for minimizing smoothness vs. χ^2. A value of 1 is a good starting point, but it can be varied up or down to observe the effect (e.g., by factors of 10). Additional optional parameters are available to further define the scaling model and refinement procedure (see `mdx2.scale --help`). By default, 5 cycles of alternating least-squares are performed and the scaling model is written to the file `scales.nxs`. To plot the refined scale factors (Fig. 11), open the file in *NeXpy* and double-click `scaling_model` in the file tree.

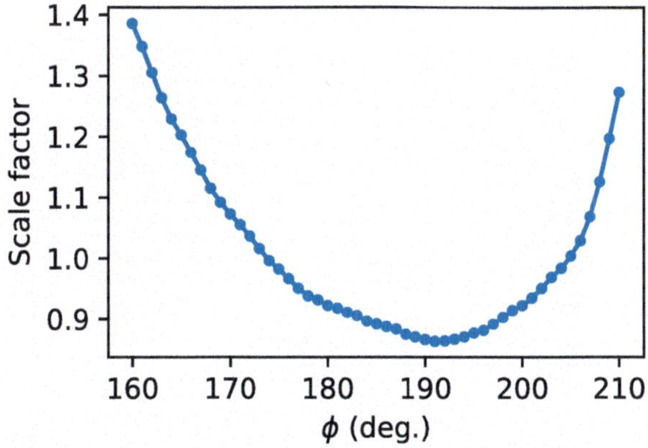

Fig. 11 Scale factor per image in a rotation set after refinement using *mdx2*.

3.5.3 Merging

After scaling, the equivalent observations can be merged using Eqs. (17) and (18) as follows:

```
mdx2.merge corrected.nxs --scale scales.nxs
```

The file `merged.nxs` is produced with a table of intensities and error estimates.

3.6 Visualization

The processed diffuse scattering data from Section 3.5 is stored as a table of values covering the asymmetric unit of reciprocal space. In order to visualize the data, it is necessary to convert from a table to an array and symmetry-expand the data. The function *mdx2.map* can be used for generating slices and volumes. First, generate a slice as follows:

```
mdx2.map geometry.nxs merged.nxs --limits -50 50 -50 50 0 0 \
  --outfile slice.nxs
```

The `--limits` argument defines the boundaries of the slice in fractional (Miller index) units. In this case, the slice is a central section (*hk0*) with h from -50 to 50 (fractional units), k from -50 to 50 (fractional units), and $l = 0$. With the $4 \times 4 \times 4$ sampling used, the slice will have 201×201 voxels.

Open the file `slice.nxs` in *NeXpy* and double-click `intensity` in the file tree to display a slice through the reciprocal space map. Adjust the sliders in the signal tab to make the background scattering variations clear (as in Fig. 12).

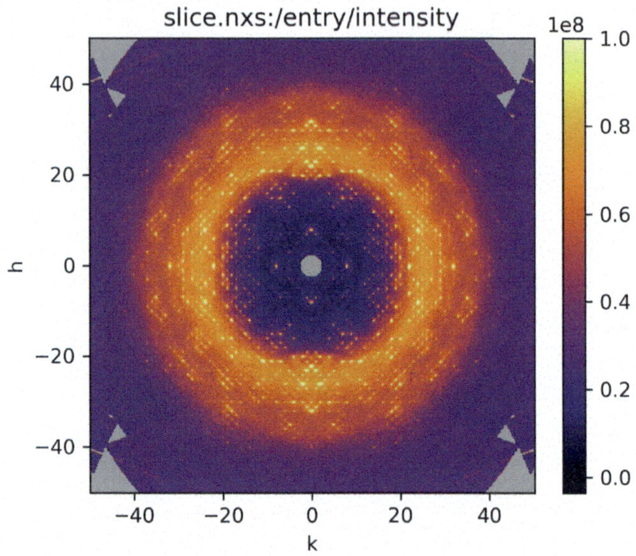

Fig. 12 Slice through the reciprocal space map of insulin visualized using *NeXpy*.

A similar command can be used to generate reciprocal space volumes:

```
mdx2.map geometry.nxs merged.nxs --limits 0 50 0 50 0 50 \
   --outfile map.nxs
```

To conserve computing resources, we have restricted the map to the positive octant of reciprocal space ($h \geq 0$, $k \geq 0$, $l \geq 0$), which is the same as the other seven by symmetry. Open `map.nxs` using *NeXpy* and double-click `intensity` in the file tree. Scroll through z-slices. It helps to unselect "autoscale" in the "z" tab so that intensity scale is fixed, and then adjust it in the "signal" tab.

3.7 Map statistics

In this section, we use *mdx2* with *pandas* to calculate statistical properties of the diffuse map. *Pandas* is a Python library commonly used in data science to manipulate tabular data. It will be used here without explanation; excellent tutorials are available elsewhere (VanderPlas, 2016).

In a Python session or Jupyter notebook, first import the needed libraries and define convenient functions as follows:

```
from mdx2.utils import loadobj, saveobj
from mdx2.data import HKLTable
import pandas as pd
import numpy as np

Crystal = loadobj('geometry.nxs','crystal')

def hkl2s(h,k,l):
    """Compute the magnitude of s from Miller indices."""
    UB = Crystal.ub_matrix
    s = UB @ np.stack((h,k,l))
    return np.sqrt(np.sum(s*s,axis=0))
```

3.7.1 Intensity vs. resolution

The following Python program calculates the mean and standard deviation of intensity in shells of constant resolution (Fig. 13A):

```
# load table as mdx2 object and convert to pandas dataframe
tab = loadobj('merged.nxs','hkl_table')
tab.s = hkl2s(tab.h,tab.k,tab.l)
df = tab.to_frame().set_index(['h','k','l']).sort_index()

# divide into 100 equal sized bins
s_bins = pd.cut(df['s'],100)

# compute statistics in each bin
df_isoavg = df.groupby(s_bins).agg({
    's':'mean',
    'intensity':['mean','std'],
    'intensity_error':'mean'})

# plot statistics vs. s
df_isoavg.set_index(('s','mean')).plot();
```

3.7.2 Correlation of half-datasets

To quantify the precision of the final diffuse map, the dataset can be split into equivalent halves and the correlation coefficient computed between them in shells of constant resolution [see Chapter 1 (Pei et al., 2023)].

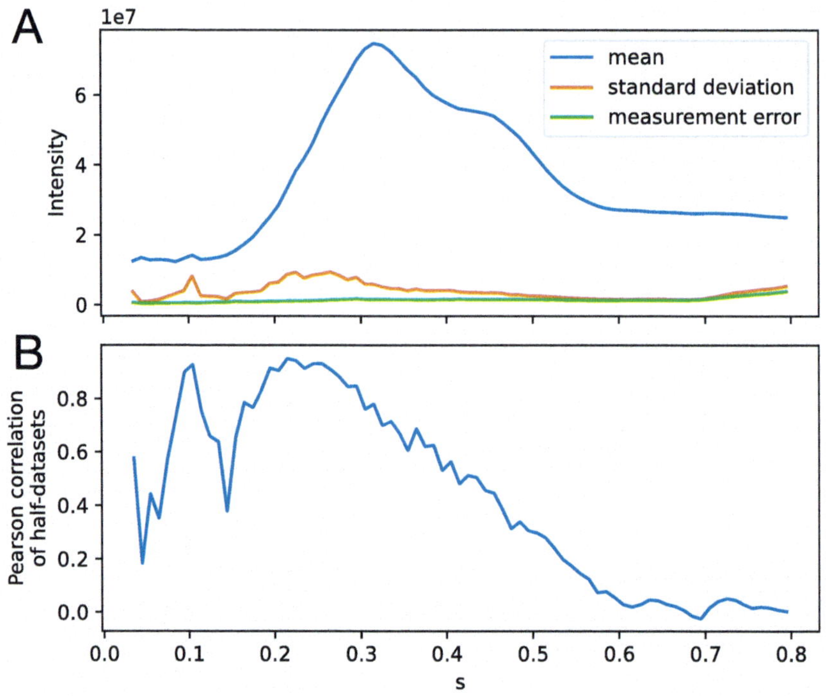

Fig. 13 Intensity statistics for the insulin diffuse scattering dataset after scaling. (A) The mean (blue) and standard deviation (orange) of intensity were computed within each resolution shell for the merged reciprocal space map. The variations in the diffuse pattern are much smaller than the isotropic component. However, they are larger than the noise in the data (average measurement error, propagated from Poisson statistics), shown in green. (B) The dataset was split into equivalent halves and merged separately, and the Pearson correlation was computed between these half datasets as a function of resolution (1/*s*). By this measure, the variations in the diffuse pattern contain significant signal up to ~2 Å resolution. The dip at low resolution is related to the strength of the halo features relative to the background in this dataset.

Mdx-lib has built-in tools for splitting unmerged datasets randomly with similar statistical weight. This method is not yet available in *mdx2* version 0.3.1. Here, we use *pandas* to split the data by assigning asymmetric units to each half (splitting on the `op` column of the data table). The following python script reads the unmerged data from `corrected.nxs` and saves two files: `split_1.nxs` and `split_2.nxs`.

```
tab = loadobj('corrected.nxs','hkl_table')
df = tab.to_frame()

# split into two halves based on the symmetry operator
grouped = df.groupby(pd.cut(df['op'],2))

# save as nexus files for merging with mdx2.merge
for grp,filename in
zip(grouped,('split_1.nxs','split_2.nxs')):
    print('interval:',grp[0])
    _tmp = HKLTable.from_frame(grp[1])
    _tmp.ndiv = tab.ndiv # add back the grid subdivision info
    saveobj(_tmp,filename)
```

Next, scale and merge each half dataset using *mdx2* on the command line:

```
mdx2.merge split_1.nxs --scale scales.nxs \
   --outfile merged_1.nxs
mdx2.merge split_2.nxs --scale scales.nxs \
   --outfile merged_2.nxs
```

Finally, load the merged datasets as *pandas* dataframes and compute correlation coefficients in shells of constant resolution (Fig. 13B):

```
# load the tables and convert to pandas dataframe
t1 = loadobj('merged_1.nxs','hkl_table')
t2 = loadobj('merged_2.nxs','hkl_table')
t1.s = hkl2s(t1.h,t1.k,t1.l)
df1 = t1.to_frame().set_index(['h','k','l'])
df2 = t2.to_frame().set_index(['h','k','l'])

# merge the two tables
df12 = df1.merge(df2,on=('h','k','l'),suffixes=('_1','_2'))

# compute pearson correlation within resolution shells
s_bin = pd.cut(df12.s,100) # 100 bins
int_grp = df12.groupby(s_bin)['intensity_1','intensity_2']
corr_binned = int_grp.corr(method='pearson').unstack()
corr_binned['s'] = df12.groupby(s_bin)['s'].mean()

# plot the results
corr_binned.plot(x='s',y=('intensity_1','intensity_2'))
```

4. Conclusions

This chapter has introduced a modern approach to macromolecular diffuse scattering based on low dose, high redundancy data collection and reciprocal space mapping. The promise of diffuse scattering for studying correlated motion was made clear by Caspar and colleagues 35 years ago, however it has only recently become feasible to acquire the accurate and detailed data needed to account for all of the motions occurring in a protein crystal. Much of the recent work on diffuse scattering has focused on data processing, and we expect a continued focus on methods development in the near future. However, the field has reached a turning point. As outlined in Chapter 1 (Pei et al., 2023), now that the physical basis for diffuse scattering from protein crystals is understood, the demands on sample quality and data processing have been clarified. Although methods will continue to evolve, we expect increased standardization of software and workflows going forward. Here, we have made the first steps toward standardization in the form of *mdx2*. Furthermore, by introducing user-friendly software tools and rigorous tutorials, we have reduced the burden of expertise required to enter the field of diffuse scattering. With these advances, the goal of moving the field beyond proof-of-concept experiments to biological applications is within reach.

Acknowledgments

The *mdx2* software and tutorial in Section 3 are based on materials developed for the Data Reduction Workshop at the 2022 Erice School on Diffuse Scattering. We are grateful to the organizers and participants for providing both the motivation to develop *mdx2* and valuable feedback. We thank Kevin Dalton and members of the *DIALS* development team for answering questions and helping us get started with *dxtbx*. The insulin dataset was collected at the Center for High-Energy X-ray Sciences (CHEXS), which is supported by the at National Science Foundation (BIO, ENG and MPS Directorates) under award DMR-1829070, and the Macromolecular Diffraction at CHESS (MacCHESS) facility, which is supported by award P30GM124166 from the National Institute of General Medical Sciences, National Institutes of Health, and by New York State's Empire State Development Corporation (NYSTAR). This work was supported by National Institutes of Health grant GM124847 (to N.A.).

Appendix A

A.1. Obtaining and running the tutorial in Jupyter notebooks

The tutorials in Section 3 are available on the *mdx2* GitHub page as separate Jupyter notebooks (https://github.com/ando-lab/mdx2). The notebooks can be downloaded individually or obtained by cloning the *mdx2* repository.

The code is designed to run on a personal computer with ~20 Gb of free disk storage and at least 4 Gb of RAM. A unix-like operating system is assumed (Linux, OSX, or Windows Subsystem for Linux).

A.2. Downloading the tutorial dataset

The dataset from insulin is available on Zenodo (https://dx.doi.org/10.5281/zenodo.6536805). First, download `insulin_2_1.tar` and extract the tar archive. The file will expand to a directory called `images` with subfolders `insulin_2_1`, `insulin_2_bkg`, and `metrology`. Place `images` in the same directory as the Jupyter notebooks.

A.3. Setting up the python environment

Install *miniconda3* with Python version *3.10*. Installers and instructions are available on the web: https://docs.conda.io/en/latest/miniconda.html. To prevent miniconda from interfering with existing conda installations, it should be prevented from modifying the users' bash scripts or running "conda init" (`-b` flag). We also recommend choosing a non-default install location; `~/miniconda-mdx2` is assumed in the following examples, but can be modified if needed.

Example (MacOS x64):

```
wget https://repo.anaconda.com/miniconda/Miniconda3-
py310_22.11.1-1-MacOSX-x86_64.sh -O miniconda-installer.sh
bash miniconda-installer.sh -b -p ~/miniconda-mdx2
```

Next, activate the *mdx2* conda environment and install *mdx2*'s dependencies.

```
source ~/miniconda-mdx2/bin/activate
conda install -c conda-forge dxtbx nexusformat pandas numexpr
```

If the conda environment is de-activated (by typing `conda deactivate` or starting a new terminal session), it can be re-activated by typing `source ~/miniconda-mdx2/bin/activate`. For convenience, the following line can be added to the shell startup script:

Example (~/.bash_profile).

```
alias activate_mdx2="source ~/miniconda-mdx2/bin/activate"
```

Then, the mdx2 environment can be activated by typing `activate_mdx2`.

A.4. Installing mdx2

First, activate the *mdx2* conda environment (see above). Then install *mdx2* from the GitHub repository using pip.

```
pip install git+https://github.com/ando-lab/mdx2.git@v0.3.1
```

The tag `@v0.3.1` at the end of the repository address specifies version 0.3.1 for consistency with this tutorial. If this tag is omitted, the most up-to-date version will be installed (see Version notes at https://github.com/ando-lab/mdx2).

The *mdx2* tools should now be available at the command line. Check the version as follows:

```
mdx2.version
```

The version number `0.3.1` should be displayed.

A.5. Installing DIALS, NeXpy, and JupyterLab

First, activate the *mdx2* conda environment (see above). Then install the programs using conda.

```
conda install -c conda-forge dials nexpy jupyterlab
```

With the *mdx2* conda environment active, *nexpy* can be launched from the command line by typing `nexpy`. Similarly, Jupyter Lab can be launched by typing `jupyter lab`. The *DIALS* command-line tools are also available. For instance `dials.version` prints the version information and install location.

References

Ayyer, K., Yefanov, O. M., Oberthür, D., Roy-Chowdhury, S., Galli, L., Mariani, V., ... Chapman, H. N. (2016). Macromolecular diffractive imaging using imperfect crystals. *Nature, 530*(7589), 202–206.

Busing, W. R., & Levy, H. A. (1967). Angle calculations for 3- and 4-circle X-ray and neutron diffractometers. *Acta Crystallographica, 22*(4), 457–464. https://doi.org/10.1107/S0365110×67000970

Caspar, D., Clarage, J., Salunke, D., & Clarage, M. (1988). Liquid-like movements in crystalline insulin. *Nature, 332*(6165), 659–662.

Chapman, H. N., Fromme, P., Barty, A., White, T. A., Kirian, R. A., Aquila, A., ... Spence, J.C.H. (2011). Femtosecond x-ray protein nanocrystallography. *Nature, 470*(7332), 73–77.

Clarage, J. B., & Phillips, G. N., Jr. (1997). *Analysis of diffuse scattering and relation to molecular motion. Methods in Enzymology*. Elsevier, 407–432.

De Klijn, T., Schreurs, A., & Kroon-Batenburg, L. (2019). Rigid-body motion is the main source of diffuse scattering in protein crystallography. *IUCrJ, 6*(2), 277–289.

Evans, P. R., & Murshudov, G. N. (2013). How good are my data and what is the resolution. *Acta Crystallographica. Section D, 69*(7), 1204–1214. https://doi.org/10.1107/S0907444913000061

Glover, I., Harris, G., Helliwell, J., & Moss, D. (1991). The variety of x-ray diffuse scattering from macromolecular crystals and its respective components. *Acta Crystallographica. Section B, Structural Science, 47*(6), 960–968.

Grosse-Kunstleve, R. W., Sauter, N. K., Moriarty, N. W., & Adams, P. D. (2002). The computational crystallography toolbox: Crystallographic algorithms in a reusable software framework. *Journal of Applied Crystallography, 35*(1), 126–136.

Hamilton, W. C., Rollett, J. S., & Sparks, R. A. (1965). On the relative scaling of X-ray photographs. *Acta Crystallographica, 18*(1), 129–130.

Kabsch, W. (2010). XDS. *Acta Crystallographica. Section D, Biological Crystallography, 66*(2), 125–132.

Könnecke, M., Akeroyd, F. A., Bernstein, H. J., Brewster, A. S., Campbell, S. I., Clausen, B., ... Wuttke, J. (2015). The NeXus data format. *Journal of Applied Crystallography, 48*(1), 301–305.

Meisburger, S. P., Case, D. A., & Ando, N. (2020). Diffuse X-ray scattering from correlated motions in a protein crystal. *Nature Communications, 11*(1), 1271. https://doi.org/10.1038/s41467-020-14933-6

Meisburger, S. P., Case, D. A., & Ando, N. (2023). Robust total X-ray scattering workflow to study correlated motion of proteins in crystals. *Nature Communications, 14*(1), 1228. https://doi.org/10.1038/s41467-023-36734-3

Meisburger, S. P., Thomas, W. C., Watkins, M. B., & Ando, N. (2017). X-ray scattering studies of protein structural dynamics. *Chemical Reviews, 117*(12), 7615–7672.

Moore, P. B. (2009). On the relationship between diffraction patterns and motions in macromolecular crystals. *Structure (London, England: 1993), 17*(10), 1307–1315.

Morgan, A. J., Ayyer, K., Barty, A., Chen, J. P. J., Ekeberg, T., Oberthuer, D., ... Chapman, H. N. (2019). *Ab initio* phasing of the diffraction of crystals with translational disorder. *Acta Crystallographica. Section A, Foundations and Advances, 75*(1), 25–40. https://doi.org/10.1107/S2053273318015395

Neviani, V., Lutz, M., Oosterheert, W., Gros, P., & Kroon-Batenburg, L. (2022). Crystal structure of the second extracellular domain of human tetraspanin CD9: Twinning and diffuse scattering. *IUCrData,* 7(9), x220852.

Ourmazd, A., Moffat, K., & Lattman, E. E. (2022). Structural biology is solved—Now what? *Nature Methods, 19*(1), 24–26.

Parkhurst, J. M., Brewster, A. S., Fuentes-Montero, L., Waterman, D. G., Hattne, J., Ashton, A. W., ... Winter, G. (2014). *dxtbx*: The diffraction experiment toolbox. *Journal of Applied Crystallography, 47*(4), 1459–1465. https://doi.org/10.1107/S1600576714011996

Peck, A., Poitevin, F., & Lane, T. J. (2018). Intermolecular correlations are necessary to explain diffuse scattering from protein crystals. *IUCrJ, 5*(2), 211–222.

Pei, X., Bhatt, N., Wang, H., Ando, N., & Meisburger, S. (2023). Introduction to diffuse scattering and data collection [submitted]. *Methods in Enzymology.*

Rowicka, M., Kudlicki, A., Zelinka, J., & Otwinowski, Z. (2004). Coordinate transformations in modern crystallographic computing. *Acta Crystallographica. Section A, Foundations and Advances, 60*(6), 542–549. https://doi.org/10.1107/S0108767304017398

Rupp, B. (2009). *Biomolecular crystallography: Principles, practice, and application to structural biology*. Garland Science.

Schreurs, A. M., Xian, X., & Kroon-Batenburg, L. M. (2010). EVAL15: A diffraction data integration method based on ab initio predicted profiles. *Journal of Applied Crystallography, 43*(1), 70–82.

Shoemaker, S. C., & Ando, N. (2018). X-rays in the cryo-electron microscopy era: Structural biology's dynamic future. *Biochemistry, 57*(3), 277–285.
Simonov, A., De Baerdemaeker, T., Boström, H. L., Rios Gomez, M. L., Gray, H. J., Chernyshov, D., ... Goodwin, A. L. (2020). Hidden diversity of vacancy networks in prussian blue analogues. *Nature, 578*(7794), 256–260.
Su, Z., Dasgupta, M., Poitevin, F., Mathews, I. I., Bedem, H., van den, Wall, M. E., ... Wilson, M. A. (2021). Reproducibility of protein x-ray diffuse scattering and potential utility for modeling atomic displacement parameters. *Structural Dynamics, 8*(4), 044701.
The NeXpy Development Team. (2023). *NeXpy* (Version v1.0.0) [Computer software]. GitHub. ⟨https://github.com/nexpy/nexpy⟩.
Van Benschoten, A. H., Liu, L., Gonzalez, A., Brewster, A. S., Sauter, N. K., Fraser, J. S., & Wall, M. E. (2016). Measuring and modeling diffuse scattering in protein x-ray crystallography. *Proceedings of the National Academy of Sciences of the United States of America, 113*(15), 4069–4074.
VanderPlas, J. (2016). *Python data science handbook: Essential tools for working with data.* O'Reilly Media, Inc.
Wall, M. E. (1996). *Diffuse features in x-ray diffraction from protein crystals* [PhD thesis, Princeton University]. http://public.lanl.gov/mewall/Wall-Princeton-1996.pdf.
Wall, M. E. (2009). Methods and software for diffuse X-Ray scattering from protein crystals. In R. Foote, & J. Lee (Vol. Eds.), *Micro and nano technologies in bioanalysis. Methods in Molecular Biology™. Vol. 544.*
Wall, M. E., Clarage, J. B., & Phillips, G. N. (1997). Motions of calmodulin characterized using both bragg and diffuse x-ray scattering. *Structure, 5*(12), 1599–1612.
Wall, M. E., Ealick, S. E., & Gruner, S. M. (1997). Three-dimensional diffuse x-ray scattering from crystals of staphylococcal nuclease. *Proceedings of the National Academy of Sciences, 94*(12), 6180–6184.
Waterman, D. G., Winter, G., Gildea, R. J., Parkhurst, J. M., Brewster, A. S., Sauter, N. K., & Evans, G. (2016). Diffraction-geometry refinement in the *DIALS* framework. *Acta Crystallographica. Section D, 72*(4), 558–575. https://doi.org/10.1107/S2059798316002187
Winter, G., Beilsten-Edmands, J., Devenish, N., Gerstel, M., Gildea, R. J., McDonagh, D., ... Evans, G. (2022). DIALS as a toolkit. *Protein Science: A Publication of the Protein Society, 31*(1), 232–250.
Xu, D., Meisburger, S. P., & Ando, N. (2021). Correlated motions in structural biology. *Biochemistry, 60*(30), 2331–2340. https://doi.org/10.1021/acs.biochem.1c00420

CHAPTER THREE

Molecular-dynamics simulations of macromolecular diffraction, part I: Preparation of protein crystal simulations

David C. Wych[a,b] and Michael E. Wall[a,*]
[a]Computer, Computational and Statistical Sciences Division, Los Alamos, NM, United States
[b]Center for Nonlinear Studies, Los Alamos National Laboratory, Los Alamos, NM, United States
*Corresponding author. e-mail address: mewall@lanl.gov

Contents

Abstract

Molecular-dynamics (MD) simulations of protein crystals enable the prediction of structural and dynamical features of both the protein and the solvent components of macromolecular crystals, which can be validated against diffraction data from X-ray crystallographic experiments. The simulations have been useful for studying and predicting both Bragg and diffuse scattering in protein crystallography; however, the preparation is not yet automated and includes choices and tradeoffs that can impact the results. Here we examine some of the intricacies and consequences of the choices involved in setting up MD simulations of protein crystals for the study of diffraction data, and provide a recipe for preparing the simulations, packaged in an

ISSN 0076-6879, https://doi.org/10.1016/bs.mie.2023.06.008

accompanying Jupyter notebook. This article and the accompanying notebook are intended to serve as practical resources for researchers wishing to put these models to work.

1. Introduction

Molecular-dynamics (MD) simulations of biomolecules are prepared using an atomistic structural model, usually sourced from the Protein Data Bank (PDB) (Berman et al., 2000). The dynamics are simulated using numerical integration over small time steps subject to the potentials defined by the parameters of a force field that defines how the atoms interact. It has long been appreciated that MD simulations are useful for providing finer detail, context, or interpretation to experimental observations (see, e.g., Karplus & McCammon, 2002). Comparison with experimental observations also is critical for assessing the level of confidence that can be placed in an MD simulation (see, e.g., van Gunsteren & Mark, 1998).

Most MD simulations are prepared using structures obtained from macromolecular crystallography (MX) experiments. Over 85% of the structures in the PDB are MX structures (see https://www.rcsb.org/stats/summary). It is common for the simulations to be compared to the crystal structure, but a direct comparison can be tricky: MD simulations are almost always performed at room temperature (about 300 K), but over 95% of crystal structures are collected at cryogenic temperatures (Keedy et al., 2015) to protect against radiation damage (Garman, 2010; Holton, 2009; Warkentin et al., 2013). The difference is important because studies have shown that cryo-temperature MX structures differ significantly from their room-temperature counterparts, in protein structure and structural heterogeneity (Ebrahim et al., 2022; Fraser et al., 2011; Keedy et al., 2014), ordered water structure (Stachowski, Vanarotti, Seetharaman, Lopez, & Fischer, 2022), ligand binding (Mehlman et al., 2023), and even in the detailed network of interactions that appear to contribute to allosteric signaling (Keedy, 2019). Additionally, the structure obtained by MX is in the crystalline state, whereas most MD simulations are performed in the solution state. This difference is also important as experimental and simulation-based studies have shown that there can be differences between crystalline and solution state in terms of protein structure (Janowski, Liu, Deckman, & Case, 2016; Lautz, Kessler, Kaptein, & van Gunsteren, 1987; Terada & Kidera, 2012; Van Gunsteren & Berendsen, 1984; Verli, Calazans, Brindeiro, Tanuri, & Guimarães, 2007), structural heterogeneity

(Ahlstrom, Vorontsov, Shi, & Miyashita, 2017; Cerutti, Trong, Stenkamp, & Lybrand, 2009), and interactions with ligands (Cerutti et al., 2009; Vorontsov & Miyashita, 2009) albeit sometimes quite subtle (Stocker, Spiegel, & van Gunsteren, 2000). Especially when compared with room-temperature MX structures, crystalline-state MD simulations are unique in that they can be prepared and processed to allow for direct comparison to experimental data with no major caveats. The comparison of crystalline MD simulations and MX data can be revealing from both the simulation side and the experimental side: crystalline MD simulations can be used to enhance crystallographic modeling (Mikhailovskii, Xue, & Skrynnikov, 2022; Wych et al., 2023) and crystallographic data can be used to enhance MD simulation—in particular, comparison with X-ray crystallographic data is an emerging technique for studying and improving force fields (Cavender et al., 2023; Cerutti & Case, 2019; Hu & Jiang, 2010; Krieger, Darden, Nabuurs, Finkelstein, & Vriend, 2004; Schnieders, Fenn, Pande, & Brunger, 2009).

It is possible to compute a wide variety of features of interest in X-ray crystallography from an MD simulation of a protein crystal, including:

- average protein structure (Avbelj, Moult, Kitson, James, & Hagler, 1990; Cerutti, Le Trong, Stenkamp, & Lybrand, 2008; Janowski et al., 2016; Janowski, Cerutti, Holton, & Case, 2013; Lautz et al., 1987; Meinhold & Smith, 2005a; York et al., 1994)
- multiple conformations for residues (e.g. side chains, ligands—even water molecules) and their occupancies/populations (Cerutti et al., 2009; Cerutti, Freddolino, Duke, & Case, 2010; Janowski et al., 2013; Wych et al., 2023) or full protein structural ensembles (Wych et al., 2023)
- structure factors and crystallographic densities (total, or decomposed into, e.g., protein, solvent, or ion components) (Janowski et al., 2013; Janowski et al., 2016;; Wall, Calabró, Bayly, Mobley, & Warren, 2019; Wych et al., 2023)
- root-mean-square fluctuations and/or temperature factors, also known as B-factors or atomic displacement parameters (ADPs) (Bond, Faraldo-Gómez, Deol, & Sansom, 2006; Krieger et al., 2004; Cerutti et al., 2009; Cerutti et al., 2010; Lautz et al., 1987; Vitkup, Ringe, Karplus, & Petsko, 2002; York et al., 1994; Meinhold & Smith, 2005a; Héry, Genest, & Smith, 1997; Janowski et al., 2013; Hu & Jiang, 2010)
- diffuse scattering intensities (total, isotropic, and anisotropic) (Meinhold & Smith, 2005b; Meisburger, Case, & Ando, 2020; Meisburger, Case, & Ando, 2023; Wall, 2018; Wall, Clarage, & Phillips, 1997; Wall, Van

Benschoten, et al., 2014; Wych, Fraser, Mobley, & Wall, 2019; Clarage, Romo, Andrews, Pettitt, & Phillips, 1995).

In some cases, there is more than one way to do the calculation: e.g., B-factors can be estimated either by direct analysis of atomic fluctuations or by refinement of a structure against the MD-simulated Bragg intensities (Cerutti & Case, 2019); similarly, average or representative structures can be measured by, e.g., free energy landscape, principle component, or structural cluster analysis of the conformations sampled in the trajectory (Papaleo, Mereghetti, Fantucci, Grandori, & De Gioia, 2009) or by refinement of a single-structure or ensemble structural model against the MD-simulated Bragg intensities (Wall, 2018; Chen & Chapman, 2001). Using MD simulations of protein crystals, one can perform controlled studies using both the simulated crystallographic data and the ensemble of atomistic structures which produced them, and this can be used to validate MD simulations of water structure (Van Gunsteren & Berendsen, 1984; Wall et al., 2019) and to aid in interpretation of crystallographic data and enhance crystallographic modeling and refinement techniques (Vitkup et al., 2002; Wych et al., 2023).

Simulating a protein in the crystalline state can come at the cost of additional computational time and/or resources as compared to the more common solution state, due to the increased number of copies of the protein involved; however, the potential extra cost does not necessarily result in a decrease in the efficiency of obtaining the desired results. The additional copies of the protein enable increased statistics for studying the internal motions of proteins. In addition, the solvent fraction in the simulation box is typically much higher in solution simulations than in crystalline simulations: simulating in solution requires placing the protein in a box surrounded by a large enough water layer such that the protein is shielded from interaction with its identical neighbor(s) in adjacent periodic boxes, which usually results in solvent occupying most of the simulated box volume. A crystalline simulation, by contrast, typically has about 50% water by volume (Weichenberger, Afonine, Kantardjieff, & Rupp, 2015). Crystalline simulations therefore can improve the efficiency of simulating multiple copies of a protein, compared to solution simulations.

Efficiency arguments aside, one of main benefits of simulating in the crystalline state is the ability to calculate structure factors and scattering intensities which can be directly compared with crystallographic data. Most MD simulations start with the protein in a state determined by crystallographic experiments, and then simulate in the solution state—it is only in the crystalline MD case that the protein is being simulated in a state designed to reproduce the real conditions during experimental data collection (the solution state is appropriate for comparison to NMR experiments). Although structure factors can be calculated from any atomistic simulation (as we will discuss, all molecular-dynamics simulations using periodic boundary conditions are in some sense crystalline) it is only in the crystalline MD case that the simulated crystal corresponds with the actual physical crystal being studied; in the solution state case, the "crystal" created by the periodic boundary conditions is unphysical, and the structure factors calculated from the simulations cannot be compared directly with structure factors from experiment, though the densities can be aligned and compared (and indeed, have been used to investigate water density in single-protein solution state simulations) (Ge et al., 2022).

To help lower the barrier to entry for performing crystalline MD simulations, and to encourage improvement and innovation in the space, here we provide a practical guide for preparing MD simulations of protein crystals. Our goal is not to present a comprehensive review or definitive reference, but to help researchers who are interested in getting started with performing the simulations. We cover some of the background on the many intricacies of preparing and running a crystalline MD simulation, based on our experience in putting these models to practice. To increase the accessibility of the methods, we make use of a suite of free and extensible software tools which are standard in the computational chemistry and crystallography communities. We also provide example commands to carry out the various steps and distribute an accompanying Jupyter notebook containing code that readers can use to produce these models for themselves. A co-published article in this volume (with its own Jupyter notebooks) is dedicated to analysis of the simulations with a focus on comparison to both Bragg and diffuse diffraction data (Wych & Wall, in press).

2. Methods and discussion

2.1 Jupyter notebook

This article presents step-by-step instructions for preparing MD simulations of protein crystals. The Jupyter notebook that accompanies this article is an interactive document containing both exposition and interactive code blocks which can be examined, edited, and run. The notebook contains sections and code blocks corresponding to each of the methods laid out in the sections below, to provide the reader with a practical, interactive experience applying the methods to an example system (staphylococcal nuclease) as they read through this article. The notebook and the files necessary to make use of it are distributed with *Lunus* diffuse scattering software (Wall, 2009), in the *GitHUB* repository (*https://github.com/lanl/lunus/tree/master/examples/tutorials*), along with detailed instructions for how to download and use them. In short, one will need to first install a *conda* distribution such as *miniconda* (*https://docs.conda.io/en/latest/miniconda.html*) and, from the terminal, set up a *conda* environment (*conda create -n crystalMD*; *conda activate crystalMD*) in which one should install *git* and *git-lfs* (*conda install git git-lfs*). Git can then be used to clone the *lunus* repository (*git clone https://github.com/lanl/lunus.git*) and download all the files necessary to work through the notebook. From there, one can traverse to the *tutorials* directory (*cd lunus/examples/tutorials/*) and follow the instructions laid out in the *README.md* and *INSTALL.md* files to install all of the necessary dependencies for the notebooks to run correctly. The notebook for this guide is in the *crystalline_MD_prep/* subdirectory, and is called *MDPreparation.ipynb*. It should be noted that the notebook and the installation guidelines have been prepared and tested on Mac/UNIX operating systems and have not been tested on Windows systems. However, one should be able to make use of them in the Windows Subsystem for Linux (WSL) environment available on Windows version 10 or later (*https://ubuntu.com/wsl*).

2.2 Preparing a structural model: Missing atoms and residues

Structural models deposited in the PDB may be missing atoms if there is not sufficient support for them in the density. This is often due to structural heterogeneity or disorder, with residues at the N- and C-termini among the most common to be left out of structures, but residues on flexible loops or even atoms of individual side chains may also be left out if the density does not support them. To build a crystalline MD model, these atoms must be built back in, to ensure that the content of the model closely resembles what is in the crystal, and to increase the accuracy of the crystal packing, protein–protein, and protein–solvent interactions.

In principle, adding back in missing atoms is a simple procedure. Although their specific conformations are unknown, the chemical composition of missing residues and side chains is typically known, and methods for building computational models of molecular structures are well-developed in the computational chemistry toolkit. However, the space of conformations for even a single side chain can be large: a lysine side chain, e.g., has nine degrees of freedom, corresponding to the nine choices of bond angle and dihedral angle which determine the specific shape of the molecule. For longer stretches of amino acids, with additional independent degrees of freedom for the backbone structure, the conformational state space can be large in the absence of additional information (Levinthal, 1969). Fortunately, the positions of atoms in the crystal structure that have well-resolved experimental density provide valuable information, and we have found that commonly available methods can produce a reasonable starting structure for simulation.

One possible approach to completing the protein model at the N- and C-termini is to extend the secondary structure that already exists in adjacent residues which have well-resolved density (Wych et al., 2019). For example, if the nearest resolved residues in the structure are part of an alpha-helix, the missing residues can be appended to continue the alpha-helix secondary structure. Another option, if it is available, is to model the missing residues using homology modeling [copying the conformation of residues in other, solved structures with the same sequence (Topham et al., 1990)]. Both options are available in *UCSF ChimeraX* (Pettersen et al., 2004; Pettersen et al., 2021) (in *Tools > Structure Editing > Build Structure* or *Model/Refine Loops*) as well as *coot* (Emsley, Lohkamp, Scott, & Cowtan, 2010). Another option for modeling missing regions or generating structures for whole proteins is to use *AlphaFold2* (Jumper et al., 2021), a popular machine learning model which predicts protein structures based on sequence information alone. Initial studies indicate that *AlphaFold2* can reproduce many crystal structures reasonably well, and that poorly predicted regions can be improved using experimental data (Terwilliger et al., 2022).

Care must be taken to ensure that atoms which are added to a crystal structure neither clash with existing atoms nor clash with atoms in symmetry-related copies of the protein. In *coot*, this is quite easy, as one can simply turn on symmetry-related copies, and examine the added residues one by one. However, if the added region is too large to inspect each residue individually, one can also create copies of the protein using *cctbx* tools (using the example below as a guide) and calculate the pairwise distances between the added atoms and all atoms in other symmetry-related copies. If a clash is detected (usually, defined as a pairwise distance less than 3 Å), one can fine-tune the structure in that region. Another option is to use *ChimeraX* to build a unit cell (in *Tools > Higher-order Structure > UnitCell* and click the button *make copies*) and then select the modeled-in residues and check for clashes (in *Tools > Structure Analysis > Find Clashes/Contacts*).

AmberTools (Case et al., 2022) includes yet another option to prepare a protein structure *.pdb* file with missing atoms added, in the simplest case (a protein with only standard amino acid residue types, and missing atoms only on residues already included in the model). Code block 1 from the notebook illustrates how to use this method. It makes use of a standard tool for processing PDB structure files such that they will be ready for simulation, *pdb4amber*, with the *--add-missing-atoms* option. This tool is experimental as of *AMBER21*, but in our experience does a reasonable job of adding heavy atoms if they are missing from PDB structures.

```
!pdb4amber -i protein_calcium.pdb -o protein_calcium_amber.pdb --add-missing-atoms
```

Code block 1 Example of adding missing atoms using *pdb4amber*, taken from the notebook. *pdb4amber* takes in a *.pdb* structure file from the PDB and outputs a *.pdb* file with a format more suitable for simulation. The *--add-missing-atoms* flag tells the program to add heavy atoms missing from the structure, though it will not add entire residue groups (e.g. N- and C-termini), only atoms missing on existing residues in the structure; it is experimental as of *AMBER21*, so may not produce the desired outcome; in this case, one may use an alternative method such as one of the ones discussed here and input the resulting structure to *pdb4amber* with the flag removed.

Another important consideration is the need to add hydrogen atoms to the MD model: structures in the PDB often do not contain hydrogens, if only because X-ray crystallographic experiments at typical resolutions cannot resolve them. Advances in joint neutron- and X-ray crystallographic refinement led to the development of the riding hydrogen method in *cctbx* and *Phenix* which adds hydrogens based on geometric constraints (Liebschner, Afonine, Urzhumtsev, & Adams, 2020). Accurate placement of hydrogens in the structure is important as it can have an appreciable effect on the physical interactions which determine the protein's structure and dynamics. In MD system preparation, adding hydrogens to most residues is straightforward and is performed automatically by the built-in system preparation tools in *AMBER* and *GROMACS* (*pdb4amber* and *pdb2gmx*, respectively). For titratable side chains, it is necessary to inspect the PDB header or the publication associated with the PDB entry for information about the pH of the crystallization buffer. Although the pK_a values in proteins differ from the free amino acid values, this can guide choices for the system preparation. If there are neutron diffraction data available, this information can potentially help guide which specific proton to add, as was done, e.g., in Wall et al. (2019), Wych et al. (2023). It also might be possible to make educated guesses informed by the chemical environment in the region (i.e., is there a negatively charged residue nearby that would be reasonably associated with a partially positively charged nitrogen?); however, in the absence of experimental information, the model will need to be interpreted in light of the uncertainty. In reality, side chains do not exist in one protonation/tautomerization state, but standard MD models do not allow for a dynamically changing protonation state, so one must make compromises [though hybrid-solvent explicit water and Generalized Born/Poisson–Boltzmann-based constant pH simulations are available in both *AMBER* (Case et al., 2022) and *GROMACS* (Berendsen, van der Spoel, & van Drunen, 1995; Abraham et al., 2015), if capturing the effects of (de-) protonation events is important to the study at hand (Baptista, Teixeira, & Soares, 2002; Machuqueiro & Baptista, 2006; Swails, York, & Roitberg, 2014)]. A future simulation paradigm, in which quantum effects such as charge equilibration and protonation reactions are modeled explicitly, would allow for more accurate treatment of these residues, and the effects of these protonation changes on protein structure and dynamics.

2.3 Force fields: Choices and parametrization

The force field refers to the precise potential energy function for the atoms in MD simulation given their configuration, which in turn determines the dynamics. The total potential energy is the sum of terms describing general interactions that are applied to the system being studied through the selection of the force-field parameters. The velocities for the atoms are typically initialized randomly using a Boltzmann distribution. At each step of MD simulation, the positions and velocities of the atoms are fed into the potential energy function defined by the force field, and the dynamics are calculated through iterative rounds of numerical integration of the equations of motion over a small time step [$O(1)$ fs, with the exact value depending on the methods used].

The force field parameters include: the resting geometry and springs constants for bonds and bond angles; parameters to define multimodal torsion angle profiles; atomic partial charges which define the Coulomb interaction; and nonbonded van der Waals interaction parameters. For protein atoms, many force fields have been well-studied and tested and should be suitable for crystalline MD simulations (Best et al., 2012; Cerutti, Swope, Rice, & Case, 2014; Feller & MacKerell, 2000; Hornak et al., 2006; Kaminski, Friesner, Tirado-Rives, & Jorgensen, 2001; Lindorff-Larsen et al., 2010; MacKerell, Banavali, & Foloppe, 2000; Maier et al., 2015; Schuler, Daura, & Van Gunsteren, 2001).

These force fields have been developed and improved over many decades, making heavy use of crystal structures for comparisons. (Historical comparisons to crystal structures for force field development have been performed mostly using solution-state simulations rather than crystalline simulations, which does raise concerns about whether the potentials might contain biases that would be inappropriate for crystalline simulations.) Additional sources of information for force fields include quantum chemistry calculations, NMR data, and free energy measurements. There are also various force fields for simulating explicit water molecules (rather than using continuum or implicit solvent models) which have been refined to reproduce bulk phase structure, pair distribution functions, and thermodynamic data. The most well-studied and commonly used of these are the SPC (Berendsen, Postma, van Gunsteren, & Hermans, 1981), SPC/E (Berendsen, Grigera, & Straatsma, 1987), and TIP3P and TIP4P (Jorgensen, Chandrasekhar, Madura, Impey, & Klein, 1983) models. The MD community is also turning its attention to using crystallography data for improving force fields [see, e.g., (Cavender et al., 2023)]. For example, an expansive investigation of the differences between different various force fields, in the context of crystalline MD simulations can be found in (Janowski et al., 2016). Various choices are reasonable; in our hands, satisfactory crystalline MD simulations have been obtained using CHARMM27 (MacKerell et al., 2000), AMBER99SB-ildn (Lindorff-Larsen et al., 2010), and AMBER14SB (Maier et al., 2015) with TIP3P (Jorgensen et al., 1983) water parameters. Code block 2 contains a line from the notebook that performs a standard parametrization of a protein system with bound calcium ions, using the AMBER99SB-ildn force field and the SPCE water model. The *-ignh* flag replaces the hydrogens in the input *.pdb* file with ones added by *GROMACS*, which facilitates compatibility with the force field naming conventions.

```
!gmx pdb2gmx -f protein_calcium_amber.pdb -o protein_calcium_gmx.pdb \
     -p protein.top --water spce --ff amber99sb-ildn -ignh
```

Code block 2 *pdb2gmx* takes in a *.pdb* structure file and outputs another structure file (in the *.pdb* or *.gro* format; or others, if desired) along with a *.top* topology file consistent with the choice of force field. The *–water* option specifies the desired water model, in this case SPCE (others include tip3p, tips3p, tip4p, tip5p, and SPC). The *–ff* option specifies the desired force field, in this case *amber99sb-ildn*; *GROMACS* comes with several built-in force field directories, but others can be easily added. (If one wishes to use a force field that is unavailable in the distribution of *GROMACS* (or test a new or edited force field parameter set) one can simply place the force field directory in the *GROMACS* directory tree at the location *$GROMACS_HOME/share/gromacs/top/*. Many user-contributed options, including AMBER14SB, are available for download at *https://www.gromacs.org/user_contributions.html*.)

Developing general small molecule force field parameters is challenging because the space of molecules is combinatorially large. Compiling a single set of atom types and their parameters that will work for every input molecule is extremely difficult—a grand challenge in the field of computational chemistry. The Open Force Field Initiative (Mobley et al., 2018), for example, is a collaboration of researchers organized for this exact purpose: automating the tasks of parametrizing small molecules by direct chemical perception. At present, most small molecules are modeled by classifying atoms into atom types that correspond to an element in a particular chemical environment. The atom types are used to obtain the force field parameters. The difficulty arises because the atom type system is an imperfect generalization, and the free energy landscape for small molecules, with even very few degrees of freedom, can still be rather complicated, requiring quantum chemistry calculations of the partial charges, van der Waals parameters, and geometric parameters necessary to correctly model the energy landscape. Researchers often use programs like *AmberTools' antechamber*, *parmchk2*, and *tleap* to generate parameters and then test them in energy minimization and/or simulation. Similar results can be achieved with, e.g., the *CHARMM-GUI* (Jo, Kim, Iyer, & Im, 2008) or the *SwissParam* web server (Zoete et al., 2011). Researchers must be on the guard for unphysical geometries or interactions; if problems are observed in testing, these must be corrected by hand. For example, despite their ubiquity *in vivo*, bioorganic phosphates can assume poor geometries in MD simulations, requiring focused research and new parameter sets for working with (poly-)phosphorylated amino acids and small molecules like ADP and ATP (Meagher, Redman, & Carlson, 2003; Steinbrecher, Latzer, & Case, 2012; Stoppelman, Ng, Nerenberg, & Wang, 2021). One *ad hoc* fix is to add Leonard-Jones (LJ) parameters to the hydrogen(s) (which often do not have LJ parameters). Such a fix may be adequate for many purposes; for example, in the case of a restrained bound ligand, the local geometry may be preserved even if there are deficiencies in the force field. As force fields improve, we can hope that such issues will become increasingly rare.

2.4 The Computational Crystallography Toolbox (cctbx)

In the section that follows, and in the accompanying paper covering the analysis of MD simulations of protein crystals (Wych & Wall, in press), we will rely heavily on the Computational Crystallography Toolbox (*cctbx*) (Grosse-Kunstleve, Sauter, Moriarty, & Adams, 2002). *cctbx* is a software suite for the processing, analysis, and modeling of crystallographic data which [together with *CCP4* (Winn et al., 2011)] forms an open source foundation for *Phenix* (Adams et al., 2010), a widely used crystallographic refinement program. *cctbx* has tools for processing and analyzing molecular structure files and crystallographic data files. Below, we will use *cctbx* to load in a protein structure file and make use of the crystallographic symmetry operations and unit cell box geometry to construct a supercell (2×2×2 tiling of unit cells) from the asymmetric unit structure.

2.5 Expanding the system using the crystalline symmetry

In a standard protein solution MD simulation, one would prepare a box with the protein in the center and space between the protein and the edges of the box into which solvent will be placed. This box is most often rectangular but can also have other suitable shapes, including triclinic, octahedral, or dodecahedral. The edges of the box are commonly treated as a periodic boundary so that, e.g., molecules that exit through the right edge enter through the left; similarly for the top and bottom, and back and front edges (i.e., for a cubic box, the system is topologically a three-dimensional torus). This means that even a solution MD system with periodic boundary conditions effectively corresponds to an infinite, perfect crystal, even if it does not correspond to a physically realized system.

For the specific case of MD simulations of protein crystals, we construct the simulated box such that it is as close as possible to the real protein crystal, using the experimentally determined unit cell information. The simplest approach is to place a copy of the crystal structure at each symmetry-related location. Another approach, used in Wych et al. (2023), is to draw randomly from a set of structures obtained, e.g., using ensemble refinement (Burnley, Afonine, Adams, & Gros, 2012). To handle cases where there are multiple conformations for individual side chains or extended structural regions, conformations may be randomly sampled to be statistically consistent with the occupancies in the crystal structure. In adopting the latter two approaches, it is important to avoid building models with overlapping components. In practice, we have found that this can be done by rejecting individual choices that result in clashes with the parts of the model that already have been built; however, it remains to be seen how generally applicable this technique will be.

The notebook provides an example of how to build a crystalline system using the defined symmetry, beginning with the line shown at the top of Code block 3. Before creating the crystalline MD model, some preprocessing of the *.pdb* file obtained from the PDB is performed. At present there is no general purpose tool for doing this; the notebook shows some steps that are taken in the case of staphylococcal nuclease. The detailed explanation for these steps in this case is included in the notebook, above the cell containing the top line in Code block 3. Similar steps might be required/desired in other cases.

The bottom line of Code block 3 from the notebook runs the *propagate.py* script, included in the *code/* subdirectory, which implements the simplest approach to building the model. The *CRYST1* record in the header of a *.pdb* crystal structure file from the PDB contains the information necessary to expand the asymmetric unit into a full model of the unit cell, using the rotational and translational operations of the space group (Fig. 1). The unit cell is then replicated along each axis to produce a supercell system of the desired size. The more unit cells (N) along a given axis, the finer the sampling ($1/N$) of reciprocal space is with respect to the Bragg lattice for comparisons to diffuse scattering data. In a single unit cell simulation box, atoms separated by distances greater than ½ a unit cell dimension within the box can be much closer, due to the periodic boundary conditions. Therefore, to avoid artificially high correlations between atoms separated by up to a lattice vector along any dimension, a supercell of at least 2×2×2 unit cells is recommended for diffuse scattering simulations (Wall, 2018). Fig. 2A shows such a 2×2×2 supercell created for crystalline staphylococcal nuclease simulations.

```
with open("protein_calcium_gmx.pdb", 'r') as input_prot:
.
.
.
!python ./code/propagate.py -au asymmetric_unit.pdb -o supercell.pdb -x 2 -y 2 -z 2
```

Code block 3 At the bottom of this code block is a line from the notebook that runs *propagate.py*, which uses *cctbx* tools to expand a *.pdb* structure file to the unit cell, and to tile multiple copies of the unit cell creating a supercell. In this case, a 2×2×2 unit cell supercell system is created. Above this line, the structure is preprocessed to create the asymmetric unit *.pdb* file.

2.6 Solvent modeling

Experimentally, finding the correct buffer or mother liquor for crystallization is a key step in obtaining high-quality crystals. Small changes in the composition of the buffer solution can have substantial effects on the stability of protein crystals (or, even, whether crystals form at all). Indeed, for hen egg-white lysozyme, experiments have shown that differences in concentration of NaCl on the order of 1 mM or less can have a large effect on the crystallization success rate (Chen, Yin, Shi, & Ma, 2012).

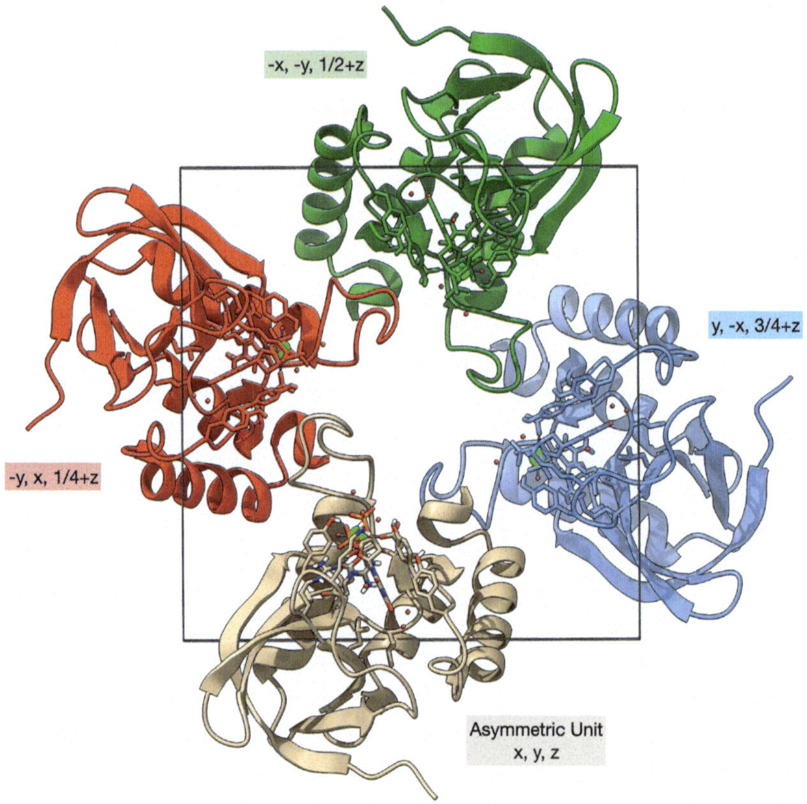

Fig. 1 Protein crystal unit cell of staphylococcal nuclease (PDB: 4WOR), with asymmetric unit in beige, and symmetry-related copies in red, blue, and green, with their associated $P4_1$ space group symmetry operations. All transformations are performed after conversion to fractional coordinates. Unit cell bounding box is indicated using gray lines.

Although faithful modeling of the solvent would no doubt be important for accurate modeling of the protein crystallization process, we do not yet know which details are most critical for simulating diffraction experiments. Current best practice involves explicitly modeling each individual water molecule, as well as trying to incorporate as many small molecules from the crystal buffer as is possible. For many purposes it might be sufficient to neutralize the system with ions (to satisfy the requirements for Ewald summation of electrostatic contributions to the forces and energies), and then to approximate the correct concentrations of

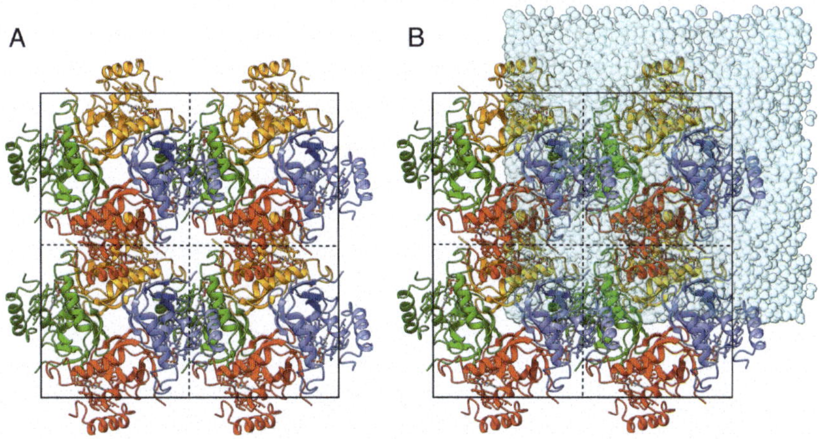

Fig. 2 (A) MD simulation model of crystalline staphylococcal nuclease, viewed along the c-axis. The system is a supercell of 2×2×2 unit cells. This view shows a 2 × 2 configuration of unit cells within the ab-plane, separated by dashed black lines; a second 2 × 2 layer lies beneath the visible layer. Each unit cell contains four proteins configured according to the $P4_1$ symmetry (red, blue, green, and orange cartoons; equivalent proteins in adjacent unit cells are colored the same). (B) The same system as in (A), with waters added (blue transparent spheres). Periodic boundary conditions imply that atoms lying outside the simulation box (black solid lines) are equivalently located at their symmetry-equivalent positions within the box. For example, atoms above the top of the box equivalently may be shifted down by a vertical box length, and atoms to the right of the box similarly may be translated left.

charged particles, by adding additional elemental ions or ionic molecules on top of those needed for neutralization. (In the absence of experimental information, the concentrations of buffer molecules in the crystal is generally assumed to be the same as in the crystal mother liquor, so long as they fit in the solvent cavities. However, the crystalline environment can influence the chemical potential for the buffer constituents, and their concentrations are generally unknown.)

A major advantage of using explicit waters in crystalline MD simulations is that the resulting trajectories can be used to yield information about ordered waters. Various choices made in preparation can have consequences for the model's ability to reproduce the ordered waters from crystallographic models accurately, including the modeling of structural heterogeneity, protonation states on titratable residues, and applying harmonic restraints to bias the protein toward the crystal structure. To assess the ability of the simulations to reproduce the crystallographic waters, all ordered waters from the experimentally determined structure can be removed prior to solvation. However, in some studies, it may be advantageous to maintain selected ordered waters in the structure (e.g., if they are important components of a catalytic mechanism, or an active site chemical environment). One must then choose whether to restrain these waters in their experimentally-determined positions. It is possible, for both types of system preparation, to observe the dynamics/kinetics of ordered water molecules by analysis of the simulation trajectory, e.g., by calculating water density or keeping track of the water occupancy of the region of interest.

Code block 4 from the notebook is the first of a series of *GROMACS* commands that add water to a processed supercell *.pdb* file, and then neutralize the system with ions. In *GROMACS*, solvation of an MD system begins with a tiling of the entire simulated system with a smaller, prefabricated box of water molecules (the default box is *spc216.gro*, a coordinate file describing a 1.86 nm cube filled with 216 water molecules in the conformation specified by the SPC force field parameter set, simulated at 300 K). Water molecules clashing with atoms already present in the simulated system are deleted, leaving water in the cavities between molecules. Ions, or other small molecules, are placed into the system by deleting water molecules and adding ions to the void left behind, to avoid clashes. A solvated system for crystalline staphylococcal nuclease MD simulations is illustrated in Fig. 2B.

```
!gmx editconf -f supercell_no_chains_renumbered.pdb -o supercell_corrected_box.gro
-box 96.998 96.998 126.86 -angles 90 90 90 -noc
```

Code block 4 *GROMACS editconf* is used to ensure that the supercell simulation box has the correct geometry, with the -box and -angles options specifying the lengths of the sides of the box and the angles between the axes. Subsequent steps in the notebook are: *gmx solvate* to fill the supercell system with waters in prefabricated boxes (*spc216.gro*), producing a structure file in the *.gro* format, with waters added, and a new version of the topology file (*supercell.top*) which incorporates information about the newly added waters; *gmx grompp* using the topology and structure files to create a portable binary run input (*.tpr*) file; and *gmx genion* to randomly replace waters with ions. The *-neutral* flag tells *genion* to add as many ions as are necessary to eliminate the net charge. The *-rmin* option of *genion* might need to be set lower than the default to fit in the correct number of ions (0.4 in the notebook example). This option decreases the minimum distance allowed between protein atoms and ions added in. For a standard solution system, there is usually enough water volume surrounding the protein that there will rarely be an issue finding space to adding ions. However, for protein crystal systems, the relative solvent volume is much lower, so it can be harder to find the space to add ions.

2.7 Minimization

Immediately after the MD model is first constructed, the atoms typically are not in a low-energy configuration according to the force field. For instance, a water molecule might be placed with its partially-negatively-charged oxygen adjacent to a negatively charged atom on an amino acid side chain (e.g., aspartate), whereas a more energetically favorable configuration might have the water oriented differently. Energy minimization is the process by which the system is allowed to relax into a state that is more favorable, according to the force field. This is an important first step, as initial configurations with very high energy can cause unphysical dynamics once the simulation is set in motion.

Code block 5 from the notebook marks the beginning of the initial energy minimization step. In the notebook, energy minimization is performed for the solvated and neutralized system, and the portable energy file produced during minimization is analyzed to plot the potential energy as a function of the minimization time step. During minimization, the gradients of the energy with respect to the atomic coordinates are computed and are used to make small adjustments the coordinates in a step-wise manner, moving the system downhill into a local minimum in the energy landscape. Energy minimization normally proceeds until the gradient drops beneath a defined threshold. This is a sign that the system has relaxed into the lowest energy state one can reasonably expect, subject to the force field and initial coordinates. The progress of minimization can be assessed with *gmx energy*, which takes the portable energy (*.edr*) file as input, and outputs the selected analysis data in the *.xvg* format. It can be useful to plot the *y*-axis on a log scale, as the change in energy for the first few steps of minimization can be quite large (orders of magnitude). A useful reference point to keep in mind is that the energy of a hydrogen bond in the secondary structure of peptides in water is ~10 kJ/mol (~2 kcal/mol) (Sheu, Yang, Selzle, & Schlag, 2003)—if changes in energy from one step of energy minimization to the next are on the order of only a few hydrogen bonds over the entire supercell system, one can probably consider the system well relaxed.

```
!gmx grompp -f mdp_files/min.mdp -c supercell_solvated_neutral.gro -p supercell.top
-o min_1.tpr -maxwarn 2
```

Code block 5 Line from the notebook that creates a *GROMACS* input file for energy minimization. *grompp* is used to create a portable binary run input (*.tpr*) file, using the structure (*.gro*) and topology (*.top*) files, according to the simulation parameters specified in the MD parameter (*.mdp*) file. After this step in the notebook, *gmx mdrun* is used to run the minimization, with the *-deffnm* flag specifying the default file name for all of the output files from the simulation (again, specified in the*.mdp* file). Then *gmx energy* is used to output potential energy data for the minimization, for analysis. [Tip: Energy minimization may fail (in *GROMACS*) due to the appearance of an "unsettled water," or a water whose position is such that its contribution to the overall energy is extremely large. A common, reliable fix is to increase the minimum distance to the solute for placing waters during solvation (the *-radius* parameter in *GROMACS' solvate*) so that only waters that are further from the protein are kept in the filtering step, after the system box is tiled.]

2.8 Adjusting the pressure via iterative solvation and equilibration

In standard solution protein MD simulations in the NPT ensemble (constant particle number and controlled pressure and temperature), the pressure is maintained using a barostat which changes the lengths of the simulated box edges on the fly to achieve the desired pressure: decreasing the box side lengths lowers the volume and increases the pressure at constant temperature; vice versa for increasing the box side lengths. For comparison of simulations to experimental diffraction data, it is beneficial to keep the dimensions of the box fixed, and to perform a NVT simulation instead (constant particle number and volume, and controlled temperature). Code block 6 from the notebook prepares a solvated, neutralized, and minimized crystalline MD system for equilibration in the NVT ensemble. Subsequent cells in the notebook run the equilibration simulation, and process the resulting portable energy file to output the pressure as a function of the simulation time step.

```
!gmx grompp -f mdp_files/equil.mdp -c minimization_1/min_1.gro -p supercell.top -r
supercell_solvated_neutral.gro -o equil_1.tpr -maxwarn 2
```

Code block 6 Line from the notebook that creates a *GROMACS* input file for equilibration. *gmx grompp* is used to create a portable binary run input (*.tpr*) file for equilibration, according to the simulation parameters specified in the MD parameter (*.mdp*) file, using the structure (*.gro*) file from the end of minimization and the topology (*.top*) file as input.

After minimization, the first steps of MD simulations enable the system to further relax and sample configurations that are representative of what is favored by the force field. It is common for the pressure in the NVT simulation to be negative after this relaxation (e.g., −1000 bar or below), and for the addition of more waters to be required to increase the system to atmospheric pressure. (A rationale for the negative pressure is that the waters in the predefined boxes used for solvation are packed in a way that might not be energetically favorable near the solute. After solvation, upon energy minimization and equilibration, the waters may pack more favorably, opening voids in the bulk, leading to the negative pressure. These voids may then be filled with additional waters, bringing the system near atmospheric pressure.) This may be accomplished by successive rounds of solvation and equilibration. Our practice has been to continue the iterative solvation and equilibration procedure until the average pressure during equilibration (plus or minus standard error) is within the range from (−100, 100) bar (Fig. 3). Pressure fluctuations during simulation are typically larger than this range, as shown in Fig. 3. Estimates of the number of waters to add in successive rounds of solvation can be arrived at by linear interpolation: that is, tracking the increase in pressure as a function of the number of waters added, and extrapolating the number of waters that would be required to bring the system to the desired pressure of one bar if this trend continued.

2.9 Harmonic restraints

Although MD simulations of protein crystals can remain reasonably close to the crystal structure [the root mean squared distance from the crystal structure for protein atoms on the order of 0.75–2 Å (Janowski et al., 2016; Wall, Adams, Fraser, & Sauter, 2014)], they can deviate enough (especially in the packing) to interfere with comparisons to crystallographic data, e.g., between simulated and experimental structure factors or densities. The long-term solution to this issue is to improve the MD models and force fields. However, in the interim, when fidelity to the crystal structure is critical, one may restrain the simulation to the crystal structure by attaching the heavy atoms to their crystallographic coordinates with springs. Such so-called harmonic restraints are readily applied in common biomolecular MD software; another possible approach, yielding softer restraints, is to fold the copies of the protein back into the asymmetric unit and to restrain the mean coordinates of each atom (computed by averaging over all of the copies) to the crystal structure coordinates (Xue & Skrynnikov, 2014).

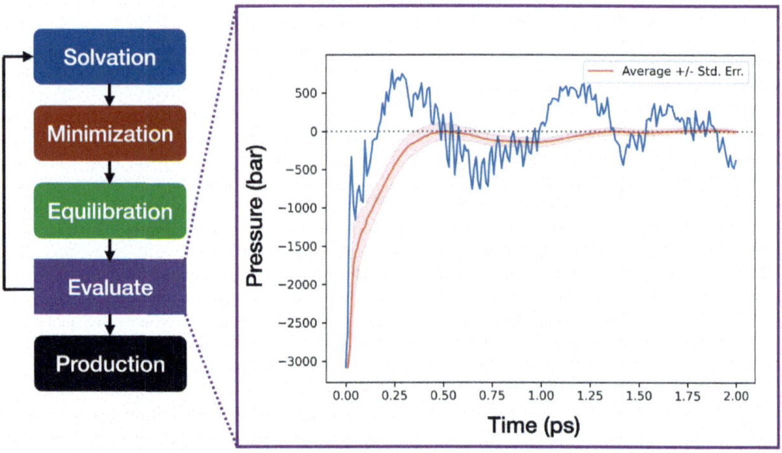

Fig. 3 Left: The flow-chart of the process of iterative solvation and equilibration: waters are added in the solvation step, and the system is energy minimized and equilibrated before the pressure is analyzed in the evaluation step; if the pressure is above or below the range (−100,100) bar, the number of waters and the change in pressure are analyzed to determine a new number of waters needed, which are added in a new solvation step, before the process is repeated; if the average pressure after equilibration is within the range (−100,100) bar, the system is considered ready for production simulation. Right: Fluctuations in pressure during a 2 ps equilibration for a 2×2×2 unit-cell crystalline MD simulation of staphylococcal nuclease (PDB: 4WOR) after two rounds of iterative solvation and equilibration from the Jupyter notebook accompanying this paper. Red line and filled regions represent the running average +/− standard error of the mean. Pressure equilibrates to 2.08 +/− 34.49 bar, indicating that the system is ready for production simulation.

Without restraints, in addition to deviations in the protein structure and packing, the ordered water structure can be disrupted, with elimination of restraints leading to a very significant drop off in the ability of the simulations to reproduce the positions of crystallographic waters (Wall et al., 2019).

There is an inherent tradeoff in the choice of the strength of harmonic restraints: restrain too strongly, and the MD model will not provide useful information about structural heterogeneity; restrain too weakly, and one is susceptible to the biases in the force field that can make the average structure deviate too much from the crystal structure. We have found empirically that a reasonable harmonic restraint strength in *GROMACS* for recovering water structure is about 200 kJ/mol/nm^2, which, although the motions are greatly decreased, still allows for local regions of the protein to deviate substantially from the experimentally determined structure at room temperature (300 K) (Wall et al., 2019). On the other hand, another tradeoff of restraints concerns accurate reproduction of experimentally determined B-factors: here, restraining too strongly will limit conformational heterogeneity, as each atom will behave like a spring with a well-defined spring constant that is the same for all atoms. Restraint strengths are often set equivalently for all heavy atoms; one can of course set the strength of each atom's harmonic restraint independently, but this is not commonly done. Multiple studies using crystalline MD simulations have found that, although the atomic coordinates in unrestrained MD simulations deviate substantially from the crystal structure, they can reproduce the B-factors quite accurately (Cerutti & Case, 2019; Janowski et al., 2013; Janowski et al., 2016; Wall, 2018). We also note that prior studies that have satisfactorily reproduced diffuse scattering data have not used restraints (Meisburger et al., 2020; Meisburger et al., 2023; Wall, 2018; Wall, Van Benschoten, et al., 2014; Wych et al., 2019; Meinhold & Smith, 2005a, b). We have seen that, compared to unrestrained simulations, the ability to reproduce diffuse scattering data can decrease using spring constants of 200 kJ/mol/nm^2 (unpublished); however, it is possible that using weaker restraints might improve diffuse scattering simulations, by preserving the crystal packing while allowing for sufficient structural heterogeneity.

2.10 Production simulation

Production refers to the stretch of simulation that will be used for analysis. During the equilibration step, restraints are often kept quite strong (1000 kJ/mol/nm^2) to ensure that the simulations remain close to the crystal structure. For production, however, one might wish to weaken or even relax restraints to allow for more accurate simulation of B-factors, correlated dynamics, etc. As an example of how to do this, Code block 7 shows a line from the notebook that generates weaker restraints for us in production. The notebook shows how to include these restraints when reading in the topology files and an example of how to use them is provided in a *GROMACS.mdp* parameter file (*prot.mdp*).

```
!echo "Protein-H" | gmx genrestr -f protein_calcium_amber.pdb \
    -o posre_Protein_weaker.itp -fc 200 200 200
```

Code block 7 Line from the notebook that uses *gmx genrestr* to generate weaker restraints for the non-hydrogen atoms of the protein (200 kJ/mol/nm^2).

Production simulations of protein diffraction are usually on the order of 100 ns up to many microseconds. If restraints are completely relaxed, a preproduction stretch of simulation (something on the order of 10 ns or so, possibly longer for large systems) might be useful to ensure that the pressure is still in a desired range, before committing to a longer run; one may add or remove waters as needed to bring the pressure back into range. The simulation duration depends on whether or not restraints are used, and what information is being sought. If restraints are used, a simulation of about 100 ns might be sufficient. If one wishes to sample large scale molecular motions such as conformational changes on the scale of a whole protein, as may be important for modeling the cloudy, large-scale features in diffuse scattering, a simulation duration of a microsecond or longer might be desired. For Bragg analysis, a sampling interval of 10 ps is sufficient from a subset of the trajectory covering the final 10 ns of production simulation produced simulated Bragg data with information content sufficient to improve structure modeling and interpretation (Wych et al., 2023). For diffuse scattering, in studies so far, a sampling interval of 2 ps or below has enabled experimentally observed diffuse scattering features such as streaks/arcs, cloudy features, speckles, and halos to be reproduced (Meisburger et al., 2020; Meisburger et al., 2023; Wall, 2018; Wall, Adams, et al., 2014; Wych et al., 2019). The choice of sampling and averaging strategies can influence the accuracy of diffuse scattering simulations (Wall, 2018); future studies should yield insights into what the optimal strategies might be.

3. Summary

This guide has walked through steps involved in the preparation of MD simulations of protein crystals. First is the preparation of the protein structure and the parametrization of the system: atoms or residues which are missing from the protein structure must be added back in before the crystalline MD model is constructed. Next, the crystal structure is expanded to the unit cell and tiled into a supercell system using *cctbx* tools. Solvation and charge neutralization follows, with iterative rounds of solvation and equilibration to bring the system near atmospheric pressure. In performing a production simulation, there is a choice to be made about whether to use restraints to the crystal structure, with resulting tradeoffs in the fidelity of the simulations to the average structure or the dynamics. In particular, because crystalline MD simulations do not yet simultaneously reproduce the experimental crystal structure and dynamics with high fidelity, unrestrained or very weakly restrained simulations should be preferred if one is interested in studying the dynamics of the system

(B-factors, covariances of atomic displacements, diffuse scattering, etc.), and more strongly restrained simulations should be preferred if one wishes to keep the average structure of the protein closer to the crystal structure, to compare directly to experimental density or structure factor amplitudes, or to study the solvent structure, including ordered water networks.

To accompany this guide, we have also prepared a Jupyter notebook, in which the reader can actively prepare and simulate a system by reading through the notebook and running the code blocks inline. We hope that these resources will allow researchers to make use of these methods more easily. The many connections that can be made to crystallographic observables, including computation of structure factors (densities) from the full system as well as the protein and solvent, separately, the average structure, the B-factors, and the isotropic and anisotropic diffuse scattering, in addition to the analysis of the covariance of atomic displacements and more, are covered in another chapter in this volume (Wych & Wall, in press), which is also accompanied by Jupyter notebooks providing the reader with a similar opportunity for active engagement with the methods.

Acknowledgments

We are grateful to David Case and David Mobley for discussions about crystalline MD simulations. Many thanks to the attendees of the tutorials on molecular-dynamics simulations of biological diffuse scattering at the 57th course of the International School of Crystallography in Erice, Italy, who tested early versions of the Jupyter notebook. David C. Wych and Michael E. Wall were supported by the Exascale Computing Project (grant No. 17-SC-20-SC), a collaborative effort of the US Department of Energy Office of Science and National Nuclear Security Administration. David C. Wych was supported by the Center for Nonlinear Studies at Los Alamos National Laboratory (Technical Release Number: LA-UR-23-23654; under US Department of Energy contract No. 89233218CNA000001 to Triad National Security).

References

Adams, P. D., Afonine, P. V., Bunkoczi, G., Chen, V. B., Davis, I. W., Echols, N., ... Zwart, P. H. (2010). PHENIX: A comprehensive Python-based system for macromolecular structure solution. *Acta Crystallographica Section D, 66*, 213–221.

Ahlstrom, L. S., Vorontsov, I. I., Shi, J., & Miyashita, O. (2017). Effect of the crystal environment on side-chain conformational dynamics in cyanovirin-N investigated through crystal and solution molecular dynamics simulations. *PLoS One, 12*, e0170337.

Avbelj, F., Moult, J., Kitson, D., James, M., & Hagler, A. (1990). Molecular dynamics study of the structure and dynamics of a protein molecule in a crystalline ionic environment, Streptomyces griseus protease A. *Biochemistry, 29*, 8658–8676.

Baptista, A. M., Teixeira, V. H., & Soares, C. M. (2002). Constant-p H molecular dynamics using stochastic titration. *The Journal of Chemical Physics, 117*, 4184–4200.

Berendsen, H. J. C., Grigera, J. R., & Straatsma, T. P. (1987). The missing term in effective pair potentials. *The Journal of Physical Chemistry, 91*, 6269–6271.

Berendsen, H. J. C., Postma, J. P. M., van Gunsteren, W. F., & Hermans, J. (1981). *Interaction models for water in relation to protein hydration in intermolecular forces*. Reidel Publishing Company, 331–342.

Berendsen, H. J. C., van der Spoel, D., & van Drunen, R. (1995). GROMACS: A message-passing parallel molecular dynamics implementation. *Comp Phys Comm, 91*, 43–56.

Berman, H. M., Westbrook, J., Feng, Z., Gilliland, G., Bhat, T. N., Weissig, H., ... Bourne, P. E. (2000). The protein data bank. *Nucleic Acids Research, 28*, 235–242.

Best, R. B., Zhu, X., Shim, J., Lopes, P. E. M., Mittal, J., Feig, M., & MacKerell, A. D., Jr. (2012). Optimization of the additive CHARMM all-atom protein force field targeting improved sampling of the backbone ϕ, ψ and side-chain $\chi1$ and $\chi2$ dihedral angles. *Journal of Chemical Theory and Computation, 8*, 3257–3273.

Bond, P. J., Faraldo-Gómez, J. D., Deol, S. S., & Sansom, M. S. (2006). Membrane protein dynamics and detergent interactions within a crystal: A simulation study of OmpA. *Proceedings of the National Academy of Sciences, 103*, 9518–9523.

Burnley, B. T., Afonine, P. V., Adams, P. D., & Gros, P. (2012). Modelling dynamics in protein crystal structures by ensemble refinement. *Elife, 1*, e00311.

Case, D. A., Aktulga, H. M., Belfon, K., Ben-Shalom, I. Y., Berryman, J. T., Brozell, S. R., ... Kollman, P. A. (2022). *Amber 2022*. San Francisco: University of California.

Cavender, C. E., Case, D. A., Chen, J. C.-H., Chong, L. T., Keedy, D. A., Lindorff-Larsen, K., ... Gilson, M. K. (2023). *Structure-based experimental datasets for benchmarking of protein simulation force fields*. ⟨https://doi.org/10.48550/arXiv.2303.11056⟩.

Cerutti, D. S., & Case, D. A. (2019). Molecular dynamics simulations of macromolecular crystals. *WIREs Computational Molecular Science, 9*, e1402.

Cerutti, D. S., Freddolino, P. L., Duke, R. E., Jr, & Case, D. A. (2010). Simulations of a protein crystal with a high resolution X-ray structure: Evaluation of force fields and water models. *The Journal of Physical Chemistry. B, 114*, 12811–12824.

Cerutti, D. S., Le Trong, I., Stenkamp, R. E., & Lybrand, T. P. (2008). Simulations of a protein crystal: Explicit treatment of crystallization conditions links theory and experiment in the streptavidin–biotin complex. *Biochemistry, 47*, 12065–12077.

Cerutti, D. S., Swope, W. C., Rice, J. E., & Case, D. A. (2014). ff14ipq: A self-consistent force field for condensed-phase simulations of proteins. *Journal of Chemical Theory and Computation, 10*, 4515–4534.

Cerutti, D. S., Trong, I. L., Stenkamp, R. E., & Lybrand, T. P. (2009). Dynamics of the streptavidin–biotin complex in solution and in its crystal lattice: Distinct behavior revealed by molecular simulations. *The Journal of Physical Chemistry. B, 113*, 6971–6985.

Chen, R.-Q., Yin, D.-C., Shi, J.-Y., & Ma, X.-L. (2012). Sensitivity of lysozyme crystallization to minute variations in concentration. *Acta Crystallographica Section D: Biological Crystallography, 68*, 584–591.

Chen, Z., & Chapman, M. S. (2001). Conformational disorder of proteins assessed by real-space molecular dynamics refinement. *Biophysical Journal, 80*, 1466–1472.

Clarage, J. B., Romo, T., Andrews, B. K., Pettitt, B. M., & Phillips, G. N. (1995). A sampling problem in molecular dynamics simulations of macromolecules. *Proceedings of the National Academy of Sciences, 92*, 3288–3292.

Ebrahim, A., Riley, B. T., Kumaran, D., Andi, B., Fuchs, M. R., McSweeney, S., & Keedy, D. A. (2022). The temperature-dependent conformational ensemble of SARS-CoV-2 main protease (Mpro). *IUCrJ, 9*, 682–694.

Emsley, P., Lohkamp, B., Scott, W. G., & Cowtan, K. (2010). Features and development of Coot. *Acta Crystallographica Section D: Biological Crystallography, 66*, 486–501.

Feller, S. E., & MacKerell, A. D. (2000). An improved empirical potential energy function for molecular simulations of phospholipids. *The Journal of Physical Chemistry. B, 104*, 7510–7515.

Fraser, J. S., van den Bedem, H., Samelson, A. J., Lang, P. T., Holton, J. M., Echols, N., & Alber, T. (2011). Accessing protein conformational ensembles using room-temperature X-ray crystallography. *Proceedings of the National Academy of Sciences, 108*, 16247–16252.
Garman, E. F. (2010). Radiation damage in macromolecular crystallography: What is it and why should we care? *Acta Crystallographica Section D: Biological Crystallography, 66*, 339–351.
Ge, Y., Wych, D. C., Samways, M. L., Wall, M. E., Essex, J. W., & Mobley, D. L. (2022). Enhancing sampling of water rehydration on ligand binding: A comparison of techniques. *Journal of Chemical Theory and Computation, 18*, 1359–1381.
Grosse-Kunstleve, R. W., Sauter, N. K., Moriarty, N. W., & Adams, P. D. (2002). The Computational Crystallography Toolbox: Crystallographic algorithms in a reusable software framework. *Journal of Applied Crystallography, 35*, 126–136.
Héry, S., Genest, D., & Smith, J. C. (1997). Fluctuation and correlation in crystalline lysozyme. *Journal of Chemical Information and Computer Sciences, 37*, 1011–1017.
Holton, J. M. (2009). A beginner's guide to radiation damage. *Journal of Synchrotron Radiation, 16*, 133–142.
Hornak, V., Abel, R., Okur, A., Strockbine, B., Roitberg, A., & Simmerling, C. (2006). Comparison of multiple Amber force fields and development of improved protein backbone parameters. *Proteins: Structure, Function, and Bioinformatics, 65*, 712–725.
Hu, Z., & Jiang, J. (2010). Assessment of biomolecular force fields for molecular dynamics simulations in a protein crystal. *Journal of Computational Chemistry, 31*, 371–380.
Janowski, P. A., Cerutti, D. S., Holton, J., & Case, D. A. (2013). Peptide crystal simulations reveal hidden dynamics. *Journal of the American Chemical Society, 135*, 7938–7948.
Janowski, P. A., Liu, C., Deckman, J., & Case, D. A. (2016). Molecular dynamics simulation of triclinic lysozyme in a crystal lattice. *Protein Science, 25*, 87–102.
Jo, S., Kim, T., Iyer, V. G., & Im, W. (2008). CHARMM-GUI: A web-based graphical user interface for CHARMM. *Journal of Computational Chemistry, 29*, 1859–1865.
Jorgensen, W. L., Chandrasekhar, J., Madura, J. D., Impey, R. W., & Klein, M. L. (1983). Comparison of simple potential functions for simulating liquid water. *The Journal of Chemical Physics, 79*, 926–935.
Jumper, J., Evans, R., Pritzel, A., Green, T., Figurnov, M., Ronneberger, O., ... Potapenko, A. (2021). Highly accurate protein structure prediction with AlphaFold. *Nature, 596*, 583–589.
Kaminski, G. A., Friesner, R. A., Tirado-Rives, J., & Jorgensen, W. L. (2001). Evaluation and reparametrization of the OPLS-AA force field for proteins via comparison with accurate quantum chemical calculations on peptides. *The Journal of Physical Chemistry. B, 105*, 6474–6487.
Karplus, M., & McCammon, J. A. (2002). Molecular dynamics simulations of biomolecules. *Nature Structural Biology, 9*, 646–652.
Keedy, D. (2019). Journey to the center of the protein: Allostery from multitemperature multiconformer X-ray crystallography. *Acta Crystallographica Section D, 75*, 123–137.
Keedy, D. A., Kenner, L. R., Warkentin, M., Woldeyes, R. A., Hopkins, J. B., Thompson, M. C., ... Uervirojnangkoorn, M. (2015). Mapping the conformational landscape of a dynamic enzyme by multitemperature and XFEL crystallography. *Elife, 4*, e07574.
Keedy, D. A., Van Den Bedem, H., Sivak, D. A., Petsko, G. A., Ringe, D., Wilson, M. A., & Fraser, J. S. (2014). Crystal cryocooling distorts conformational heterogeneity in a model Michaelis complex of DHFR. *Structure (London, England: 1993), 22*, 899–910.
Krieger, E., Darden, T., Nabuurs, S. B., Finkelstein, A., & Vriend, G. (2004). Making optimal use of empirical energy functions: Force-field parameterization in crystal space. *Proteins: Structure, Function, and Bioinformatics, 57*, 678–683.
Lautz, J., Kessler, H., Kaptein, R., & van Gunsteren, W. F. (1987). Molecular dynamics simulations of cyclosporin A: The crystal structure and dynamic modelling of a structure in apolar solution based on NMR data. *Journal of Computer-Aided Molecular Design, 1*, 219–241.

Levinthal, C. (1969). In J. T. P. De Brunner, & E. Munck (Eds.). *Mossbauer spectroscopy in biological systems* (pp. 22–24). Monticello, Illinois: University of Illinois Press.

Liebschner, D., Afonine, P. V., Urzhumtsev, A. G., & Adams, P. D. (2020). Implementation of the riding hydrogen model in CCTBX to support the next generation of X-ray and neutron joint refinement in Phenix. *Methods in Enzymology, 634*, 177–199.

Lindorff-Larsen, K., Piana, S., Palmo, K., Maragakis, P., Klepeis, J. L., Dror, R. O., & Shaw, D. E. (2010). Improved side-chain torsion potentials for the Amber ff99SB protein force field. *Proteins, 78*, 1950–1958.

Abraham, M. J., Murtola, T., Schulz, R., Páll, S., Smith, J. C., Hess, B., & Lindahl, E. (2015). GROMACS: High performance molecular simulations through multi-level parallelism from laptops to supercomputers. *SoftwareX, 1–2*, 19–25.

Machuqueiro, M., & Baptista, A. M. (2006). Constant-pH molecular dynamics with ionic strength effects: Protonation–conformation coupling in decalysine. *The Journal of Physical Chemistry. B, 110*, 2927–2933.

MacKerell, A. D., Jr, Banavali, N., & Foloppe, N. (2000). Development and current status of the CHARMM force field for nucleic acids. *Biopolymers: Original Research on Biomolecules, 56*, 257–265.

Maier, J. A., Martinez, C., Kasavajhala, K., Wickstrom, L., Hauser, K. E., & Simmerling, C. (2015). ff14SB: Improving the accuracy of protein side chain and backbone parameters from ff99SB. *Journal of Chemical Theory and Computation, 11*, 3696–3713.

Meagher, K. L., Redman, L. T., & Carlson, H. A. (2003). Development of polyphosphate parameters for use with the AMBER force field. *Journal of Computational Chemistry, 24*, 1016–1025.

Mehlman, T., Biel, J. T., Azeem, S. M., Nelson, E. R., Hossain, S., Dunnett, L. E., ... Keedy, D. A. (2023). Room-temperature crystallography reveals altered binding of small-molecule fragments to PTP1B. *Elife, 12*, e84632.

Meinhold, L., & Smith, J. C. (2005a). Fluctuations and correlations in crystalline protein dynamics: A simulation analysis of staphylococcal nuclease. *Biophysical Journal, 88*, 2554–2563.

Meinhold, L., & Smith, J. C. (2005b). Correlated dynamics determining X-ray diffuse scattering from a crystalline protein revealed by molecular dynamics simulation. *Physical Review Letters, 95*, 218103.

Meisburger, S. P., Case, D. A., & Ando, N. (2020). Diffuse X-ray scattering from correlated motions in a protein crystal. *Nature Communications, 11*, 1–13.

Meisburger, S. P., Case, D. A., & Ando, N. (2023). Robust total X-ray scattering workflow to study correlated motion of proteins in crystals. *Nature Communications, 14*, 1228.

Mikhailovskii, O., Xue, Y., & Skrynnikov, N. R. (2022). Modeling a unit cell: Crystallographic refinement procedure using the biomolecular MD simulation platform Amber. *IUCrJ, 9*, 114–133.

Mobley, D. L., Bannan, C. C., Rizzi, A., Bayly, C. I., Chodera, J. D., Lim, V. T., ... Eastman, P. K. (2018). Escaping atom types in force fields using direct chemical perception. *Journal of Chemical Theory and Computation, 14*, 6076–6092.

Papaleo, E., Mereghetti, P., Fantucci, P., Grandori, R., & De Gioia, L. (2009). Free-energy landscape, principal component analysis, and structural clustering to identify representative conformations from molecular dynamics simulations: The myoglobin case. *Journal of molecular graphics and modelling, 27*, 889–899.

Pettersen, E. F., Goddard, T. D., Huang, C. C., Couch, G. S., Greenblatt, D. M., Meng, E. C., & Ferrin, T. E. (2004). UCSF Chimera–A visualization system for exploratory research and analysis. *Journal of Computational Chemistry, 25*, 1605–1612.

Pettersen, E. F., Goddard, T. D., Huang, C. C., Meng, E. C., Couch, G. S., Croll, T. I., ... Ferrin, T. E. (2021). UCSF ChimeraX: Structure visualization for researchers, educators, and developers. *Protein Science: A Publication of the Protein Society, 30*, 70–82.

Schnieders, M. J., Fenn, T. D., Pande, V. S., & Brunger, A. T. (2009). Polarizable atomic multipole X-ray refinement: Application to peptide crystals. *Acta Crystallographica Section D, 65*, 952–965.

Schuler, L. D., Daura, X., & Van Gunsteren, W. F. (2001). An improved GROMOS96 force field for aliphatic hydrocarbons in the condensed phase. *Journal of Computational Chemistry, 22*, 1205–1218.

Sheu, S.-Y., Yang, D.-Y., Selzle, H., & Schlag, E. (2003). Energetics of hydrogen bonds in peptides. *Proceedings of the National Academy of Sciences, 100*, 12683–12687.

Stachowski, T. R., Vanarotti, M., Seetharaman, J., Lopez, K., & Fischer, M. (2022). Water networks repopulate protein–ligand interfaces with temperature. *Angewandte Chemie International Edition, 61*, e202112919.

Steinbrecher, T., Latzer, J., & Case, D. (2012). Revised AMBER parameters for bioorganic phosphates. *Journal of Chemical Theory and Computation, 8*, 4405–4412.

Stocker, U., Spiegel, K., & van Gunsteren, W. (2000). On the similarity of properties in solution or in the crystalline state: A molecular dynamics study of hen lysozyme. *Journal of Biomolecular NMR, 18*, 1–12.

Stoppelman, J. P., Ng, T. T., Nerenberg, P. S., & Wang, L.-P. (2021). Development and validation of AMBER-FB15-compatible force field parameters for phosphorylated amino acids. *The Journal of Physical Chemistry. B, 125*, 11927–11942.

Swails, J. M., York, D. M., & Roitberg, A. E. (2014). Constant pH replica exchange molecular dynamics in explicit solvent using discrete protonation states: Implementation, testing, and validation. *Journal of Chemical Theory and Computation, 10*, 1341–1352.

Terada, T., & Kidera, A. (2012). Comparative molecular dynamics simulation study of crystal environment effect on protein structure. *The Journal of Physical Chemistry. B, 116*, 6810–6818.

Terwilliger, T. C., Leibschner, D. L., Croll, T., Williams, C. J., McCoy, A. J., Poon, B. K., ... Read, R. J. (2022). AlphaFold predictions: Great hypotheses but no match for experiment. *bioRxiv, 2011*(2021), 517405.

Topham, C. M., Thomas, P., Overington, J. P., Johnson, M. S., Eisenmenger, F., & Blundell, T. L. (1990). An assessment of COMPOSER: a rule-based approach to modelling protein structure. In J. Kay, G. Lunt, & D. Osguthorpe (Vol. Eds.), *Vol 57. Protein Structure, Prediction and Design, Biochemical Society Symposium* (pp. 81–90). Portland Press, London.

Van Gunsteren, W., & Berendsen, H. (1984). Computer simulation as a tool for tracing the conformational differences between proteins in solution and in the crystalline state. *Journal of Molecular Biology, 176*, 559–564.

van Gunsteren, W. F., & Mark, A. E. (1998). Validation of molecular dynamics simulation. *The Journal of Chemical Physics, 108*, 6109–6116.

Verli, H., Calazans, A., Brindeiro, R., Tanuri, A., & Guimarães, J. A. (2007). Molecular dynamics analysis of HIV-1 matrix protein: Clarifying differences between crystallographic and solution structures. *Journal of Molecular Graphics and Modelling, 26*, 62–68.

Vitkup, D., Ringe, D., Karplus, M., & Petsko, G. A. (2002). Why protein R-factors are so large: A self-consistent analysis. *Proteins: Structure, Function, and Bioinformatics, 46*, 345–354.

Vorontsov, I. I., & Miyashita, O. (2009). Solution and crystal molecular dynamics simulation study of m4-cyanovirin-N mutants complexed with di-mannose. *Biophysical Journal, 97*, 2532–2540.

Wall, M. E. (2009). Methods and software for diffuse X-ray scattering from protein crystals. *Methods in Molecular Biology, 544*, 269–279.

Wall, M. E. (2018). Internal protein motions in molecular-dynamics simulations of Bragg and diffuse X-ray scattering. *IUCrJ, 5*, 172–181.

Wall, M. E., Adams, P. D., Fraser, J. S., & Sauter, N. K. (2014). Diffuse X-ray scattering to model protein motions. *Structure (London, England: 1993), 22*, 182–184.

Wall, M. E., Calabró, G., Bayly, C. I., Mobley, D. L., & Warren, G. L. (2019). Biomolecular solvation structure revealed by molecular dynamics simulations. *Journal of the American Chemical Society, 141*, 4711–4720.

Wall, M. E., Clarage, J. B., & Phillips, G. N., Jr (1997). Motions of calmodulin characterized using both Bragg and diffuse X-ray scattering. *Structure (London, England: 1993), 5*, 1599–1612.

Wall, M. E., Van Benschoten, A. H., Sauter, N. K., Adams, P. D., Fraser, J. S., & Terwilliger, T. C. (2014). Conformational dynamics of a crystalline protein from microsecond-scale molecular dynamics simulations and diffuse X-ray scattering. *Proceedings of the National Academy of Sciences, 111*, 17887–17892.

Warkentin, M., Hopkins, J. B., Badeau, R., Mulichak, A. M., Keefe, L. J., & Thorne, R. E. (2013). Global radiation damage: Temperature dependence, time dependence and how to outrun it. *Journal of Synchrotron Radiation, 20*, 7–13.

Weichenberger, C. X., Afonine, P. V., Kantardjieff, K., & Rupp, B. (2015). The solvent component of macromolecular crystals. *Acta Crystallographica Section D: Biological Crystallography, 71*, 1023–1038.

Winn, M. D., Ballard, C. C., Cowtan, K. D., Dodson, E. J., Emsley, P., Evans, P. R., ... Wilson, K. S. (2011). Overview of the CCP4 suite and current developments. *Acta Crystallographica Section D, 67*, 235–242.

Wych, D. C. & Wall, M. E. (in press). Molecular-dynamics simulations of macromolecular diffraction, part II: Analysis of protein crystal simulations. *Methods in Enzymology*.

Wych, D. C., Aoto, P. C., Vu, L., Wolff, A. M., Mobley, D. L., Fraser, J. S., ... Wall, M. E. (2023). Molecular-dynamics simulation methods for macromolecular crystallography. *Acta Crystallographica Section D: Structural Biology, 79*, 50–65.

Wych, D. C., Fraser, J. S., Mobley, D. L., & Wall, M. E. (2019). Liquid-like and rigid-body motions in molecular-dynamics simulations of a crystalline protein. *Structural Dynamics, 6*, 064704.

Xue, Y., & Skrynnikov, N. R. (2014). Ensemble MD simulations restrained via crystallographic data: Accurate structure leads to accurate dynamics. *Protein Science: A Publication of the Protein Society, 23*, 488–507.

York, D. M., Wlodawer, A., Pedersen, L. G., & Darden, T. A. (1994). Atomic-level accuracy in simulations of large protein crystals. *Proceedings of the National Academy of Sciences, 91*, 8715–8718.

Zoete, V., Cuendet, M. A., Grosdidier, A., & Michielin, O. (2011). SwissParam: A fast force field generation tool for small organic molecules. *Journal of Computational Chemistry, 32*, 2359–2368.

CHAPTER FOUR

Molecular-dynamics simulations of macromolecular diffraction, part II: Analysis of protein crystal simulations

David C. Wych[a,b] and Michael E. Wall[a,*]
[a]Computer, Computational and Statistical Sciences Division, Los Alamos, NM, United States
[b]Center for Nonlinear Studies, Los Alamos National Laboratory, Los Alamos, NM, United States
*Corresponding author. e-mail address: mewall@lanl.gov

Contents

Abstract

Molecular-dynamics (MD) simulations have contributed substantially to our understanding of protein structure and dynamics, yielding insights into many biological processes including protein folding, drug binding, and mechanisms of protein–protein interactions. Much of what is known about protein structure comes from macromolecular crystallography (MX) experiments. MD simulations of protein crystals are useful in the study of MX because the simulations can be analyzed to calculate almost any crystallographic observable of interest, from atomic coordinates to structure factors and densities, B-factors, multiple conformations and their populations/occupancies, and diffuse scattering intensities. Computing resources and software to support crystalline MD simulations are now readily available to many researchers studying protein structure and dynamics and who may be interested in advanced interpretation of MX data, including diffuse scattering. In this work, we outline methods of analyzing MD simulations of protein crystals and provide accompanying Jupyter notebooks as practical resources for researchers wishing to perform similar analyses on their own systems of interest.

Methods in Enzymology, Volume 688 2023 Published by Elsevier Inc.
ISSN 0076-6879, https://doi.org/10.1016/bs.mie.2023.06.012

1. Introduction

Molecular-dynamics (MD) simulations of protein crystals can be processed to calculate a number of crystallographic observables of interest in macromolecular crystallography (MX). MX crystal structures in the Protein Data Bank [PDB, *rcsb.org* (Berman et al., 2000)] consist of coordinates for each atom and their associated B-factors [also known as atomic displacement parameters (ADPs)], which represent the degree of small-scale harmonic disorder for the atoms, and occupancies, which represent the fraction of unit cells containing the atom at the coordinates specified. These properties are not directly read out from the experimental data but are refined altogether as model parameters until the predictions match the experimental data as closely as possible (see, e.g., Refs. Afonine et al., 2012; Murshudov et al., 2011). Together, these parameters define the structural model and a corresponding density for the crystal's unit cell. Using crystallographic structural models as input, MD simulations of protein crystals can provide an independent source of information about each one of these parameters, providing insight in to the structural and dynamical features of the protein and solvent which the model attempts to capture through refinement.

MD simulations provide information about protein crystal structure and dynamics that can be useful for enhancing crystallography studies. For example, variations in protein atom positions cannot always adequately be modeled with small-scale harmonic oscillations about a single average position, even at cryogenic temperatures (García, Krumhansl, & Frauenfelder, 1997; Keedy et al., 2014; Keedy et al., 2015), leading to methods for obtaining crystal structures with multiple conformations of selected side chains (Keedy et al., 2015; Riley et al., 2021) or ensemble models with multiple copies of the entire protein (Burnley, Afonine, Adams, & Gros, 2012; Kuriyan et al., 1991; Ploscariu, Burnley, Gros, & Pearce, 2021; Wall, Clarage, & Phillips, 1997). In addition, beyond the ordered waters in the first solvation layer, solvent density is usually modeled as uniform within the bulk regions, motivating methods to better characterize the solvent density (Afonine, Adams, Sobolev, & Urzhumtsev, 2021; Badger & Caspar, 1991). MD simulations can provide independent information for these tasks by modeling the full protein structural ensemble, including anharmonic motions and structural heterogeneity, and by including a full atomic-detail model of the solvent, with each water molecule, ion, and constituent of the crystallographic buffer solution represented explicitly. Early protein crystal MD simulations compared the simulated water structure to experimental crystal structures (van Gunsteren, Berendsen, Hermans, Hol, & Postma, 1983), and recent work shows that the simulations are capable of accurately reproducing the positions of ordered waters (Wall, Calabró, Bayly, Mobley, & Warren, 2019). Crystalline MD simulations can be used to recover the B-factors from refinement (Cerutti & Case, 2019; Janowski, Cerutti, Holton, & Case, 2013; Janowski, Liu, Deckman, & Case, 2016; Wall, 2018), and can aid in interpretation of crystallographic data by computing densities from any individual component of the MD model; for example, a recent study leveraged this strategy to develop an alternative ordered solvent model and to discover alternative conformations in crystallographic refinement (Wych et al., 2023).

MD simulations also have the potential to play a role in understanding the differences between cryogenic and room-temperature protein crystals. Over 95% of the MX structures in the PDB were collected at cryogenic temperatures (Keedy et al., 2015), to protect against radiation damage [protein crystals are thirty to fifty times less sensitive to radiation at 100 K versus 300 K (Warkentin et al., 2013)]. These cryo-temperature structures are often used to make inferences about protein structure and function at room temperature. However, a growing literature suggests that protein and water structure change quite dramatically between cryogenic and room temperature (Keedy et al., 2015; Russi et al., 2017; Stachowski, Vanarotti, Seetharaman, Lopez, & Fischer, 2022)—not to mention changes in protein structural heterogeneity (Fraser et al., 2011; Keedy et al., 2014; Keedy, 2019) and ligand binding (Mehlman et al., 2022). Controlled studies comparing MD simulations of cryogenic vs room-temperature conditions potentially can be used to better understand the differences between these states (Nagai, Tama, & Miyashita, 2019; Vitkup, Ringe, Petsko, & Karplus, 2000). In addition, in an effort to obtain room-temperature data without the effects of radiation damage, researchers have begun to make use of X-ray free electron lasers (XFELs), which produce very brilliant, fast pulses of X-rays (Barends, Stauch, Cherezov, & Schlichting, 2022)—pulses so brilliant that they cause the crystal to explode, but so brief that diffraction measurements can be made before they do, even at room temperature. These quick pulses also allow for time-resolved experiments with resolutions from microseconds down to as low as femtoseconds. Such time-resolved experiments are an especially attractive target for MD simulations of protein crystals, which use femtosecond-scale time steps and can readily probe the relevant time scales.

Researchers also have been looking to expand the scope of X-ray crystallographic measurement and modeling to include information about correlated dynamics: to this end, there has been increased interest in the study of diffuse scattering in macromolecular crystallography (see, e.g., reviews Benoit & Doucet, 1995; Clarage, Phillips, 1997; Meisburger, Thomas, Watkins, & Ando, 2017; Wall, Adams, Fraser, & Sauter, 2014; Xu, Meisburger, & Ando, 2021). Standard crystallographic modeling is concerned with only the Bragg data—a lattice of bright spots in the diffracted X-ray intensity, whose amplitudes encode information about the average electron density in the crystal's unit cell. The diffuse scattering is the intensity that lies between the Bragg peaks and is associated with correlated variations in the electron density (Meisburger et al., 2017), which are especially interesting when they are due to protein dynamics. It is much weaker, pixel-by-pixel, than the Bragg peaks in diffraction images, but it accounts for most of the scattered X-rays in macromolecular crystallography experiments (Clarage, Phillips, 1997). The diffuse scattering encompasses both relatively sharp haloes around Bragg peaks, which result from large-scale phonon-like motions (Meisburger, Case, & Ando, 2020; Polikanov & Moore, 2015), and cloudy, patterned features that result from correlations on a shorter length scale (see, e.g., review Meisburger et al., 2017). The diffuse scattering is more challenging to model than the Bragg data, as there are far too many independent parameters for all two-point correlations of atomic displacements to be refined directly.

One route to progress in diffuse scattering studies is to develop phenomenological models with a limited number of parameters (Ayyer et al., 2016; Caspar, Clarage, Salunke, & Clarage, 1988; Clarage, Clarage, Phillips, Sweet, & Caspar, 1992; Doucet & Benoit, 1987; Meisburger et al., 2020; Moore, 2009; Peck, Poitevin, & Lane, 2018; Phillips, Fillers, & Cohen, 1980; Polikanov & Moore, 2015; Wall et al., 1997; Wych, Fraser, Mobley, & Wall, 2019). Such models so far have been concerned with the diffuse scattering from the protein only, and so ignore contributions from the solvent (or interactions of the protein with the solvent). These models are parametrized using variables which are tuned to optimize agreement with the diffuse scattering data, similar to structural refinement.

A complementary route to progress has been the development of crystalline MD simulation models of diffuse scattering (Clarage, Romo, Andrews, Pettitt, & Phillips, 1995; Faure et al., 1994; Hery, Genest, & Smith, 1998; Meinhold & Smith, 2005a; Meisburger et al., 2020; Meisburger, Case, & Ando, 2023; Wall et al., 2014; Wall, 2018; Wych et al., 2019). The crystalline MD model is the only one, currently, which includes contributions to the diffuse scattering from both the protein and the solvent (Meinhold & Smith, 2005a; 2005b; Meisburger et al., 2020; Meisburger et al., 2023; Wall et al., 1997; Wall et al., 2014, 2018). In addition, the MD model is not refined: the predictions of the model are determined by the force field, the choices made in the preparation of the system, and the character of the production simulation. As such, presently, the crystalline MD model provides independent information about dynamics that can be validated using the experimental data. The agreement that has been achieved between the MD simulations and the total diffuse scattering data is high (Meinhold & Smith, 2005a; 2005b; Meisburger et al., 2020; Wall et al., 2014; Wall, 2018; Wych et al., 2019) however, the total includes a large isotropic component that includes roughly equal contributions from both the solvent and protein in MD simulations (Meinhold & Smith, 2005b; Wall et al., 2014). The agreement of the MD simulations with the anisotropic component, which accounts for ~10% of the total signal and in MD simulations is dominated by the contribution from the protein (Wall et al., 2014), is not yet as high as that obtained by fitting the phenomenological models [particularly the elastic network/normal modes models and liquid-like motions models (Meisburger et al., 2020)]. Improving the agreement of the crystalline MD with the diffuse scattering is a major goal in the field and not only will increase our understanding of protein crystal dynamics but might eventually benefit MD simulations more broadly by improving force fields (Cavender et al., 2023).

Despite the increasing utility of MD simulations in MX experiments, the number of researchers making use of these simulations for this purpose is currently limited by the lack of automated tools, making the barrier to entry relatively high. In this work, which is a companion to the MD preparation paper in this volume (Wych & Wall, in press), we provide examples and discussion of the methods used to calculate structure factors and densities (decomposed into protein and solvent components), refine a structural model against the data computed from simulation, calculate the diffuse scattering (isotropic and anisotropic components), analyze the covariance matrix of atomic displacements (which is intimately related to the diffuse scattering), produce a protein structural ensemble model from each frame of the simulation, and predict the temperature factors by both fluctuation analysis and refinement. We hope this chapter, and the accompanying Jupyter notebook (https://www.github.com/lanl/examples/tutorials), will serve as an on-ramp for researchers in structural biology or computational chemistry wishing to analyze MD simulations in conjunction with crystallographic data.

2. Methods and discussion

2.1 Jupyter notebooks

This article is accompanied by Jupyter notebooks, which are interactive documents containing both exposition and interactive code blocks that can be examined, edited, and run. The notebooks are divided according to their application to Bragg or diffuse analysis. We have organized them to follow the order below, facilitating their use side-by-side while reading the article. The Bragg notebook provides examples using cAMP-dependent protein kinase (PKA), and covers the subject matter of Ref. (Wych et al., 2023); the diffuse notebook uses staphylococcal nuclease (SNase) and covers the subject matter of Ref. (Wych et al., 2019). Numbered Code blocks in this review are copied from the notebooks to serve as cross-references. The notebooks and the files necessary to make use of them are distributed with *Lunus* software for diffuse scattering (Wall, 2009), in the *GitHub* repository (https://github.com/lanl/lunus/tree/master/examples/tutorials/). In short, one will need to first install a *conda* distribution such as *miniconda* (https://docs.conda.io/en/latest/miniconda.html) and, from the terminal, set up a *conda* environment (*conda create -n crystalMD*; *conda activate crystalMD*) in which one should install *git* and *git-lfs* (*conda install git git-lfs*). Git can then be used to clone the *lunus* repository (*git clone* https://github.com/lanl/lunus.git) and download all the files necessary to work through the notebook. From there, one can traverse to the *tutorials* directory (*cd lunus/examples/tutorials/*) and follow the instructions laid out in the *README.md* and *INSTALL.md* files to install all of the necessary dependencies for the notebooks to run correctly. The notebook for the Bragg analysis is found in the *crystalline_MD_analysis_bragg/* subdirectory, named *MDAnalysisBragg.ipynb*. The notebook for the diffuse analysis is found in the *crystalline_MD_analysis_diffuse/* subdirectory, named *MDAnalysisDiffuse.ipynb*. Instructions for downloading and using the notebooks are in the *README.md* and *INSTALL.md* files in the *tutorials/* directory. It should be noted that these notebooks and the installation guidelines have been prepared and tested on Mac/UNIX operating systems, and have not been tested on Windows systems. However, one should be able to make use of them in the Windows Subsystem for Linux (WSL) environment available on Windows version 10 or later (https://ubuntu.com/wsl).

2.2 Preparation of the MD trajectory for analysis

In the complementary chapter in this volume (Wych & Wall, in press), we reviewed methods for preparing protein crystalline systems for MD simulations. Following the instructions in that chapter (with examples provided in the associated Jupyter notebook) one should be able to produce a trajectory for a protein crystalline system, which can then be analyzed using any of the methods outlined in this paper. There are a few basic methods one must use to process the raw trajectory from MD simulation to prepare it for analysis: extracting a subset of the trajectory, down-sampling in time, trajectory fitting, and periodic boundary correction.

One often needs to extract a subset of the full trajectory for analysis, as the simulation takes time to arrive at a stable state; whereas unrestrained simulations can take microseconds to arrive at an equilibration state (measured by e.g., root-mean-squared deviation from the initial structure) (Case, in press), harmonically restrained simulations can arrive at an equilibrated state in less than 100 ns (see, e.g., Wych et al., 2023). Diffuse scattering calculations have been computed from trajectory subsets from hundreds of nanoseconds (Wych et al., 2019) to many microseconds (Meisburger et al., 2020; Wall et al., 2014; Wall, 2018). Short duration simulations (e.g., shorter than 1 ns) have insufficient sampling for reproducible calculations of the cloudy diffuse features that vary about the isotropic signal (Clarage et al., 1995), but microsecond-scale simulations overcome this limitation sufficiently to yield satisfactory calculations (Wall et al., 2014). The difficulty in sampling is due to the lack of convergence of pairwise atom displacement correlations on short time scales (Meinhold & Smith, 2005a). Another chapter in this volume demonstrates that, for a 5 μs simulation of a 7 × 7 × 7 crystalline system of triclinic lysozyme, decreasing the length of the processed trajectory from 5 μs to 400 ns leads to only a small differences in calculated total diffuse maps that include the isotropic features (with a correlation coefficient of 0.9859 between the two) (Case, in press). For Bragg analysis, convergence times are much shorter, as the signal is only sensitive to the mean density; studies have used trajectories as long as many microseconds (Janowski et al., 2013), but a subset as short as 10 ns has proven sufficient when analyzing simulations that are restrained to the crystal structure (Wall et al., 2019; Wych et al., 2023).

Down-sampling refers to selecting frames from the trajectory separated by time steps longer than that of the original trajectory; this can be necessary because trajectory files are quite large (tens to hundreds of gigabytes) and/or analysis may not require the processing of every frame. For analysis of diffuse scattering, sampling every 2 ps is reasonable for simulating the cloudy features that vary about the isotropic signal (Wall et al., 2014). When focusing on the entire signal, which is dominated by the isotropic component, or on the relatively bright diffuse haloes in the neighborhood of the Bragg peak, coarser sampling may be possible: for a 5 μs simulation of a 7 × 7 × 7 crystalline system of triclinic lysozyme, changing from sampling every 0.4 ns to sampling 40 ns only introduces small differences to the calculated diffuse scattering maps (with a correlation coefficient 0.9775 between the two) (Case, in press). For Bragg analysis, one can sample frames every 10 ps and achieve satisfactory results using restrained simulations (Wych et al., 2023).

Trajectory fitting refers to translating selected atoms in each frame of the simulation such that the distance between the atoms in the frames and in a reference structure is minimized. This is necessary, as the center of mass of the supercell can drift concertedly from its positions at the start, due to the finite precision of the numerical integration (Case, in press). In addition to making trajectories harder to interpret and visualize, this drift can introduce errors in the phases of the structure factors computed from the trajectory if it is not corrected.

Periodic boundary correction refers to the process of mapping coordinates from the trajectory to equivalent coordinates, subject to the periodic boundary conditions (PBCs), such that the trajectory is physically consistent over time: for example, atoms in a protein which exit the periodic box on one side re-appear on the opposite side; however, this jump from one side of the box to the other is unphysical, it is simply a result of the periodic mapping. Examples of periodic boundary correction include keeping molecules whole (mapping back atoms of proteins which cross the periodic boundary, so the protein is kept intact) and eliminating jumps in the trajectory when the center of mass of the protein crosses the periodic boundary. Such corrections are mainly important for visualization but may also be important for alignment of frames prior to computations, depending on the method of alignment employed.

All of these methods of trajectory processing can be accomplished with *GROMACS*' *trjconv* method, as described in the Bragg analysis notebook. Using *GROMACS* (and potentially other software) for alignment, it is important to carry out the periodic boundary correction (options *-pbc mol* and *-pbc nojump,* in that order) in advance, to correct for the effect of atoms jumping to the opposite side of the box during the simulation, which can introduce errors. The upper right panel of Fig. 1 shows a visualization of a snapshot after processing, in the context of the MD–MX procedure (Wych et al., 2023), a set of methods using Bragg analysis of MD simulations to enhance MX protein structure determination.

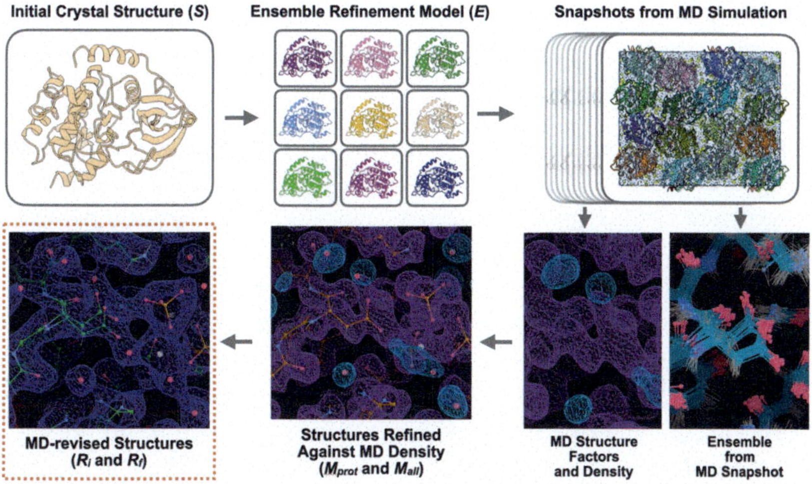

Fig. 1 The focus of the Bragg analysis of crystalline MD simulations in this review is the MD–MX procedure to revise macromolecular crystal structures (Wych et al., 2023), illustrated here. In this procedure, an initial crystal structure (top left) is used to prepare an MD simulation model: in (Wych et al., 2023) this was done by using *phenix.ensemble_refinement* (Burnley et al., 2012) to generate an ensemble of 32 structures (top center, illustrated using 9 differently colored cartoon ribbons) that were randomly placed at each symmetry-related location in a 2×2×2 supercell model; however, as described in (Wych & Wall, in press), a single-structure or multiconformer model may also be used to build the supercell. Snapshots from the MD simulation (top right) are used to generate simulated data and MD ensemble visualizations (bottom right; protein density in purple and water density in cyan). The initial structure is refined against the simulated data, leading to a revised protein structure and a new water network (bottom center; densities on bottom right). The revised structure is refined against the experimental data to produce the initial MD-revised structure R_i, and manual improvements are made guided by the MD density and snapshots, leading to a final MD-revised structure R_f (bottom left; $2F_o$–F_c density in blue at 1σ with traces of positive F_o–F_c density at visible in green at 3σ). Reproduced from *Acta Crystallographica Section D* (Wych et al., 2023) under the Creative Commons Attribution License.

2.3 Bragg analysis

Once the trajectories have been processed as described above, both the Bragg and diffuse analysis may be performed. We begin by describing the Bragg analysis. The MD–MX procedure (Wych et al., 2023) provides the context for the Bragg analysis methods described in this review, and the accompanying Bragg analysis notebook (see Jupyter notebooks) contains code to carry out elements of the analysis. Working through the examples in the notebook will give readers experience with key steps in the

MD–MX procedure that should facilitate the application of the methods to other systems. The organization of this section follows that in the notebook, so that they can be used side-by-side.

2.3.1 Calculating mean structure factors and intensities from the MD trajectory

The structure factor $F(\boldsymbol{q})$ is the Fourier Transform of the electron density of the simulation volume, where $\boldsymbol{q}$ is the scattering vector. The intensity of diffracted X-rays measured by a detector is proportional to $|F(\boldsymbol{q})|^2$ meaning that information about the density itself is hard to directly extract because the phases of the structure factors are lost ($|Ae^{-ix}|^2 = Ae^{-ix} Ae^{ix} = A^2$). Despite this, it is still possible to extract information out of these diffraction experiments, because the magnitudes of the structure factors are still measured: crystallographers model the electron density by building an atomistic structural model and compare the magnitude of the calculated structure factors with the magnitude of structure factors from experiment.

Using standard crystallography methods, structure factors can be calculated from each frame of an MD simulation as

$$F(\boldsymbol{q}) = \sum_i f_i(\boldsymbol{q}) e^{i\boldsymbol{q}\cdot\boldsymbol{r}_i}$$

where $f_i(\boldsymbol{q})$ is the atomic form factor of atom i, and $\boldsymbol{r}_i$ is its position. The *Lunus* software package (https://www.github.com/lanl/lunus) contains many tools designed for the study of diffuse scattering, including software for working with, processing, and analyzing experimentally collected diffraction data as well as calculating simulated scattering data from MD simulation trajectories. The method *xtraj.py* is a Python script which computes structure factors, intensities, and diffuse scattering from MD simulation trajectories in the *GROMACS .xtc* format. It uses the Computational Crystallography Toolbox (Grosse-Kunstleve, Sauter, Moriarty, & Adams, 2002), or *cctbx*, and can be run on multiple machines in parallel with MPI. Mean structure factors are output by *xtraj.py* in the *.mtz* file format (by default, in a file named *fcalc.mtz*) and computed using

$$\langle F(\boldsymbol{q})\rangle = \frac{1}{T}\sum_t F_t(\boldsymbol{q}).$$

where $F_t(\boldsymbol{q})$ is the structure factor computed from the frame at time t and T is the total number of frames. The mean intensity (output by *xtraj.py*, by default, in the file named *icalc.mtz*) is computed using

$$\langle |F(\boldsymbol{q})|^2\rangle = \frac{1}{T}\sum_t |F_t(\boldsymbol{q})|^2.$$

Once *Lunus* is installed, *xtraj.py* can be called with a simple command-line call. An example of an *xtraj.py* command from the notebook is shown in Code block 1. The trajectory is from a simulation of a system of 2×2×2 unit cells in GROMACS *.xtc* format [methods to analyze crystalline MD trajectories to produce information about RMS deviations, B-factors, individual protein trajectories (*XtalAnalyze.sh* and *XtalPlot.sh*), and electron density maps (*md2map.sh*) are also available in the *XtalAnalysis* package in the *AMBER18* software suite (Case, in press)]. The density of the asymmetric unit is desired for the subsequent analysis; this can be obtained by specifying the unit cell and space group of the original crystal structure, which folds the density into the asymmetric unit; placing the density on an absolute scale then requires dividing the result by the number of copies of the asymmetric unit present in the simulation volume (e.g., 32 copies for a 2×2×2 system of PKA in space group $P2_12_12_1$). In this example, the entire system is selected; the notebook also shows how to select just the water molecules or just the rest of the system, without the water molecules, which is dominated by the protein. The lower right panel of Fig. 1 shows an example of such a decomposition of the density into water and the protein. Structure factors can be calculated from any component of the simulation—one could even calculate the density of a single water molecule or atomic ion as it wanders throughout the crystal's solvent cavities. Such a decomposition of the density into protein and solvent components cannot be done directly using experimental data, as the experimental density does not come with labels that would allow the components to be distinguished. The ability to decompose the MD density into components provides unique information that can be leveraged to enhance MX structure determination, e.g., by helping to distinguish between the protein and the solvent (Wych et al., 2023).

```
!python {xtraj} \
        traj="./data/PKA_sample_trajectory.xtc" \
        top="./data/PKA_crystal_structure_supercell.pdb" \
        unit_cell="58.976,79.7436,99.058,90.0,90.0,90.0" \
        space_group="P212121" d_min=1.6 first=0 last=50 \
        diffuse=diffuse_full_PKA_sample.hkl selection="all" \
        fcalc=fcalc_full_PKA_sample.mtz icalc=icalc_full_PKA_sample.mtz
```

Code block 1 Example *xtraj.py* execution to compute structure factors and intensities for the asymmetric unit from the trajectory in the trajectory file, copied from the Bragg analysis notebook (the full path to *xtraj.py* is provided using the xtraj variable in braces). The trajectory is in the *GROMACS .xtc* format, and the topology information is specified using the reference structure file (in the *.pdb* format). The *unit_cell*, and *space_group* parameters tell *xtraj.py* how to fold the density of the P_1 supercell into the asymmetric unit. The command specifies a trajectory beginning at frame 0 (the first frame) and ending at frame number 50 (51 frames total). Results are truncated at a resolution of 1.6 Å.

To visualize the MD structure factors as densities, a straightforward method is to use *coot* (Emsley, Lohkamp, Scott, & Cowtan, 2010), which is available in a CCP4 installation. If one opens coot, and opens a structure file (e.g. the crystal structure) by clicking *File > Open Coordinates…* one can visualize the densities, superimposed on the structure, by selecting *File > Open MTZ…*, and selecting the *fcalc.mtz* file(s) produced by *xtraj.py* (Fig. 1, bottom center panel). The *File > Auto Open MTZ* option will only work for.*mtz* files containing both density and difference density information with the correct formatting and will not produce anything if an *xtraj.py*-outputted *fcalc.mtz* file is selected (it is designed for.*mtz* files outputted by phenix refinement). The color and RMSD (sigma) value for the density can be selected in the *Display Manager* by clicking the *Properties* button associated with a structure factor file. [A few other important details for using *coot* to visualize information from crystalline MD simulations: (i) the default sigma value for *coot* is 1.5 sigma for densities and 3.0 sigma for difference densities; we recommend using a sigma (RMSD) value of 1.0 for the *fcalc.mtz* files produced by *xtraj.py* as it allows one to see more of the details present in the MD density; (ii) any ensemble structural model needs to start with a model number of 1 for *coot* to open and display it correctly (e.g., ensembles starting with model number 0 will not load properly); (iii) to make sure that the densities will be superimposed on top of the crystal structure when visualizing, one must ensure that the supercell is propagated starting from the crystal structure coordinates, the trajectory is translationally fitted to the supercell, and the correct *unit_cell* and *space_group* information is provided to *xtraj.py* when calculating structure factors from the MD simulation trajectory.]

2.3.2 Using the MD structure factors to refine an MD structural model

Using a crystalline MD model, it is possible to refine a structural model using the MD structure factors. To do this, the MD structure factors need to be processed to produce simulated diffraction data. This involves using the *sftools* method from *CCP4* to process the MD data and produce an *.mtz* file with *Iobs* and *SIGIobs* columns that correspond to the intensity and the variance of the intensity at each Miller index. An example of how to do this using the *generate_iobs.sh* script provided with the notebook is shown in

Code block 2. The resulting.*mtz* file then can be used in refinement just as one would use crystallographic data.

```
!bash ./code/generate_iobs.sh data/fcalc_full.mtz iobs_full.mtz
```

Code block 2 This line in the Bragg notebook calls an example script to generate fake Bragg intensities from MD structure factors. The script scales the simulated data so the values are within an acceptable range for downstream analysis, here divided by 100. It also adds an error column, required for downstream analysis (arbitrarily defined as the square root of the intensity, assuming they are due to photon counting noise). The resulting *iobs_full.mtz* file can be used in crystallographic structural refinement software as if it were experimental data, to obtain a structural model derived from the MD structure factors.

Such a strategy has been used to develop the MD–MX procedure for enhancing crystallographic refinement (Fig. 1, Wych et al., 2023). A key step in this procedure is protein-first refinement, in which the isolated protein model is first refined using the MD structure factors computed from only the protein component of the system. Ordered waters are then added using the total MD structure factors, and the resulting structure is refined using the total MD structure factors to produce a synthetic MD crystal structure (Fig. 1, bottom center). The resulting structural model is then refined against the experimental data to obtain a revised crystal structure (Fig. 1, bottom left). In a final stage, as described below, the ensemble of proteins in MD snapshots may be overlaid and used to suggest local regions where the crystal structure might be better modeled using multiple conformations. In an application to PKA, the MD–MX procedure yielded lower experimental R-factors; an MD-derived water model that is complementary to that obtained by water picking in the experimentally derived density map, including more ordered waters; and alternative protein side chain and backbone conformations that were not identified using standard methods.

Protein-first refinement can be accomplished using *Phenix* with the general outline of commands in the Bragg notebook beginning with the line shown in Code block 3. The refinement strategy is carried out for the example system of PKA.

```
!phenix.refine ./files_for_refinement/single_structure_refinement.eff --unused_ok
```

Code block 3 The first line in the protein-first refinement section of the Bragg notebook. In this step, the initial structure is refined against only the intensities calculated from the protein component of simulation. The *.eff* file contains all of the specifications for refinement, and the details of each step of the process are outlined in the notebook.

2.3.3 MD simulation B-factors

B-factors are parameters of a crystallographic structural model that are refined alongside the coordinates and occupancies for each atom, representing the degree of small-scale harmonic disorder about each atom's average position. This small-scale harmonic disorder can be modeled as isotropic or anisotropic. These parameters are found in every PDB file in the *ATOM, HETATM,* or associated *ANISOU* records and are important for providing context to atomic positions which might otherwise be viewed as completely precise and static. Assuming the atomic displacements follow a 3D Gaussian distribution (which is true for harmonic disorder), the mean structure factor is given by

$$\langle F(\boldsymbol{q})\rangle = \sum_i f_i(\boldsymbol{q})\, e^{-\frac{1}{2}\boldsymbol{q}^T\mathbf{U}_i\boldsymbol{q}} e^{i\boldsymbol{q}\cdot\boldsymbol{r}_i}$$

where $\mathbf{U}_i = \langle \boldsymbol{u}_i\boldsymbol{u}_i^T\rangle$ is the variance–covariance matrix of the displacements $\boldsymbol{u}_i$ of atom i from the mean. In this equation, the effect of the disorder is to multiply the atomic form factor $f_i(\boldsymbol{q})$ of each atom by the Debye–Waller factor $T_i(\boldsymbol{q})$, defined as (see, e.g., chapter 2 of James, 1948)

$$T_i(\boldsymbol{q}) = e^{-\frac{1}{2}\boldsymbol{q}^T\mathbf{U}_i\boldsymbol{q}}$$

The anisotropic displacement parameters that appear in the PDB *ANISOU* records are the six unique parameters of the matrix $\mathbf{U}_i$. The anisotropic B factor is defined as $\mathbf{B} = 8\pi^2\mathbf{U}_i$, and the effective isotropic B-factor is given by $B_i = \frac{8}{3}\pi^2\mathrm{Tr}(\mathbf{U}_i)$, which is the value that appears in the PDB records. (The factor of 3 comes from the fact that the displacement being averaged to compute the B-factor is projected along the direction of $\boldsymbol{q}$ and leaves out components along two other orthogonal directions.) In the case where the displacements are isotropically distributed, the B-factor may be computed as $B_i = \frac{8}{3}\pi^2\langle \boldsymbol{u}_i^2\rangle$. The quantity $\langle \boldsymbol{u}_i^2\rangle$ is the mean squared displacement of atom i, which is the square of another commonly calculated quantity in MD simulations: the root-mean squared fluctuation, or RMSF ($RMSF_i^2 = \langle \boldsymbol{u}_i^2\rangle$).

In addition to yielding atomic coordinates corresponding to the MD structure factors, protein-first refinement yields MD B-factors in the*.pdb* file. This is our recommended way to obtain B-factors for comparison with experimental data. With the definitions above, one also can calculate the anisotropic (isotropic) B-factors directly from an MD simulation by simply calculating $\mathbf{U}_i$ (or $\langle \boldsymbol{u}_i^2 \rangle$), averaging over the snapshots for each atom, and multiplying by $8\pi^2$ (or $\frac{8}{3}\pi^2$). Naively, one might think that these two methods for calculating B-factors should produce the same results. However, even if the actual fluctuations from the MD are exactly known (not the case for crystallographic refinement), the simulations include anharmonic motions, which can introduce changes in the atom displacement distribution that break the underlying assumptions of the Debye–Waller factor model. Even in the case of refinement against the MD simulated data the resulting B-factors might not be equivalent to the values calculated using the atomic coordinates from MD snapshots. Moreover, the detailed method of calculating the B-factors in this way can have a significant impact on the agreement with the B-factors from refinement (Janowski et al., 2016). For example, calculating the RMSF by first rotationally and translationally fitting each protein from the supercell trajectory to the coordinates of the asymmetric unit prior to calculating the $\langle \boldsymbol{u}_i^2 \rangle$ values generates MD B-factors that disagree more with the experimental B-factors than calculating the RMSF by first aligning each supercell snapshot by center of mass and performing crystal symmetry and translation operations to map each protein back on to the asymmetric unit, before computing the averages. In particular, the second analysis retains the influence of overall rotations and translations of proteins, which are present in the MD simulations, inspire translation/libration/screw-axis (TLS) models of B-factors in Bragg analysis (Schomaker & Trueblood, 1968; Urzhumtsev, Afonine, & Adams, 2013), and, as supported by diffuse scattering studies (Meisburger et al., 2020; Polikanov & Moore, 2015), might actually be an important contributor to vibrations in real protein crystals.

2.3.4 Visualizing crystalline MD snapshots as ensembles of symmetry-related proteins

Each frame of a crystalline MD simulation trajectory includes an ensemble of symmetry-related proteins. To visualize an ensemble from a frame of the MD trajectory, we need to reverse the process of expanding the asymmetric unit to the supercell: the proteins need to be sent from their positions in the supercell (Fig. 1) back on to the orientation and position of the asymmetric unit from the initial structural model (Fig. 2). This can be accomplished with the *reverse_propagate.py* script which makes use of *cctbx* tools (Code block 4).

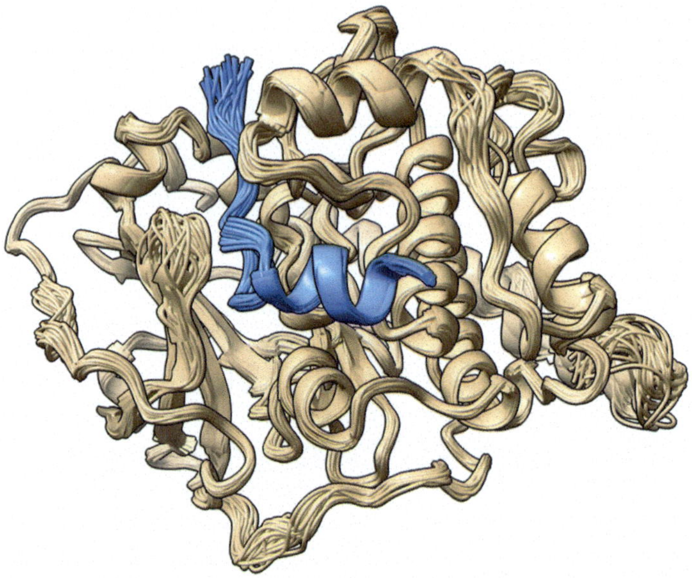

Fig. 2 Protein structural ensemble mapped back to the asymmetric unit from the supercell model of PKA, with 32 models. Main chain protein is represented in beige, with the inhibitor peptide (IP20) represented in blue.

```
!python reverse_propagate.py -xrs ./files_for_refinement/starting_structure.pdb \
      -sc ./data/supercell_sample_frame.pdb -o ensemble.pdb
```

Code block 4 This line copied from the Bragg notebook runs *reverse_propagate.py*, which uses the cctbx software suite to map proteins from their positions in the supercell back on to the asymmetric unit.

2.4 Diffuse analysis

We now turn our attention to the diffuse analysis. This analysis requires the same preprocessing of trajectories described above, with details provided in the Bragg analysis notebook. The focus for this section is the Wych et al. study (Wych et al., 2019) which includes both comparison of MD simulations to experimental data and analysis of the covariance matrix of atom displacements. The accompanying diffuse analysis notebook (see Jupyter notebooks) contains code to carry out elements of the analysis. As for the Bragg analysis, the organization of this section follows that in the notebook, so that they can be used side-by-side.

2.4.1 Calculating diffuse scattering from an MD trajectory

The diffuse intensity is calculated from a crystalline MD trajectory as the variance of the structure factors (Meisburger et al., 2017):

$$I_D(\mathbf{q}) = \langle |F(\mathbf{q})|^2 \rangle - |\langle F(\mathbf{q}) \rangle|^2$$

It may be calculated using *xtraj.py* as shown in Code block 5, copied from the diffuse notebook. This script makes use of mdtraj methods for i/o which supports a variety of trajectory formats. For use with *AMBER*, the *md2diffuse.sh* script for calculating diffuse scattering from MD trajectories is provided in the *AmberTools* package (Case, In press).

```
!python {xtraj} \
        traj="trajectory.xtc" \
        top="crystal.pdb" \
        d_min=1.6 first=0 last=50 selection="all" diffuse=diffuse_supercell.hkl \
        fcalc=fcalc_supercell.mtz icalc=icalc_supercell.mtz
```

Code block 5 In the diffuse notebook, *xtraj.py* (the full path of which is provided using the xtraj variable in braces) is used to compute structure factors and intensities from the trajectory in the *trajectory.xtc* file, using the topology information provided in the *crystal.pdb* file. In this case, the *unit_cell* and *space_group* parameters are defined by the *CRYST1* record of the *crystal.pdb* file such that *xtraj.py* considers the supercell as a P_1 unit cell with box side lengths equal to the supercell box side lengths. Because this is a supercell, reciprocal space is sampled more finely than the Bragg lattice (e.g. a 2×2×2 unit cell supercell samples reciprocal space twice as finely along each axis as a single unit cell). The *selection = "all"* argument (default) selects everything in the system. In this case, *selection = "resname SOL"* will select the water molecules, and *selection = "not resname SOL"* will select everything but the water molecules (mainly protein), using the residue name for the waters in the *.pdb* file. The selection syntax is that used in *cctbx*.

2.4.2 Comparing the simulated and experimental diffuse intensities

After running *xtraj.py* as in Code block 5, diffuse intensities will be written as a flat *h,k,l,Intensity* text file named *diffuse_supercell.hkl*. The native internal data structure for analyzing diffuse scattering maps in *Lunus* is the binary *.lat* file. An experimental dataset and the results of running *xtraj.py* on a larger MD trajectory are included in the diffuse online materials. The diffuse notebook includes a function *read_lat()* for reading the binary files in Python. Code block 6 shows an example of how this is used.

It is interesting to calculate the isotropic component of the diffuse intensity and to compare it to the experimental data. Examples of how this is done using *Lunus* methods are found in the diffuse notebook near the plots shown in Fig. 3. This analysis shows that the shape of a simulated profile is similar to that of the experimental data (Fig. 3A). The isotropic

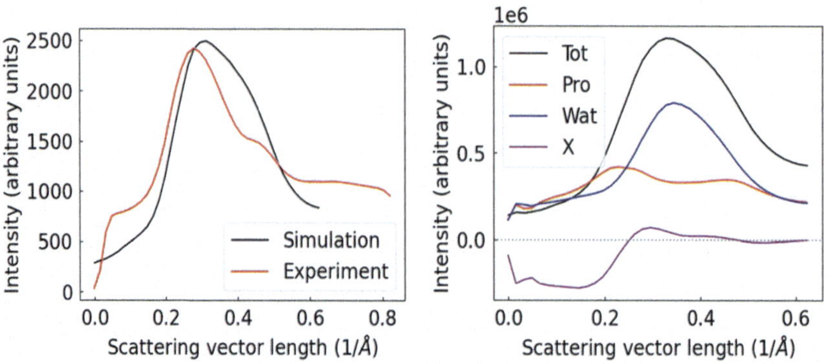

Fig. 3 Isotropic diffuse scattering profiles for staphylococcal nuclease. Left: Comparison of a simulated profile (black) with the experimental data (red). Right: A total simulated profile (Tot, black) along with isotropic profiles computed from just the protein (Pro, red) and water (Wat, blue) components, along with the residual cross-term (X, purple). The protein profile exhibits a shoulder that is similar to the one in the experimental profile.

component may be computed from the whole system or components. It is interesting to compute it from the protein and solvent components separately, by selecting the relevant parts of the system when running xtraj. The diffuse notebook shows examples of how to do this using a short trajectory included in the accompanying online materials. Code block 5 from the notebook shows how to use the selection syntax to select the non-solvent component, which is mostly protein. Decomposing the simulated profile into the protein and solvent indicates that there are substantial contributions from both (Fig. 3B).

```
header, lattice = read_lat("experimental_data.lat")
```

Code block 6 Example of how to read a *Lunus .lat* binary file as a numpy array with the *read_lat*() function in the diffuse notebook.

To perform a more detailed comparison, it is useful to first do some preprocessing to truncate the diffuse intensities, merge symmetry-equivalent values, symmetrize them using the known symmetry, and subtract the isotropic component. A helper script for this, *preprocess.sh*, is provided along with the notebook, invoked in the notebook as in Code block 7. The script takes a code for the Laue group (−3 in this case; running the script without any arguments will show a list of codes), optional lower and upper resolutions, and an optional unit cell string.

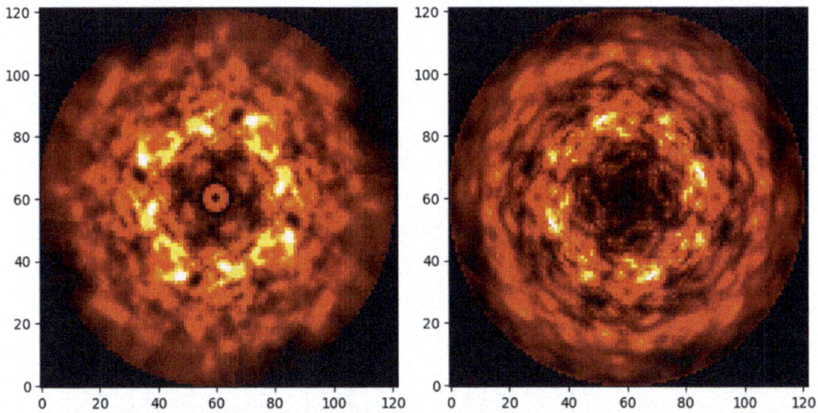

Fig. 4 Slices throughs of the experimental diffuse data (left) and simulated diffuse scattering (right) visualized in the diffuse notebook. To make the anisotropic features more visible, the 3D diffuse intensities have had the minimum intensity value subtracted in resolution shells prior to generating these images, as shown in Code block 8.

```
!bash ./code/preprocess.sh experimental_data.lat -3 200. 1.6 $(cat cell.txt)
```

Code block 7 Example from the diffuse notebook of how to preprocess diffuse intensities for comparison.

An example of the diffuse intensity calculated from an MD simulation of a protein crystalline system of SNase is shown in Fig. 4. To produce the figure, the 3D diffuse intensities have had minimum intensity value was subtracted in resolution shells using the *lunus.subminlt* method as illustrated in the line from the notebook doing this for the preprocessed experimental data in Code block 8. The simulation resembles the experimental data, with differences also apparent between them, indicating there is room for improvement in simulation of anisotropic diffuse features.

```
!lunus.subminlt experimental_data_culled_sym.lat \
     experimental_data_culled_sym_submin.lat $(cat cell.txt)
```

Code block 8 Example from the diffuse notebook of how to subtract the minimum intensity in resolution shells prior to visualization as in Fig. 4.

Code block 9 from the diffuse notebook shows how to compute the overall Pearson correlation between two *.lat* files, illustrated using the total preprocessed intensities. The computed correlation in this case is 0.91, and the correlation for the anisotropic intensities is 0.62. Nearby in the

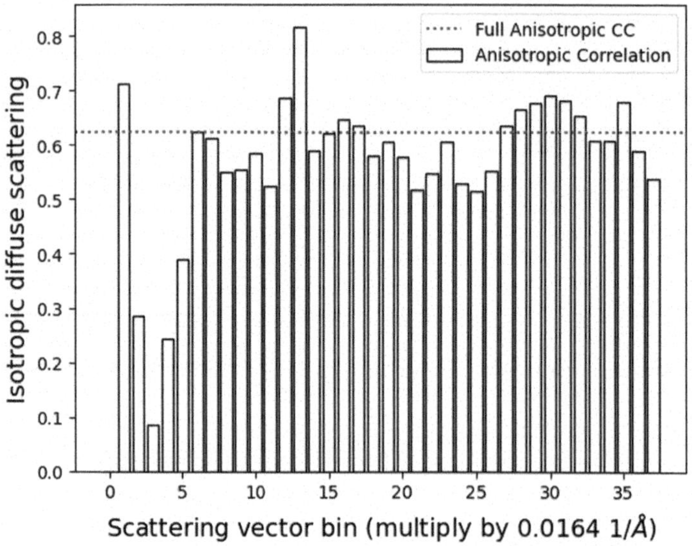

Fig. 5 Correlation of simulated and experimental diffuse intensities, as calculated in the diffuse notebook. The total correlation is 0.62, illustrated using a dotted red line.

notebook is an example of how to compute the correlation in resolution shells using *lunus.ccrlt*. Results are illustrated in Fig. 5. The use of *lunus.ccrlt* should be restricted to the anisotropic intensities, as the correlation may otherwise be largely due to the variation of the isotropic component within a resolution bin, which can dominate the overall diffuse signal.

```
!lunus.corrlt experimental_data_culled_sym.lat simulated_data_culled_sym.lat
```

Code block 9 Example from the diffuse notebook of how to use *lunus.corrlt* to compute the overall Pearson correlation between two.lat files.

2.4.3 Analysis of the covariance of atomic displacements from an MD trajectory

The diffuse scattering is of interest because it encodes information about correlated dynamics, rather than simply average structure (in the case of the Bragg data). Assuming harmonic displacements of atoms, the diffuse scattering depends on the variance–covariance matrix of atomic displacements for each atom pair (Meisburger et al., 2017):

$$\mathbf{V}_{jk} = \langle \boldsymbol{u}_j \boldsymbol{u}_k^T \rangle$$

where $\boldsymbol{u}_j$ is the atomic displacement ($\boldsymbol{u}_j = \boldsymbol{r}_j - \langle \boldsymbol{r}_j \rangle$) for atom j. The diffuse scattering can be approximated as:

$$I_D(\boldsymbol{q}) = \sum_{nj} \sum_{mk} f_j(\boldsymbol{q}) f_k^*(\boldsymbol{q}) e^{i\boldsymbol{q}\cdot(\mathbf{R}_n - \mathbf{R}_m)} e^{i\boldsymbol{q}\cdot(\langle \boldsymbol{r}_j \rangle - \langle \boldsymbol{r}_k \rangle)} T_j(\boldsymbol{q}) T_k(\boldsymbol{q}) [e^{\boldsymbol{q}^T \mathbf{V}_{njmk} \boldsymbol{q}} - 1]$$

where the sum is over the atoms j in unit cells n, and atoms k in unit cells m.

This expression can be further simplified by Taylor expanding the term in square brackets and keeping only the first term in the expansion:

$$e^{\boldsymbol{q}^T \mathrm{V}_{njmk} \boldsymbol{q}} - 1 \rightarrow \boldsymbol{q}^T \mathbf{V}_{njmk} \boldsymbol{q}$$

Everything in the expression for $I_D(\boldsymbol{q})$ is also present in the expression for Bragg diffraction except this term, containing the variance–covariance matrix of atomic displacements. This matrix has 6 unique elements for every pair of atoms in the illuminated volume of the crystal, which is too many parameters to fit directly using the data. Simpler models are therefore used to model the dynamics using diffuse scattering data. Possibilities include the liquid-like motions (LLM), independent rigid-body motions (RBM) and elastic-network/normal-modes (EN/NM) models (see Peck, Lane, & Poitevin, in press and review Meisburger et al., 2017).

The LLM model assumes that the covariance of atom pairs decays exponentially as a function of their distance apart—that is, the variance–-covariance matrix of atomic displacements for atoms j and k gets replaced by a scalar function, which decays with the pairwise distance between atoms j and k:

$$\mathbf{V}_{jk} = V_{jk} \propto \Gamma(\mathbf{\Delta r}_{jk}) = \sum_{\gamma} e^{-|\Delta \mathbf{r}_{jk}|/\gamma}$$

This means nearby atoms strongly influence each other's motions, whereas atoms that are far from each other are essentially uncorrelated in their motion—as one would expect for a liquid. By contrast, the independent RBM model assumes that the protein behaves as a rigid, solid object that can rotate and translate independently of its neighbors. Both models have a small number of parameters (O(1)) and have been used to describe the anisotropic diffuse scattering (de Klijn, Schreurs, & Kroon-Batenburg, 2019; Peck et al., 2018). The EN/NM model (Meisburger et al., 2020) is more complex [the LLM model is technically a special case of a vibrational model, not unlike the EN/NM, in which the springs are selected to yield exponentially decaying correlations (Wall, 1996). Dissertation thesis, Princeton]: to model the lattice vibrations, the proteins are treated as rigid units that are connected by springs; to model the internal vibrations, the protein is decomposed into rigid residues that also are connected by springs. The approach to fitting the spring constants [in studies so far, numbering O (100)] is hierarchical, with the lattice springs first being fit to the halo shapes, and the internal springs then being fit using the B-factors from the Bragg analysis. This model was able to reproduce both the halos around Bragg peaks (associated with acoustic phonon-like motions) and the diffuse Patterson (the Fourier transform of the diffuse scattering over reciprocal space) to high accuracy.

The elements of the $\mathbf{V}_{jk}$ matrices also may be computed directly from crystalline MD simulations. One way to do this is to use the built-in *covar* method in *GROMACS*, which takes a structure file and a trajectory file as input, and outputs the covariance matrix for each atom pair either in plain text (ASCII) format (producing a very large file in the case of a supercell), or in the *.xpm* matrix format. An example of this is shown in Code block 10 from the notebook (this line is commented out in the notebook as the trajectory file needed to produce meaningful values is too larger to be included along with the notebooks). This method can also compute the "atomic covariance" which is the trace of the covariance matrix—or the mean squared displacement for each atom, which is what we use for comparison to the LLM model in the following section. (n.b. *gmx covar* will ask the user for a selection of atoms on which to perform the covariance analysis. The covariance matrix grows as the square of the number of atoms, so one must be careful to choose a selection with a small number of atoms, otherwise the output file may be very large, and the calculation will take a very long time. For example, for a protein with 300 residues, and a supercell of 32 proteins, the full variance–covariance matrix of atomic displacements for just the alpha carbons, would have ~830 million elements, and the atomic covariance matrix would have ~92 million. With the elements stored as 16-bit floating point numbers, the full matrix would require at least 1.6 GB and the atomic covariance matrix 184 MB of memory.) The atomic covariance values for the SNase simulation trajectory are displayed in Fig. 6B.

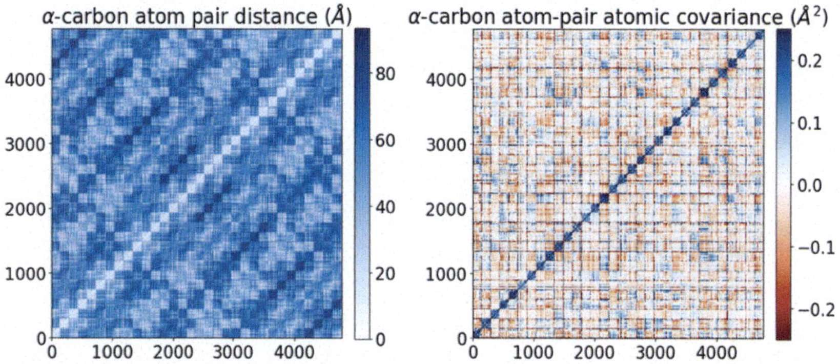

Fig. 6 (A) Backbone alpha-carbon average atom pair distance matrix computed from a supercell simulation of staphylococcal nuclease, with distances represented by color from zero (white) to 90 Å (dark blue). (B) Backbone alpha-carbon atom pair atomic covariance matrix computed from the same simulation, with zero covariance represented with white, negative atomic covariance represented in red, and positive atomic covariance represented in blue. Block matrices along the diagonal are atom pairs within proteins; off-diagonal block matrices are atom pairs across proteins. Figures are copied from the diffuse notebook.

```
# gmx covar -f traj_files/c_alpha_first_first_100ns_500_frames_fit.xtc \
    -s traj_files/c_alpha_start.pdb -av average_position.pdb -ascii \
    atomic_covariance.dat -fit
```

Code block 10 The GROMACS covar method is used to compute the diagonalized (mass-weighted) covariance matrix for the reference structure (crystal.pdb) using the trajectory (trajectory.xtc). The option *"-ascii"* will output this matrix in ASCII plain text, with elements ordered as *x1x1, x1y1, x1z1, x1x2,* … etc. The option *-xpm* will output the same, but with the matrix in *.xpm* format. The option *-xpma* will output the atomic covariance (the trace of the covariance matrix for each atom pair).

2.4.4 Associating covariance information with the LLM and RBM models of diffuse scattering

As mentioned above, the liquid-like motions model assumes that the covariance of atomic displacements decays exponentially as a function of atom-pair distance. Using the matrix of atomic covariances computed from the pairs of alpha-carbons in the command above, one can examine the relationship between the atom pair distance and the covariance to assess the validity of that assumption. However, one first needs to compute the average distance between atoms. The average structure can be outputted using the *-av* flag, which can then be used to calculate the average pairwise distance between atoms pairs. The notebook includes lines illustrating how the calculation can be done, using scripts distributed with the notebook Code block 11. Once this matrix of average pairwise distances is computed (with each element *ij* of the matrix corresponding to the average distance between atoms *i* and *j*) this matrix can be paired up, element to element, with the elements of the atomic covariance matrix, and these pairs of distances and covariances can averaged in bins to examine the relationship between atom pair distance and atomic covariance (Fig. 7). For atom pairs separated by greater than ~10 Å, the correlations closely follow an exponential decay, which is consistent with the LLM model (Wych et al., 2019).

```
# !python ./code/get_c_alphas.py
# !python ./code/pairwise_distances.py --pdb full_sys_start_calphas.pdb --out
pairwise_distances.npy
```

Code block 11 Lines in the diffuse notebook that can be used to run scripts included for calculating the distances between all Cα pairs in the system. *get_c_alphas.py* selects the alpha-carbons. *pairwise_distances.py* computes the distances, accounting for periodic boundary conditions.

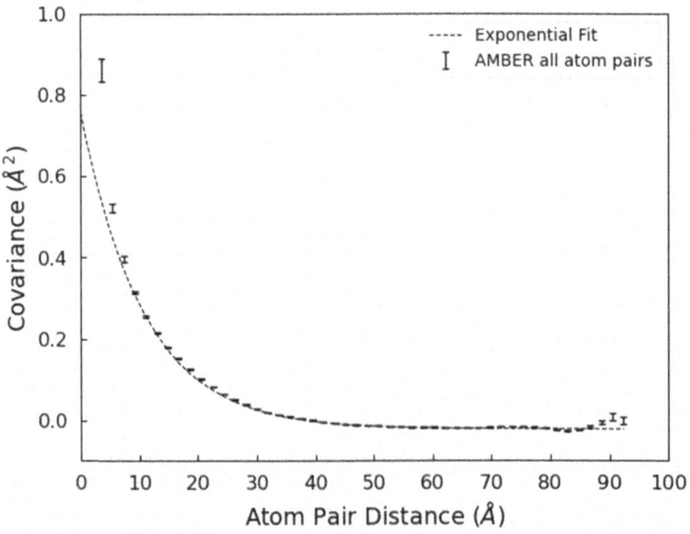

Fig. 7 Average atomic covariance of atom pairs as a function of distance for a supercell simulation of staphylococcal nuclease, with error bars representing the standard error of the mean in each bin. Exponential fit for the function $ae^{-r/g} + b$, where *r* is the atom pair distance, shown in the dashed line with parameters $a = 0.79 \pm 0.1$ Å^2, $g = 11.0 \pm 0.1$ Å, and $b = 0.0^218 \pm 0.0005$ Å^2. Figure copied from the diffuse notebook, where the analysis is performed.

Because these matrices contain elements for atom pairs across the entire supercell, we can also separate the pairs of elements according to whether the atom pairs are intra-protein or inter-protein atom pairs. When we isolate the intra-protein atom pairs, we see a different relationship—almost linear (Fig. 8A). This linear relationship between atom pair distance and covariance is consistent with small-scale rigid body translations and rotations, as demonstrated in the diffuse notebook using an included *RigidBodyMotions.py* Python module. The module generates an ensemble of structures randomly translated and rotated using values randomly drawn from normal distributions; is currently hard coded specifically for this example but can be modified for use with other proteins. The covariance of atom displacements in the ensemble is calculated; the distance-dependence is shown in red in Fig. 8A. Once the covariance of the rigid-body motions is subtracted, the residual covariance can be explained by liquid-like motions, with a correlation length similar to published values found by fitting a LLM model to the diffuse data (Fig. 8B) (Wych et al., 2019).

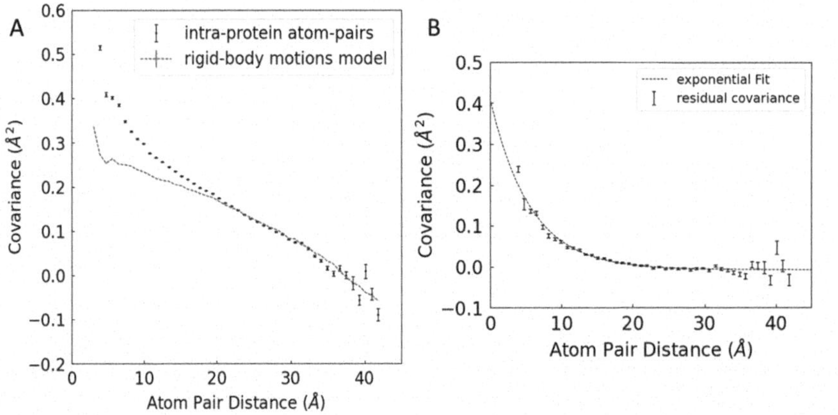

Fig. 8 Left: Average atomic covariance as a function of atom-pair separation distance for intra-protein atom pairs from a 2 × 2 × 2-unit-cell supercell simulation of staphylococcal nuclease (black points, with error bars corresponding to the standard error of the mean within each bin). The covariance estimated from an ensemble of proteins with random translations and rotations is shown in red; the standard deviation of Euler angles used to generate rotations was 0.95° and that for the translations along each axis was 0.24 Å. Right: The residual after the covariance predicted by the rigid-body motions model is subtracted (black points, with error bars corresponding to the standard error of the mean within each bin), with an exponential fit ($ae^{-r/g} + b$, where r is the atom-pair separation distance) with parameters $a = 0.42 \pm 0.2$ Å^2, $g = 5.6 \pm 0.2$ Å, and $b = -0.006 \pm 0.001$ Å^2. Figures are copied from the diffuse notebook, where the analysis is performed.

3. Summary

In this paper, we have shown how to calculate a number of X-ray crystallographic observables from MD simulations of protein crystals. We showed how to compute the structure factors from a crystalline MD simulation trajectory using *xtraj.py*, and how to calculate separate structure factors from the protein and solvent components of the trajectory. The Bragg notebook associated with this review shows how to perform key steps in the MD–MX procedure for enhancing MX structure determination (Wych et al., 2023). We then reviewed methods for folding the structures from a supercell MD snapshot into the asymmetric unit, to visualize a protein structural ensemble. The diffuse notebook includes methods for calculating the diffuse scattering from the MD simulation trajectory (again, also separated into components). We covered methods for comparing the results to experimental data including both the isotropic and anisotropic components of the diffuse scattering. We also reviewed methods for analyzing the covariance matrix of atomic displacements. In particular, we demonstrated how to analyze the dependence of the covariance on the atom-pair separation distance from a crystalline MD trajectory, and how to interpret the results in the context of both LLM and RBM models of diffuse scattering. We hope this review and the accompanying materials will serve as a jumping off point for others to use these methods in their work.

Acknowledgments

We are grateful to David Case, Steve Meisburger, and Nozomi Ando for discussions about analysis of MD simulations of diffuse scattering. Many thanks to the attendees of the tutorials on molecular-dynamics simulations of biological diffuse scattering at the 57th course of the International School of Crystallography in Erice, Italy, who tested early versions of the Jupyter notebooks. David C. Wych and Michael E. Wall were supported by the Exascale Computing Project (grant No. 17-SC-20-SC), a collaborative effort of the US Department of Energy Office of Science and National Nuclear Security Administration. David C. Wych was supported by the Center for Nonlinear Studies at Los Alamos National Laboratory (Technical Release Number: LA-UR-23-23656; under US Department of Energy contract No. 89233218CNA000001 to Triad National Security).

References

Afonine, P. V., Adams, P. D., Sobolev, O. V., & Urzhumtsev, A. (2021). A mosaic bulk-solvent model improves density maps and the fit between model and data. *bioRxiv, 2012*(2009), 471976.

Afonine, P. V., Grosse-Kunstleve, R. W., Echols, N., Headd, J. J., Moriarty, N. W., Mustyakimov, M., ... Adams, P. D. (2012). Towards automated crystallographic structure refinement with phenix.refine. *Acta Crystallographica. Section D, Biological Crystallography, 68*, 352–367.

Ayyer, K., Yefanov, O. M., Oberthur, D., Roy-Chowdhury, S., Galli, L., Mariani, V., ... Chapman, H. N. (2016). Macromolecular diffractive imaging using imperfect crystals. *Nature, 530*, 202–206.

Badger, J., & Caspar, D. (1991). Water structure in cubic insulin crystals. *Proceedings of the National Academy of Sciences, 88*, 622–626.

Barends, T. R., Stauch, B., Cherezov, V., & Schlichting, I. (2022). Serial femtosecond crystallography. *Nature Reviews Methods Primers, 2*, 59.

Benoit, J. P., & Doucet, J. (1995). Diffuse scattering in protein crystallography. *Quarterly Reviews of Biophysics, 28*, 131–169.

Berman, H. M., Westbrook, J., Feng, Z., Gilliland, G., Bhat, T. N., Weissig, H., ... Bourne, P. E. (2000). The protein data bank. *Nucleic Acids Research, 28*, 235–242.

Burnley, B. T., Afonine, P. V., Adams, P. D., & Gros, P. (2012). Modelling dynamics in protein crystal structures by ensemble refinement. *Elife, 1*, e00311.

Case, D. A. (in press). MD Simulations of macromolecular crystals: Implications for the analysis of Bragg and diffuse scattering. *Methods in Enzymology*.

Caspar, D., Clarage, J., Salunke, D., & Clarage, M. (1988). Liquid-like movements in crystalline insulin. *Nature, 332*, 659–662.

Cavender, C. E., Case, D. A., Chen, J. C.-H., Chong, L. T., Keedy, D. A., Lindorff-Larsen, K., ... Gilson, M. K. (2023). *Structure-based experimental datasets for benchmarking of protein simulation force fields*. ⟨https://doi.org/10.48550/arXiv.2303.11056⟩.

Cerutti, D. S., & Case, D. A. (2019). Molecular dynamics simulations of macromolecular crystals. *WIREs Computational Molecular Science, 9*, e1402.

Clarage, J. B., & Phillips, G. N., Jr (1997). Analysis of diffuse scattering and relation to molecular motion. *Methods in Enzymology, 277*, 407–432.

Clarage, J. B., Clarage, M. S., Phillips, W. C., Sweet, R. M., & Caspar, D. L. (1992). Correlations of atomic movements in lysozyme crystals. *Proteins, 12*, 145–157.

Clarage, J. B., Romo, T., Andrews, B. K., Pettitt, B. M., & Phillips, G. N., Jr. (1995). A sampling problem in molecular dynamics simulations of macromolecules. *Proceedings of the National Academy of Sciences of the United States of America, 92*, 3288–3292.

de Klijn, T., Schreurs, A. M. M., & Kroon-Batenburg, L. M. J. (2019). Rigid-body motion is the main source of diffuse scattering in protein crystallography. *IUCrJ, 6*, 277–289.

Doucet, J., & Benoit, J. P. (1987). Molecular dynamics studied by analysis of the X-ray diffuse scattering from lysozyme crystals. *Nature, 325*, 643–646.

Emsley, P., Lohkamp, B., Scott, W. G., & Cowtan, K. (2010). Features and development of Coot. *Acta Crystallographica Section D: Biological Crystallography, 66*, 486–501.

Faure, P., Micu, A., Perahia, D., Doucet, J., Smith, J. C., & Benoit, J. P. (1994). Correlated intramolecular motions and diffuse X-ray scattering in lysozyme. *Nature Structural Biology, 1*, 124–128.

Fraser, J. S., van den Bedem, H., Samelson, A. J., Lang, P. T., Holton, J. M., Echols, N., & Alber, T. (2011). Accessing protein conformational ensembles using room-temperature X-ray crystallography. *Proceedings of the National Academy of Sciences, 108*, 16247–16252.

García, A. E., Krumhansl, J. A., & Frauenfelder, H. (1997). Variations on a theme by Debye and Waller: From simple crystals to proteins. *Proteins: Structure, Function, and Bioinformatics, 29*, 153–160.

Grosse-Kunstleve, R. W., Sauter, N. K., Moriarty, N. W., & Adams, P. D. (2002). The Computational Crystallography Toolbox: Crystallographic algorithms in a reusable software framework. *Journal of Applied Crystallography, 35*, 126–136.

Hery, S., Genest, D., & Smith, J. C. (1998). X-ray diffuse scattering and rigid-body motion in crystalline lysozyme probed by molecular dynamics simulation. *Journal of Molecular Biology, 279*, 303–319.

James, R. (1948). *The optical principles of the diffraction of X-rays*. London: Bell.

Janowski, P. A., Cerutti, D. S., Holton, J., & Case, D. A. (2013). Peptide crystal simulations reveal hidden dynamics. *Journal of the American Chemical Society, 135*, 7938–7948.

Janowski, P. A., Liu, C., Deckman, J., & Case, D. A. (2016). Molecular dynamics simulation of triclinic lysozyme in a crystal lattice. *Protein Science, 25*, 87–102.

Keedy, D. (2019). Journey to the center of the protein: Allostery from multitemperature multiconformer X-ray crystallography. *Acta Crystallographica Section D, 75*, 123–137.

Keedy, D. A., Kenner, L. R., Warkentin, M., Woldeyes, R. A., Hopkins, J. B., Thompson, M. C., ... Uervirojnangkoorn, M. (2015). Mapping the conformational landscape of a dynamic enzyme by multitemperature and XFEL crystallography. *Elife, 4*, e07574.

Keedy, D. A., Van Den Bedem, H., Sivak, D. A., Petsko, G. A., Ringe, D., Wilson, M. A., & Fraser, J. S. (2014). Crystal cryocooling distorts conformational heterogeneity in a model Michaelis complex of DHFR. *Structure (London, England: 1993), 22*, 899–910.

Kuriyan, J., Osapay, K., Burley, S. K., Brunger, A. T., Hendrickson, W. A., & Karplus, M. (1991). Exploration of disorder in protein structures by X-ray restrained molecular dynamics. *Proteins, 10*, 340–358.

Mehlman, T. S., Biel, J. T., Azeem, S. M., Nelson, E. R., Hossain, S., Dunnett, L. E., ... Axford, D. (2022). Room-temperature crystallography reveals altered binding of small-molecule fragments to PTP1B. *Elife, 12*, e84632.

Meinhold, L., & Smith, J. C. (2005a). Fluctuations and correlations in crystalline protein dynamics: A simulation analysis of staphylococcal nuclease. *Biophysical Journal, 88*, 2554–2563.

Meinhold, L., & Smith, J. C. (2005b). Correlated dynamics determining X-ray diffuse scattering from a crystalline protein revealed by molecular dynamics simulation. *Physical Review Letters, 95*, 218103.

Meisburger, S. P., Case, D. A., & Ando, N. (2020). Diffuse X-ray scattering from correlated motions in a protein crystal. *Nature communications, 11*, 1–13.

Meisburger, S. P., Case, D. A., & Ando, N. (2023). Robust total X-ray scattering workflow to study correlated motion of proteins in crystals. *Nature Communications, 14*, 1228.

Meisburger, S. P., Thomas, W. C., Watkins, M. B., & Ando, N. (2017). X-ray scattering studies of protein structural dynamics. *Chemical Reviews, 117*, 7615–7672.

Moore, P. B. (2009). On the relationship between diffraction patterns and motions in macromolecular crystals. *Structure (London, England: 1993), 17*, 1307–1315.

Murshudov, G. N., Skubak, P., Lebedev, A. A., Pannu, N. S., Steiner, R. A., Nicholls, R. A., ... Vagin, A. A. (2011). REFMAC5 for the refinement of macromolecular crystal structures. *Acta Crystallographica. Section D, Biological Crystallography, 67*, 355–367.

Nagai, T., Tama, F., & Miyashita, O. (2019). Cryo-cooling effect on DHFR crystal studied by replica-exchange molecular dynamics simulations. *Biophysical Journal, 116*, 395–405.

Peck, A., Lane, T. J. & Poitevin, F. (in press). Modeling diffuse scattering with simple, physically interpretable models. *Methods in Enzymology*.

Peck, A., Poitevin, F., & Lane, T. J. (2018). Intermolecular correlations are necessary to explain diffuse scattering from protein crystals. *IUCrJ, 5*, 211–222.

Phillips, G. N., Jr., Fillers, J. P., & Cohen, C. (1980). Motions of tropomyosin. Crystal as metaphor. *Biophysical Journal, 32*, 485–502.

Ploscariu, N., Burnley, T., Gros, P., & Pearce, N. M. (2021). Improving sampling of crystallographic disorder in ensemble refinement. *Acta Crystallogr D Struct Biol, 77*, 1357–1364.

Polikanov, Y. S., & Moore, P. B. (2015). Acoustic vibrations contribute to the diffuse scatter produced by ribosome crystals. *Acta Crystallographica. Section D, Biological Crystallography, 71*, 2021–2031.

Riley, B. T., Wankowicz, S. A., de Oliveira, S. H. P., van Zundert, G. C. P., Hogan, D. W., Fraser, J. S., ... van den Bedem, H. (2021). qFit 3: Protein and ligand multiconformer modeling for X-ray crystallographic and single-particle cryo-EM density maps. *Protein Science: a Publication of the Protein Society, 30*, 270–285.

Russi, S., González, A., Kenner, L. R., Keedy, D. A., Fraser, J. S., & Bedem, H. v d (2017). Conformational variation of proteins at room temperature is not dominated by radiation damage. *Journal of Synchrotron Radiation, 24*, 73–82.

Schomaker, V., & Trueblood, K. N. (1968). On the rigid-body motion of molecules in crystals. *Acta Cryst, B24*, 63–76.

Stachowski, T. R., Vanarotti, M., Seetharaman, J., Lopez, K., & Fischer, M. (2022). Water Networks Repopulate Protein–Ligand Interfaces with Temperature. *Angewandte Chemie International Edition, 61*, e202112919.

Urzhumtsev, A., Afonine, P. V., & Adams, P. D. (2013). TLS from fundamentals to practice. *Crystallogr Rev, 19*, 230–270.

van Gunsteren, W. F., Berendsen, H. J., Hermans, J., Hol, W. G., & Postma, J. P. (1983). Computer simulation of the dynamics of hydrated protein crystals and its comparison with x-ray data. *Proc Natl Acad Sci U S A, 80*, 4315–4319.

Vitkup, D., Ringe, D., Petsko, G. A., & Karplus, M. (2000). Solvent mobility and the protein "glass" transition. *Nature Structural Biology, 7*, 34–38.

Wall, M. E. (1996). Diffuse features in X-ray diffraction from protein crystals. Dissertation thesis, Princeton University.

Wall, M. E. (2009). Methods and software for diffuse X-ray scattering from protein crystals. *Methods in Molecular Biology, 544*, 269–279.

Wall, M. E. (2018). Internal protein motions in molecular-dynamics simulations of Bragg and diffuse X-ray scattering. *IUCrJ, 5*, 172–181.

Wall, M. E., Adams, P. D., Fraser, J. S., & Sauter, N. K. (2014). Diffuse X-ray scattering to model protein motions. *Structure (London, England: 1993), 22*, 182–184.

Wall, M. E., Calabró, G., Bayly, C. I., Mobley, D. L., & Warren, G. L. (2019). Biomolecular solvation structure revealed by molecular dynamics simulations. *Journal of the American Chemical Society, 141*, 4711–4720.

Wall, M. E., Clarage, J. B., & Phillips, G. N., Jr (1997). Motions of calmodulin characterized using both Bragg and diffuse X-ray scattering. *Structure (London, England: 1993), 5*, 1599–1612.

Wall, M. E., Van Benschoten, A. H., Sauter, N. K., Adams, P. D., Fraser, J. S., & Terwilliger, T. C. (2014). Conformational dynamics of a crystalline protein from microsecond-scale molecular dynamics simulations and diffuse X-ray scattering. *Proceedings of the National Academy of Sciences, 111*, 17887–17892.

Warkentin, M., Hopkins, J. B., Badeau, R., Mulichak, A. M., Keefe, L. J., & Thorne, R. E. (2013). Global radiation damage: temperature dependence, time dependence and how to outrun it. *Journal of Synchrotron Radiation, 20*, 7–13.

Wych, D. C. & Wall, M. E. (In press). Molecular-dynamics simulations of macromolecular diffraction, part I: Preparation of protein crystal simulations. *Meths Enzymol.*

Wych, D. C., Aoto, P. C., Vu, L., Wolff, A. M., Mobley, D. L., Fraser, J. S., ... Wall, M. E. (2023). Molecular-dynamics simulation methods for macromolecular crystallography. *Acta Crystallographica Section D: Structural Biology, 79*.

Wych, D. C., Fraser, J. S., Mobley, D. L., & Wall, M. E. (2019). Liquid-like and rigid-body motions in molecular-dynamics simulations of a crystalline protein. *Structural Dynamics, 6*, 064704.

Xu, D., Meisburger, S. P., & Ando, N. (2021). Correlated motions in structural biology. *Biochemistry, 60*, 2331–2340.

CHAPTER FIVE

MD simulations of macromolecular crystals: Implications for the analysis of Bragg and diffuse scattering

David A. Case*
Dept. of Chemistry & Chemical Biology, Rutgers University, Piscataway, NJ, United States
*Corresponding author. e-mail address: david.case@rutgers.edu

Contents

Abstract

Some of our most detailed information about structure and dynamics of macromolecules comes from X-ray-diffraction studies in crystalline environments. More than 170,000 atomic models have been deposited in the Protein Data Bank, and the number of observations (typically of intensities of Bragg diffraction peaks) is generally quite large, when compared to other experimental methods. Nevertheless, the general agreement between calculated and observed intensities is far outside the experimental precision, and the majority of scattered photons fall between the sharp Bragg peaks, and are rarely taken into account. This chapter considers how molecular dynamics simulations can be used to explore the connections between microscopic behavior in a

Methods in Enzymology, Volume 688
ISSN 0076-6879, https://doi.org/10.1016/bs.mie.2023.06.013

crystalline lattice and observed scattering intensities, and point the way to new atomic models that could more faithfully recapitulate Bragg intensities and extract useful information from the diffuse scattering that lies between those peaks.

1. Introduction

X-ray crystallography provides a wealth of detailed information about protein structure and dynamics. (Here, as elsewhere, the term "dynamics" is used in a loose sense, describing aspects of the equilibrium conformational landscape being explored, even in the absence of any analysis of the time scales of transitions between conformers.) Experimental data is most commonly interpreted in terms of atomic models that include coordinates of the average structure, atomic displacement parameters (ADPs) that provide information about fluctuations about the mean, and information about alternate conformers that may be visible in electron density maps. These models have been extraordinarily useful in many areas of biophysical and biomedical research, but there are a number of ways in which the connections between experimental data and the underlying microscopic models of crystals might be improved:

- Most X-ray data is collected at low temperature, primarily to limit radiation damage. The cooling procedure is difficult to model at a detailed level, and is known to have systematic effects on conformational heterogeneity and on the magnitude of fluctuations (Atakisi et al., 2018; Fraser et al., 2011; Keedy et al., 2014; Stachowski et al., 2022; Woldeyes et al., 2014; Yabukarski et al., 2022). Serial crystallography, using both synchrotron and free-electron laser sources, along with careful analysis and dose-dependent studies, can reduce the dependence on low temperature, but the vast majority of protein structures are still collected at or near 100 K. Molecular dynamics approaches are of limited use there, since the time scales of even simple conformational changes become too long to be directly modeled. This chapter will consider only simulations and experiments carried out at ambient temperatures.
- Very few existing atomic models are able to reproduce the intensities of Bragg peaks to anything like the experimental uncertainties. To cite one statistic: of the more than 150,000 protein atomic models in the PDB, fewer than thirty report R_{free} less than 10%, even though typical experimental uncertainties in peak intensities are less than 5% (Holton et al., 2014).

It seems likely that most types of atomic models in current use are missing some key aspects of conformational heterogeneity, both in the protein and in the surrounding solvent.
- The majority of scattered X-rays wind up in between the intense Bragg peaks, in so-called "diffuse" scattering (Meisburger, Case, & Ando, 2020; Meisburger et al., 2017; Van Benschoten et al., 2016; Wall, Wolff, & Fraser, 2018). This data contains potential information about correlated motions in the crystal lattice, but it is not yet clear how useful this will be in practice. Other chapters in this volume deal with diffuse scattering in considerable detail; here I will discuss some connections to molecular dynamics simulations, describing computational methods to compute Bragg and diffuse scattering, and outlining some recent applications to lysozyme.

Molecular dynamics simulations of macromolecules in crystalline environments have been carried out for nearly half a century, but the relatively recent introduction of highly parallel processing via GPU acceleration has made microsecond simulations routinely accessible for the first time; see (Cerutti & Case, 2018; Wall, 2018) for recent reviews. In what follows, Section 2 briefly reviews X-ray scattering in crystals, mainly to establish notation; Section 3 presents methods for setup, equilibration and analysis of MD crystal simulations; and examples of recent applications to lysozyme are given in Sections 4 (for Bragg scattering) and 5 (for diffuse scattering). Other chapters in this volume (Meisburger & Ando, 2023; Pei, Bhatt, Wang, Ando, & Meisburger, 2023) discuss experimental data collection and analysis; this chapter discusses some ways in which comparisons to MD simulations can be made, and what might be gleaned from such comparisons; it is not intended as a review of the small but growing literature on crystalline MD simulations.

The examples discussed below are based on my experience, and largely use the *XtalTools* (https://github.com/dacase/xtaltools.git) and *cpptraj* (https://github.com/Amber-MD/cpptraj) programs. Chapters 3 and 4 of this volume (Wych & Wall, 2023a, 2023b) make use of *cctbx* and *Gromacs* to perform many of the same tasks, supporting the notion that there are many ways to proceed. In particular, the inputs and outputs of most of the workflow described here are PDB-format files, which can be used by any biomolecular MD or crystallographic refinement toolset.

2. Brief summary of X-ray scattering in crystals

To briefly summarize results that are discussed in detail in many places, (Meisburger et al., 2017; Rupp, 2010), the scattering intensity at a given point **q** in reciprocal space can be written as the square of the structure factor there:

$$|F_{crystal}(\mathbf{q})|^2 = I_{Bragg}(\mathbf{q}) + I_{diffuse}(\mathbf{q}) + I_{incoh}(\mathbf{q}) \quad (1)$$

The first two terms describe elastic scattering, and the third term describes incoherent or "Compton" scattering that creates an excited electron. This final term is independent of the structure, and depends only on the composition (numbers and types of atoms) in the unit cell. It is sometimes handled by adding this term into computed intensities, or by subtracting this term from observed values. It becomes significant at high scattering angles, but doesn't contain any structural information.

In an idealized "perfect" crystal, the electron density in each unit cell would be the same, and the only elastic term would involve Bragg scattering. This idealized situation can be described by saying that the structure factors in every cell are given by the average value in the real (experimental or simulated) system. Then scattering occurs only at discrete ("Bragg") positions:

$$I_{Bragg}(\mathbf{q}) = |F_{av}(\mathbf{q})|^2 N \nu_c^* \sum_{\mathbf{h}} \delta(\mathbf{q} - \mathbf{G_h}) \quad (2)$$

Here N is the number of unit cells, ν_c^* is the unit cell volume, and $\mathbf{G_h}$ are the positions of the Bragg peaks; the notation here follows that of (Meisburger et al., 2017). The integrated intensity depends upon additional geometric factors, but these are generally accounted for during data processing, so that what is often called the "Bragg intensity" becomes proportional to the squared magnitude of F_{av}. It is sometimes possible to place these intensities on an absolute scale, where the proportionality constant is set to unity (Lang et al., 2014; Meisburger, Case, & Ando, 2020).

The remaining term is defined as "diffuse" scattering:

$$I_{diffuse}(\mathbf{q}) = \langle |F_{crystal}(\mathbf{q}) - \overline{F_{crystal}(\mathbf{q})}|^2 \rangle \quad (3)$$

This is known as Guinier's equation, and describes scattering at arbitrary angles arising from the distribution of structure factors in different unit cells (or in the same cell at different times.) The direction of scattering is typically specified using (fractional) Miller indices in reciprocal space. The term "diffuse" is perhaps an unfortunate one, since this scattering can have sharp features as well as broader ones, as will be discussed below.

Table 1 shows the fraction of the scattered X-ray intensity into each of the three components of Eq. (1) for triclinic lysozyme. Close to two-thirds of the scattered intensity is diffuse, yet detailed examinations of such data are quite sparse.

3. MD simulations of biomolecular crystals

Biomolecular simulation codes routinely use periodic boundary conditions to represent solvated simulations, so setting up a crystal simulation is largely straightforward, as programs recognize the unit cell information in input PDB-format files. Some simulation packages can also enforce space-group symmetries, with more than one asymmetric unit (ASU) per unit cell, whereas others only allow for translational periodicity. This latter restriction can sometimes be troublesome for large systems or high symmetries (such as virus particles), but is generally not a severe problem. Although the *average* structure may be modeled as being the same in each ASU, this is not the case at any instant in time: a realistic microscopic description of the crystalline state necessarily involves conformational heterogeneity among ASUs or unit cells. The presence of alternate conformations that may be evident in electron density maps, or by the presence of diffuse scattering, are direct evidence of such variations. As a result, many biomolecular crystal simulations invoke a supercell, composed of many unit cells, as the repeating unit. This allows the resulting trajectories to accommodate alternate conformations, and to describe interactions between neighboring macromolecules. (Expansion to P1 symmetry will also lead to multiple copies of the contents of each ASU.) Having multiple chains also allows averages to be computed that should reduce inevitable statistical uncertainties in finite-length simulations.

Table 1 Fraction of X-rays appearing in Bragg reflections, and in diffuse and incoherent scattering, as a function of resolution cutoff. Data is for triclinic lysozyme (Meisburger, Case, & Ando, 2020).

Resolution range	Bragg	Diffuse	Incoherent
$\infty \rightarrow$ 3.0 Å	32%	62%	7%
$\infty \rightarrow$ 2.0 Å	20%	69%	12%
$\infty \rightarrow$ 1.2 Å	10%	63%	26%

3.1 Setup and equilibration

Most existing macromolecular crystal simulations have used an X-ray-refined average structure as the starting point, and rely on the simulated dynamics to generate the resulting heterogeneity. A common practice is to place the most highly-populated conformer (if there are alternate conformers in the refined atomic model) in a regular lattice. It is also possible to choose different alternate conformers for each chain, based on occupation estimates in a refined model. This introduces an element of randomness into the starting configuration, but helps to ensure that alternate side-chain conformations (say), which may be separated by large barriers from the most-populated conformations, are represented. As one example, *qfit* and related routines (Keedy et al., 2016; Lang et al., 2010; van Zundert et al., 2018) automate a procedure for generating alternate conformer models during crystallographic refinement; these, or other alternate conformer models can then be assembled into supercells using the *select_alts* and *XrayPrep* scripts in the *XtalTools* package.

Atomic models derived from X-ray refinement almost never describe the entire contents of the unit cell, since disordered solvent leads to a fairly flat electron density that cannot be described using localized molecular entities such as water. (It is common to describe such density as "bulk" solvent, but the solvent channels in macromolecular crystals are generally narrow enough that the solvent they contain almost certainly has properties distinct from a bulk liquid.) A thorny problem then arises, since both the composition of the solvent channels and the number of water (or other) molecules is generally unknown. In some cases, accurate measurements of crystal densities (Moreau, 2020) can provide constraints on the number of solvent molecules present. In other cases, analyses of the total scattering (including the diffuse part) can be used to obtain absolute diffraction intensities, including an estimate of the scattering amplitude in the forward direction, which can be used to determine the number of electrons in the unit cell (Meisburger, Case, & Ando, 2020).

Generally, however, theoretical considerations must be used to estimate the disordered solvent composition. Heuristic estimates of solvent content can be made from volume considerations (Lang et al., 2014). Both the composition and numbers of solvent molecules can also be estimated from integral equation theory models that posit an equilibrium between solvent channels and an exterior "mother" liquor (Gray, 2020). Alternatively, the simulations themselves can be used: one can systematically vary the number of solvent molecules (usually water) and monitor either the unit cell volume (if constant pressure simulations are used), or the internal pressure (if constant unit cell volume is enforced) (Cerutti et al., 2010). The *AddToBox* routine offers a convenient way to add given numbers of solvent molecules. A downside of this approach is that repeated test simulations are required to find the optimal solvent composition, and the result can depend more than one might like on which force field is being employed (Cerutti et al., 2010).

Once a starting structure has been constructed, equilibration generally proceeds by heating the system to the desired target temperature, with restraints on the macromolecular coordinates that are gradually released as the system adjusts to the force field being employed. One common observation is illustrated in Fig. 1, which monitors a simple metric, the deviation of the protein atoms from the starting X-ray structure. One can see that the system only "settles in" to a new structural ensemble after

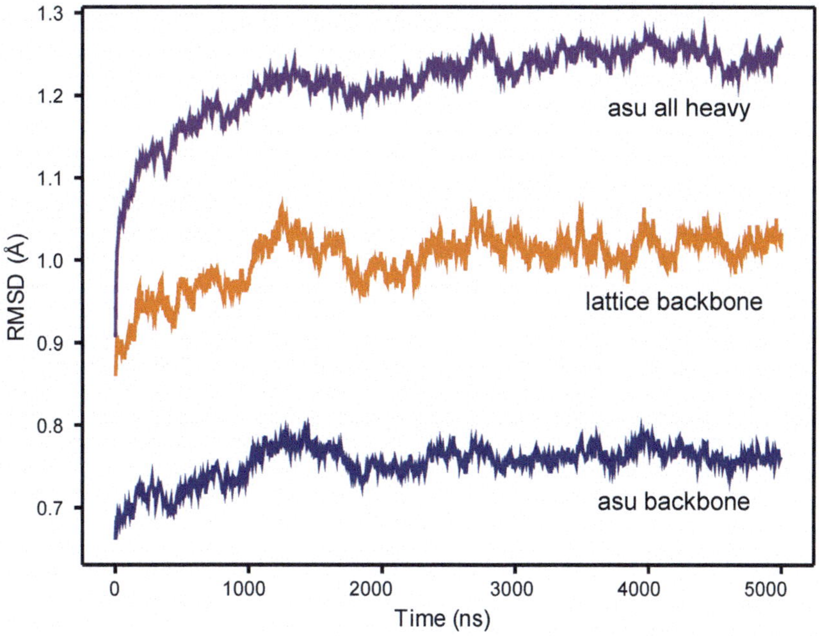

Fig. 1 Slow equilibration of structural deviation from the starting crystal structure for a 27 unit cell simulation triclinic lysozyme; simulation details are in (Meisburger, Case, & Ando, 2020). The blue and purple curves show root-mean-square deviations (RMSD) from the starting crystal structure for a single protein chain, whereas the orange curve shows results for all 27 chains as a whole.

several microseconds, which is a longer time scale than would be expected for the same protein in solution. This slow progress is surely a result of the many close contacts between protein chains, and the narrowness of the solvent channels. Solvent equilibration can be addressed by special techniques (Ross et al., 2020; Samways et al., 2020), and many other tools to accelerate sampling and equilibration are available; a discussion of these is outside the scope of this chapter.

3.2 Common analysis tasks

As with non-crystalline MD simulations, many types of analyses of the results can be made, and the choice generally depends on what sorts of information is desired. This section describes some aspects of trajectory analysis that are specific to simulations of crystals.

3.2.1 Removing unit cell "drift"

In a periodic simulation, all energies and forces should be invariant to a shift of the location of the origin of the coordinate system. This is not quite the case in practical simulations, both because of round-off errors for finite precision arithmetic, but also because the use of a grid in the particle-mesh-Ewald (PME) scheme leads to a dependence on the absolute coordinates. As a result, the unit-cell (or supercell) origin will generally undergo a random drift that can become pronounced for long simulation times. Many types of analyses are unaffected by such motion, but (in particular) the phases of computed structure factors shift as the coordinate origin moves. For this reason, it is common to begin by removing this drift by means of a translation-only structural fit of each frame to the starting coordinates. As an example, in the *cpptraj* program (Roe & Cheatham, 2013), the following command might be used:

```
rms first @C,CA,N norotate.
```

This command carries out a least-squares fit of the protein backbone atoms (named C, CA and N) to the starting structure, using only translations (via the norotate option.) All subsequent analysis can use this output.

3.2.2 Moving atoms into a principal asymmetric unit

Many types of structural analysis can be carried out in a standard fashion, but it is often convenient to use symmetry elements to translate and rotate protein chains into the positions they would have in a particular (principal) asymmetric unit (ASU) or unit cell. This creates non-physical coordinates with atoms nearly on top of each other, but simplifies visualization and other types of analysis. For example, if one has t snapshots of a supercell simulation with n asymmetric units, a different view (i.e. a new trajectory) can be created that has $n \times t$ snapshots, each with a single ASU. This makes it straightforward to take advantage of the fact that the simulation has multiple copies of the same macromolecule, and to describe (for example) the types of conformational heterogeneity present in the full ensemble. As an example, the root-mean-square (RMS) deviations shown in Fig. 1 were created using the *XtalAnalzye* package: the asu curves used results from the pseudo-trajectory with a single ASU, whereas the lattice curve treated the supercell as a single entity. (An alternative approach uses the *XtalSymm* action in cpptraj.)

3.2.3 Computing structure factors

One nice feature of crystallographic computing is that structure factors are straightforward to compute if the atomic positions in the unit cell are specified. This can be done using the *sfall* program in the CCP4 suite (https://www.ccp4.ac.uk), the *phenix.fmodel* routine in Phenix (https://phenix-online.org/), or indeed, with many other crystallographic codes, including some built-in to molecular dynamics packages. Since the simulation contains all atoms thought to be present, there is no need for any "bulk" or "disordered" solvent contribution. One complication can arise for some programs that use discrete Fourier transforms to carry out the calculation: each snapshot from a trajectory should have no atomic displacement (thermal) parameter, since the trajectory itself will describe atomic motions. But some programs are written in a way that fails if a B-factor of zero is input. A simple workaround is to assign all atoms a common B-factor (say of 15 Å^2) during the structure factor computation, and then remove this at the end by applying a B-factor of −15 to the resultant averages or standard deviations.

In practice, MD crystal simulations always have some element of true periodicity, although the repeating element might consist of many primitive unit cells. This true periodicity implies that scattering can only be computed at the Bragg positions of the supercell. For example, in the 7×7×7 supercell for triclinic lysozyme (with 343 unit cells and one chain per unit cell) described by (Meisburger, Case, & Ando, 2020), Eq. (3) allows one to describe diffuse scattering on a grid where every seventh point (in each direction) corresponds to a Bragg position for the true unit cell. This is not as fine a grid as the processed experimental data, which was processed onto a

13×11×11 grid. (An even finer experimental grid would have been possible, given the pixel size of the detector and its location relative to the crystal.) In order to compare simulated and experimental intensities, values on the experimental 13×11×11 grid were interpolated onto the 7×7×7 grid of the simulation, and a similar procedure would almost always be needed. Details of one way to do this are in (Meisburger, Case, & Ando, 2023).

The md2diffuse.sh script shows one way to carry out these trajectory analyses, using tools from the CCP4 package (Winn et al., 2011) to compute structure factors from individual snapshots. It illustrates the removal unit-cell drift, and takes pains to reduce the amount of disk space needed to store the structure factors for many frames. Since this is implemented as a simple shell script, pieces of it could be incorporated into many other workflows. The output is a list of Bragg intensities (from Eq. (2)) and diffuse intensities (from Eq. (3)), indexed with Miller indices of the supercell used in the simulation. Data processing workflows for experimental data result in a similar indexing pattern, so that direct comparisons between experiment and theory, can be made in a straightforward fashion.

One characteristic feature of such comparisons is illustrated in Fig. 2. Here the experimental intensities have been placed on an absolute scale (Meisburger, Case, & Ando, 2020), so there is no need to adjust any scaling parameters to compare to the computed values. It is immediately clear that there is very little overall correlation between computed and observed Bragg intensities, but the diffuse comparison is much more satisfying, with a strong correlation between simulated and observed data points. The origin of the large scatter for Bragg peaks is easy to understand: the average backbone structure in the simulation is about 0.4 Å; from a refined structure, such as would be deposited in the PDB. This is a small deviation in absolute terms (about one-quarter of a carbon–carbon bond length), but is well outside the expected coordinate error of about 0.03 Å; reported for this dataset (PDB ID 8dyz), which was collected to a resolution of 1.27 Å. As a result, it is difficult to learn much by comparing computed and observed Bragg intensities from unrestrained simulations, but simulations can still be of considerable use in the context of crystallographic refinement, as discussed in Section 4.

This behavior is in stark contrast to the diffuse scattering, which exhibits a correlation coefficient of 0.987 and a slope of 0.92, indicating that the computed values systematically track observations, but underestimate them by a small but systematic amount. (In Fig. 2, the "observed" values have lattice dynamics removed; but very similar plots are obtained without this

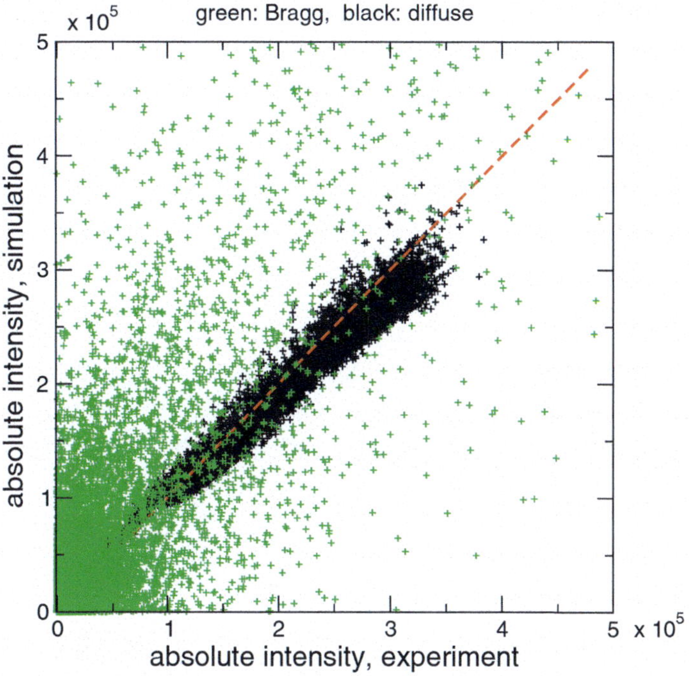

Fig. 2 Bragg intensities (green) and diffuse intensities (black), for tetragonal lysozyme. Simulation results are from a 5 μs simulation of a single unit cell. The "experimental" diffuse intensities have been modified to remove lattice vibrations, as discussed in Section 5.2. The red dashed line shows $y = x$. The PDB ID is 8dyz (Meisburger, Case, & Ando, 2023).

removal when the simulation uses supercells containing many protein chains.) But getting a good overall correlation is only a starting point. The key question, discussed in Section 5, is the extent to which simulations may guide the extraction of physically interesting information from measurements of diffuse scattering.

3.3 Convergence of simulations

A recurring practical question regarding simulations concerns how much sampling is needed to obtain converged results. There can be no single answer to this, since every protein is different, and since there has been only a limited amount of experience to provide even rules of thumb. Compounding the problem is the fact that we don't yet know how much convergence is needed in the diffuse intensities in order to extract reliable information. The sections below provide some example information for convergence in simulations of triclinic lysozyme, using a supercell of 27 unit cells, and hence 27 protein chains, each with 129 residues.

The top portion of Table 2 uses the full 5 μs of simulation, but selects progressively fewer snapshots to use compute the variance in structure factors. The widest sampling interval of 40 ns corresponds to just 125 frames, yet still gives overall correlation coefficients close to the result from averaging the full 12,500 frames, with a sampling interval of 0.4 ns. The bottom portion of the table considers the effect of the length of trajectory, with a fixed sampling interval. At least in this example, near-converged results are found after a few hundred ns of simulation. A caveat for all these comparisons is that we don't yet know how much precision is really needed to extract subtle yet interesting features of the resultant averages. The rate of convergence for such features might be slower than for the overall correlation coefficients or relative errors shown in Table 2.

The analysis here is based on the overall diffuse signal, either from experiment or simulation. Other useful comparisons give greater attention

Table 2 Convergence properties of simulations of diffuse scattering intensities, for a 27-unit cell simulation of triclinic lysozyme. The first row is treated as a standard, and all subsequent rows are compared to it. CC is the correlation coefficient for 860,000 points in reciprocal space; rel. error is the point-by-point root-mean-square difference, divided by the range of the data points, and expressed as a percent.

Sim. length	Sampling int.	CC	Rel. error
5 μs	0.4 ns	–	–
–	0.8 ns	0.9998	0.2%
–	2.0 ns	0.9992	0.3%
–	4.0 ns	0.9981	0.5%
–	20 ns	0.9893	1.2%
–	40 ns	0.9775	1.7%
4 μs	0.4 ns	0.9992	0.3%
3 μs	–	0.9978	0.5%
1 μs	–	0.9920	1.1%
0.4 μs	–	0.9859	1.5%

to the resolution dependence and anisotropy of the signal (Meisburger et al., 2023; Wall, 2018).

Quite different convergence behavior is seen in Fig. 3, which shows results for supercells of different sizes. The general shape of intensity versus scattering angle is preserved, but there are much more intense peaks as the size increases, even when plotted on a per-unit-cell basis. The basis for this behavior is discussed in Section 5.1: larger supercells permit the thermal excitation of long-range vibrational modes that correspond to the movement of many protein chains in concert. These phonon-like modes contribute to large diffuse intensities in regions near the Bragg peaks. Some recent work (Meisburger, Case, & Ando, 2023) tries to remove such

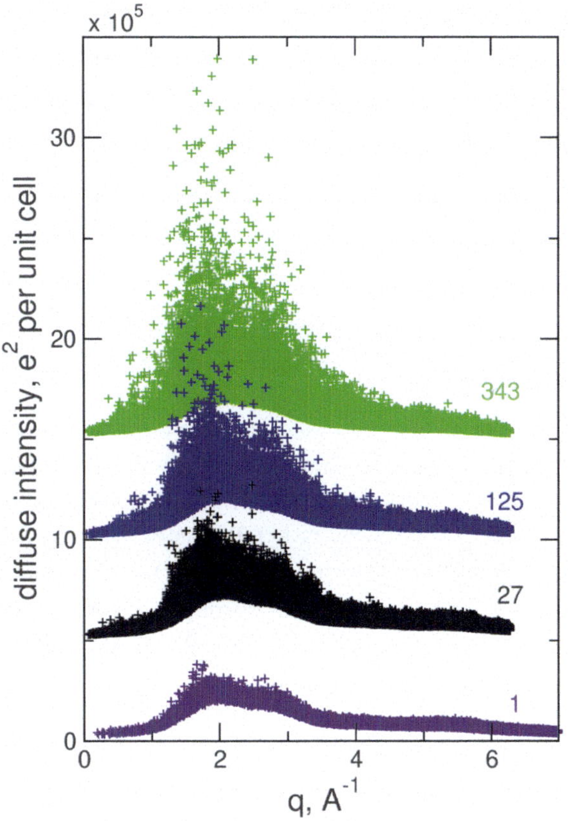

Fig. 3 Simulated diffuse intensities (for triclinic lysozyme, per unit cell) vs momentum transfer, for different numbers of unit cells. Plots for 27, 125 and 343 unit cells have been displaced upwards by 5, 10 and 15×10^5 units for visualization purposes.

behavior from the observed intensities, allowing better comparisons to simulations using only a single unit cell.

The following two sections outline ways in which MD simulations can be used for the analysis of Bragg and diffuse scattering measurements.

4. Bragg scattering and crystallographic refinement

Figure 2 suggests that unrestrained simulations are unlikely to directly show good agreement with observation for Bragg intensities. Nevertheless, simulation results can still be used to provide useful insights into what can be learned from such data.

4.1 Creating reference data sets

The use of MD simulations to create reference data to test ways in which crystallographic data is interpreted and refined has a long history (Kuriyan et al., 1986, 1991). These efforts have typically computed Bragg intensities as described in Section 3.2.3, and used that as a target dataset for conventional crystallographic refinement. Since the underlying conformational ensemble is known, the ability of refinement protocols to extract accurate information can be systematically tested. This procedure is also useful in making comparisons between simulations and published atomic models: parallel refinements against real and simulated data can make identical assumptions about how the data is processed and analyzed, and about what sorts of atomic models are used as a surrogate for the (inevitably more complex) real computational landscape (Janowski et al., 2016; Kuzmanic et al., 2014). More recent work (Liu et al., 2023) has looked at the simulation of diffraction images themselves, rather than just creating a list of Bragg intensities. This allows tests of earlier parts of the data processing protocols.

4.2 Geometry restraints and molecular force fields

There is also a long history of adding experimental restraints to MD simulations to guide the results towards conformations that are in agreement with observation. Some of the earliest implementations of X-ray refinement programs employed molecular mechanics force fields (Brünger & Karplus, 1991; Gros et al., 1990; Jack & Levitt, 1978), and this idea has been continued in recent years, including the use of quantum mechanics as well as force fields (Borbulevych et al., 2018, 2021; Caldararu et al., 2019; Fenn et al., 2011; Moriarty et al., 2020; Schnieders et al., 2009). The *phenix* refinement program (Liebschner et al., 2019) now includes an interface to Amber codes, allowing the use of gradients from force fields to replace more conventional bond and angle restraints derived from small molecules (Gray & Case, 2021; Moriarty et al., 2020; Mikhailovskii et al., 2022). This interface has been shown to guide reciprocal space refinement of protein structures with small but noticeable improvement over conventional refinement, especially at lower resolution.

4.3 Alternate conformers and ensembles

As mentioned above, crystallographic refinement against Bragg intensities can often lead to the characterization of alternate conformers, either from visualization of electron density maps, or through more automated procedures to identify likely candidates (Lang et al., 2010; Keedy et al., 2016; van Zundert et al., 2018). These are typically represented in atomic models as atoms with partial occupancy, but their very existence implies that not every unit cell is identical, even on average; this in turn suggests connections to diffuse scattering, which also depends on the spread of structure factors among unit cells.

There have also been efforts to refine crystallographic data against explicit ensembles, that is, against atomic models that contain multiple copies of entire proteins (Afonine et al., 2018; Burnley et al., 2012; Caldararu et al., 2021; Forneris et al., 2014; Ploscariu et al., 2021). Using weak X-ray restraints may also lead to improved dynamics in more conventional simulations (Xue & Skrynnikov, 2014).

4.4 Dealing with disordered solvent

Macromolecular crystals generally contain between 30% and 70% solvent, which is most often water plus ions, but may also contain alcohols (often as cryoprotectants) and other components. Some portion of the solvent can modeled with discrete (or "ordered") molecules, but the remainder is disordered. Methods to generate plausible starting configurations for MD simulations were discussed above, but few crystallographic refinement protocols attempt to account for all the solvent in this manner. The most common approach is to treat disordered solvent as having a continuous and constant density in the volume that is not occupied by discrete atoms; the constant is often treated as a refinement parameter.

One recent approach uses integral equation models (or an equivalent classical density functional approach) to model the atomic (and hence electron) density in solvent channels (Gray et al., 2022). This model (implemented in the AmberTools package) considers the solvent to be in equilibrium with a external solution with some assumed composition. The electron density is estimated at each point on a three-dimensional grid (with a typical grid spacing of 0.3–0.8 Å), with negligible densities where solvent atoms would overlap with the discrete atoms of the atomic model. In principle, this approach can determine both the amount of solvent and its composition. The latter can be of great use in nucleic acids, where only a small fraction of the cations that neutralize the DNA or RNA chains can be identified with discrete models. The general features of the predicted ion distributions are in agreement with experiment and with all-atom MD simulations, but its details are less certain (Giambasu et al., 2014, 2015, 2019).

5. Diffuse scattering

The study of diffuse scattering in proteins also has a long history, which is covered in other chapters in this volume, as well as in recent reviews (Meisburger et al., 2017; Wall et al., 2018). Modern pixel array detectors allow fine scans of reciprocal space, and have a large dynamic range that can handle weak scattering (Gruner & Lattman, 2015). These features, along with parallel advances in data processing software, make possible more complete and accurate accounts of diffuse scattering than had been available earlier (Meisburger, Case, & Ando, 2020, 2023; Wall, Wolff, & Fraser, 2018), but which reinforce and clarify many of the classic conclusions reached earlier (Clarage et al., 1992; Perez et al., 1996; Polikanov & Moore, 2015; Van Benschoten et al., 2016; Welberry, 2004).

One very visible subset of diffuse scattering is concentrated in the vicinity of Bragg reflections, and often referred to as "haloes" around the (much sharper) Bragg peaks. This sort of scattering is dominated by thermally-excited vibrations, or phonons, that describe near rigid-body motions of protein chains in the crystal lattice (Clarage et al., 1992; De Klijn et al., 2019; Perez et al., 1996; Peck et al., 2018) We look first at how recent MD simulations shed light on these motions, then on attempts to learn about other types of disorder in protein crystals.

5.1 Lattice vibrations

One structural feature of most supercell protein simulations is that the protein chains move a bit from their ideal lattice positions, even after overall unit-cell drift has been removed. Figure 4 shows some results for a

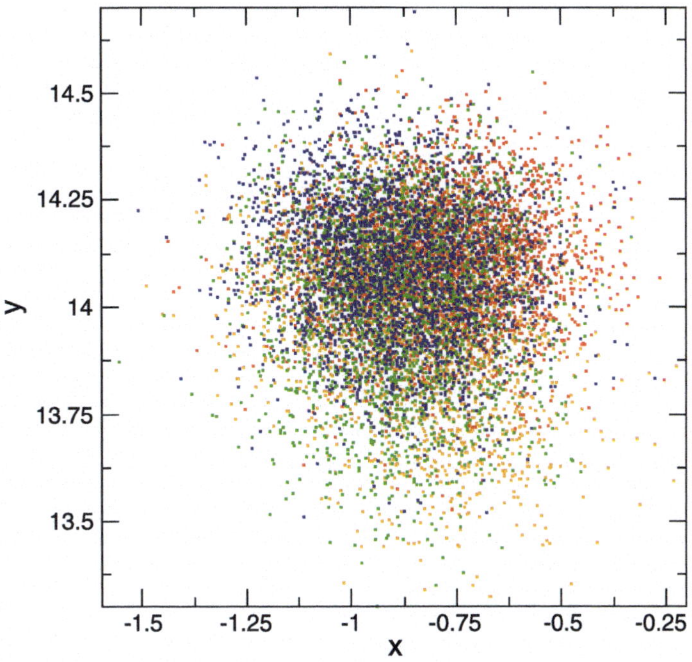

Fig. 4 Projection of the protein center of mass onto the *x* and *y* directions, from a 27-chain simulation of triclinic lysozyme (PDB Id 4lzt); simulation details are given in (Meisburger, Case, & Ando, 2020). Prior to the projection, four chains were translated back to a "primary" ASU using the unit cell parameters. Each chain is given a different color, and points represent snapshots taken from a 5 μs simulation; for simplicity, only four chains are represented, using orange, blue, green and red symbols. Distances along each direction are in Å. Similar plots of a 12 unit-cell simulation are given by (Janowski et al., 2016).

27-chain simulation of triclinic lysozyme (Meisburger, Case, & Ando, 2020). Four protein chains (chosen at random and represented by different colors) were translated back to a primary unit cell, and its center of mass projected onto the *xy* plane. This illustrates two types of disorder evident in the simulation. First, there is a "static" component, in that the center of each color is displaced from its ideal, starting, position by up to 0.2 Å; this displacement is static only on the microsecond time scale of the simulation. Second, within each color there is vibrational motion about that color's average position, by about 0.3–0.4 Å, which reflects thermal motion in the ambient-temperature simulation environment.

This distinction can serve as a model for what is surely going on in real protein crystals, where both thermally excited vibrations and more static forms of disorder both contribute to diffuse scattering. A synchrotron experiment averages over many unit cells and over long (by microscopic standards) data-collection times, and cannot distinguish between static and dynamic disorder.

It has been recognized for many years that thermally-excited lattice vibrational modes form the main component of the diffuse haloes around the Bragg peaks (Clarage et al., 1992; De Klijn et al., 2019; Perez et al., 1996; Peck et al., 2018). This component can be fit to a lattice vibrational model (Meisburger, Case, & Ando, 2020, 2023) to extract the amplitudes of the center-of-mass motions corresponding to what is seen in Figure 4. Table 3 compares fitted and simulation results for three crystal forms of lysozyme, which have different amounts of solvent content. Lattices with more solvent exhibit more thermal disorder in the protein positions, and the magnitude and trends of these RMS fluctuations are approximately reproduced in the simulations. This correspondence indicates that the lattice model used to interpret the diffuse haloes is physically reasonable, i.e., that it is more than just a data-fitting exercise.

The presence of thermally-excited lattice vibrations suggests that there should be correlations between the positions of different protein chains in the crystal lattice; indeed, it is the long-range character of these correlations that leads to the characteristic halo pattern around the Bragg positions. Figure 5 presents the center-of-mass covariance between different protein chains as a function of the distance between their centers; the zero-distance point is just the square of the RMS values shown in Table 3. It is clear from both the lattice vibrational fit, and from MD simulations, that such correlations extend to 100 Å; or more. Fits to an exponential decay have a

Table 3 RMS fluctuations (in Å) of the center of mass of a single lysozyme chain, averaged over all the 27 chains in a recent MD simulation. Experimental results and simulation details are from (Meisburger, Case, & Ando, 2023).

	Triclinic	**Orthorhombic**	**Tetragonal**
PDB Id	*6o2h*	*8dz7*	*8dxy*
Exp.	0.34	0.55	0.58
Calc.	0.36	0.58	0.52
# of chains	343	500	128
Solvent content	26.9%	44.4%	47.0%

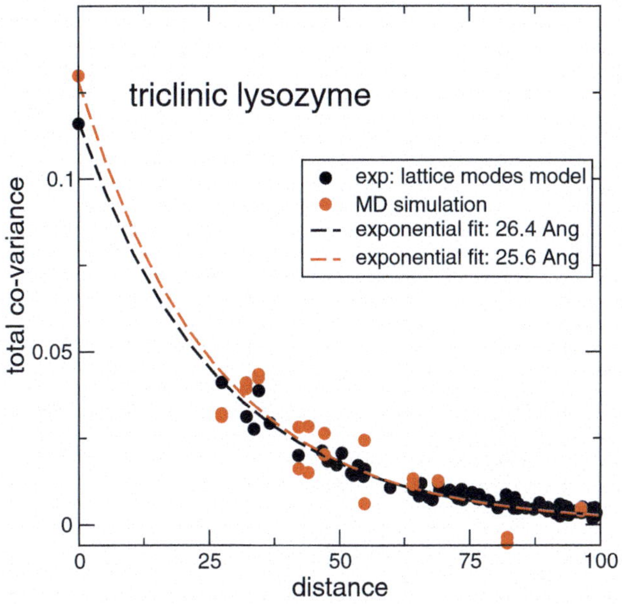

Fig. 5 Covariance of the positions of protein centers of mass versus distance for triclinic lysozyme. The *y*-axis plots the trace of the 3×3 covariance matrix of deviations from the average center of mass position for two protein chains separated by a distance show in the *x*-axis. Black circles show values derived by fitting a lattice model to experimental data (Meisburger, Case, & Ando, 2023); red circles show results from an MD simulation with 343 protein chains. Dashed lines are fits to an exponential curve, with the decay distance indicated.

length dependence of about 25 Å. The simulated results have somewhat greater scatter than those fit to experimental diffuse data, but the overall behavior is clearly almost the same. It is worth noting that this interpretation of diffuse haloes arises purely from thermally-excited phonons in a perfect lattice, and is not a result of "disorder" in either protein chains or from lattice distortions.

5.2 Internal contributions to diffuse scattering

As noted above, lattice phonons dominate the diffuse scattering near the Bragg reflection points. There has long been a hope that diffuse scattering (probably in conjunction with simulations or models) might help to characterize more interesting aspects of protein collective motions as well. Recent work describes a method to subtract the "halo" contribution from observations (Meisburger, Case, & Ando, 2023), and this may make it easier to concentrate on what remains. It is worth remembering that remaining sources of disorder that are expected to be large will include contributions from bulk solvent and from rotational disorder in surface protein side-chains. These contributions are often less interesting from a functional standpoint than are more collective (domain) motions that could contribute to binding or catalysis, and it remains to be seen how (or even if) well one can sort out all of these contributions. Simulations (Wall et al., 2018; Wych et al., 2019) have been used to analyze liquid-like and rigid-body models of internal motions (Peck et al., 2018). These models are parsimonious and general, which is both a strength and a weakness. The TLS (translation–libration–screw) model often used to model atomic displacement parameters provides a more detailed model for local rigid-body motion that has implications for diffuse scattering (Moore, 2009). MD simulations should also serve as good starting points for interpretation and refinement. Unfortunately, there is currently very little experimental data on proteins under different but similar conditions, such as comparisons between enzymes in their apo and ligand-bound states. It is likely that more information about trends in such data, in empirical models that are consistent with the data, and in the predictions of MD simulations will be required for the next big steps in interpretation of macromolecular diffuse scattering beyond the lattice phonon model.

More discussion on the potential for MD capture internal contributions to diffuse scattering can be found in Wych and Wall (2023a, 2023b).

6. Conclusions

The use of molecular dynamics simulations as an aid to interpretation of X-ray scattering is certain to grow, as increasing computer power makes simulations more affordable, and as techniques for ambient temperature experiments become more established. Time-resolved experiments (which are outside the scope of this review) offer additional prospects for connections between measurements and simulations, as do solid-state NMR results (Schanda & Ernst, 2016). Insights into atomic motions coming from solution studies and from cryo-electron microscopy will also inform the way crystal simulations are carried out and interpreted. Studies on smaller (and sometimes simpler) molecules, such as crystalline peptides and oligosaccharides, should also be fruitful (see, e.g., Janowski et al., 2013).

Acknowledgments

I thank Nozomi Ando, Steve Meisburger, James Holton and Mike Wall for many helpful discussions. This work was supported by NIH grant GM122086.

References

Afonine, P. V., Poon, B. K., Read, R. J., Sobolev, O. V., Terwilliger, T. C., Urzhumtsev, A., & Adams, P. D. (2018). Real-space refinement in PHENIX for cryo-EM and crystallography. *Acta Crystallographica. Section D, 74*, 531–544.

Atakisi, H., Moreau, D. W., & Thorne, R. E. (2018). Effects of protein-crystal hydration and temperature on side-chain conformational heterogeneity in monoclinic lysozyme crystals. *Acta Crystallographica. Section D, 74*, 264–278. https://doi.org/10.1107/S2059798318000207

Borbulevych, O., Martin, R. I., & Westerhoff, L. M. (2018). High-throughput quantum-mechanics/molecular-mechanics (ONIOM) macromolecular crystallographic refinement with PHENIX/DivCon: The impact of mixed Hamiltonian methods on ligand and protein structure. *Acta Crystallographica. Section D, 74*, 1063–1077. https://doi.org/10.1107/S2059798318012913

Borbulevych, O. Y., Martin, R. I., & Westerhoff, L. M. (2021). The critical role of QM/MM X-ray refinement and accurate tautomer/protomer determination in structure-based drug design. *Journal of Computer-aided Molecular Design, 35*, 433–451.

Brünger, A. T., & Karplus, M. (1991). Molecular dynamics simulations with experimental restraints. *Accounts of Chemical Research, 24*, 54–61.

Burnley, B. T., Afonine, P. V., Adams, P. D., & Gros, P. (2012). Modelling dynamics in protein crystal structures by ensemble refinement. *eLife, 1*, e00311.

Caldararu, O., Manzoni, F., Oksanen, E., Logan, D. T., & Ryde, U. (2019). Refinement of protein structures using a combination of quantum-mechanical calculations with neutron and X-ray crystallographic data. *Acta Crystallographica. Section D, 75*, 368–380. https://doi.org/10.1107/S205979831900175X

Caldararu, O., Ekberg, V., Logan, D. T., Oksanen, E., & Ryde, U. (2021). Exploring ligand dynamics in protein crystal structures with ensemble refinement. *Acta Crystallographica. Section D, 77*, 1099–1115. https://doi.org/10.1107/S2059798321006513

Cerutti, D. S., & Case, D. A. (2018). Molecular dynamics simulations of macromolecular crystals. *WIREs Computational Molecular Science, 8*, e1402.

Cerutti, D. S., Freddolino, P. L., Duke, R. E., Jr., & Case, D. A. (2010). Simulations of a protein crystal with a high resolution X-ray structure: Evaluation of force fields and water models. *Journal of Physical Chemistry B, 114*, 12811–12824.

Clarage, J. B., Clarage, M. S., Phillips, W. C., Sweet, R. M., & Caspar, D. L. D. (1992). Correlations of atomic movements in lysozyme crystals. *Proteins: Structure, Function, and Genetics, 12*, 145–157.

De Klijn, T., Schreurs, A., & Kroon-Batenburg, L. (2019). Rigid-body motion is the main source of diffuse scattering in protein crystallography. *IUCrJ, 6*, 277–289.

Fenn, T. D., Schnieders, M. J., Mastyakimov, M., Wu, C., Langan, P., Pande, V. S., & Brunger, A. T. (2011). Reintroducing electrostatics into macromolecular crystallographic refinement: Application to neutron crystallography and DNA hydration. *Structure, 19*, 523–533.

Forneris, F., Burnley, B. T., & Gros, P. (2014). Ensemble refinement shows conformational flexibility in crystal structures of human complement factor D. *Acta Crystallographica. Section D, 70*, 733–743. https://doi.org/10.1107/S1399004713032549

Fraser, J. S., van den Bedem, H., Samelson, A. J., Lang, P. T., Holton, J. M., Echols, N., & Alber, T. (2011). Accessing protein conformational ensembles using room-temperature X-ray crystallography. *Proceedings of the National Academy of Sciences of the United States of America, 108*, 16247–16252.

Giambasu, G., Case, D. A., & York, D. M. (2019). Predicting site-binding modes of ions and water to nucleic acids using molecular solvation theory. *Journal of the American Chemical Society, 141*, 2435–2445.

Giambasu, G. M., Luchko, T., Herschlag, D., York, D. M., & Case, D. A. (2014). Ion counting from explicit solvent simulations and 3D-RISM. *Biophysical Journal, 104*, 883–894.

Giambasu, G. M., Gebala, M. K., Panteva, M. T., Case, D. A., Herschlag, D., & York, D. M. (2015). Competitive interaction of monovalent cations with DNA predicted by 3D-RISM. *Nucleic Acids Research, 43*, 8405–8415.

Gray, J. G. (2020). Using molecular mechanics force fields and RISM densities to improve macromolecular models [Ph.D. thesis, Rutgers University].

Gray, J. G., & Case, D. A. (2021). Refinement of RNA structures using Amber force fields. *Crystals, 11*, 771.

Gray, J. G., Giambasu, G. M., Case, D. A., & Luchko, T. (2022). Integral equation models for solvent in macromolecular crystals. *Journal of Chemical Physics, 156*, 014801.

Gros, P., van Gunsteren, W. F., & Hol, W. G. J. (1990). Inclusion of thermal motion in crystallographic structures by restrained molecular dynmaics. *Science, 249*, 1149–1152.

Gruner, S. M., & Lattman, E. E. (2015). Biostructural science inspired by next-generation X-ray sources. *Annual Review of Biophysics, 44*, 33–51.

Holton, J. M., Classen, S., Frankel, K. A., & Tainer, J. A. (2014). The R-factor gap in macromolecular crystallography: An untapped potential for insights on accurate structures. *FEBS Journal, 281*, 4046–4060.

Jack, A., & Levitt, M. (1978). Refinement of large structures by simultaneous minimization of energy and R factor. *Acta Crystallographica. Section A, 34*, 931–935. https://doi.org/10.1107/S0567739478001904

Janowski, P. A., Cerutti, D. S., Holton, J., & Case, D. A. (2013). Peptide crystal simulations reveal hidden dynamics. *Journal of the American Chemical Society, 135*, 7938–7948.

Janowski, P. A., Liu, C., Deckman, J., & Case, D. A. (2016). Molecular dynamics simulation of triclinic lysozyme in a crystal lattice. *Protein Science, 25*, 87–102.

Keedy, D. A., van den Bedem, H., Sivak, D. A., Petsko, G. A., Ringe, D., Wilson, M. A., & Fraser, J. S. (2014). Crystal cryocooling distorts conformational heterogeneity in a model Michaelis complex of DHFR. *Structure, 22*, 899–910.

Keedy, D. A., Fraser, J. S., & van den Bedem, H. (2016). Exposing hidden alternative backbone conformations in X-ray crystallography using qFit. *PLoS Computational Biology, 11*, e1004507.

Kuriyan, J., Petsko, G. A., Levy, R. M., & Karplus, M. (1986). Effect of anisotropy and anharmonicity on protein crystallographic refinements. An evaluation by molecular dynamics. *Journal of Molecular Biology, 190*, 227–254.

Kuriyan, J., Ösapay, K., Burley, S. K., Brünger, A. T., Hendrickson, W. A., & Karplus, M. (1991). Exploration of disorder in protein structures by X-ray restrained molecular dynamics. *Proteins: Structure, Function, and Genetics, 10*, 340–358.

Kuzmanic, A., Pannu, N. S., & Zagrovic, B. (2014). X-ray refinement significantly underestimates the level of microscopic heterogeneity in biomolecular crystals. *Nature Communications, 5*, 3220.

Lang, P. T., Ng, H.-. L., Fraser, J. S., Corn, J. E., Echols, N., Sales, M., Holton, J. M., & Alber, T. (2010). Automated electron-density sampling reveals widespread conformational polymorphism in proteins. *Protein Science, 19*, 1420–1431.

Lang, P. T., Holton, J. M., Fraser, J. S., & Alber, T. (2014). Protein structural ensembles are revealed by redefining X-ray electron density noise. *Proceedings of the National Academy of Sciences of the United States of America, 111*, 237–242.

Liebschner, D., Afonine, P. V., Baker, M. L., Bunkoczi, G., Chen, V. B., Croll, T. I., ... Adams, P. D. (2019). Macromolecular structure determination using X-rays, neutrons and electrons: Recent developments in Phenix. *Acta Crystallographica. Section D, 75*, 861–877. https://doi.org/10.1107/S2059798319011471

Liu, N., Mikhailovskii, O., Skrynnikov, N. R., & Xue, Y. (2023). Simulating diffraction photographs based on molecular dynamics trajectories of a protein crystal: A new option to examine structure-solving strategies in protein crystallography. *IUCrJ, 10*, 16–26. https://doi.org/10.1107/S2052252522011198

Meisburger, S. P., & Ando, N. (2023). Processing macromolecular diffuse scattering data, submitted discuss experimental data collection and analysis. submitted.

Meisburger, S. P., Thomas, W. C., Watkins, M. B., & Ando, N. (2017). X-ray scattering studies of protein structural dynamics. *Chemical Reviews, 117*, 7615–7672.

Meisburger, S. P., Case, D. A., & Ando, N. (2020). Correlated motions in a protein crystal. *Nature Communications, 11*, 1271.

Meisburger, S. P., Case, D. A., & Ando, N. (2023). Robust total X-ray scattering workflow to study correlated motion of proteins in crystals. *Nature Communications, 14*, 1228. https://doi.org/10.1101/2022.08.22.504832

Mikhailovskii, O., Xue, Y., & Skrynnikov, N. R. (2022). Modeling a unit cell: Crystallographic refinement procedure using the biomolecular MD simulation platform Amber. *IUCrJ, 9*, 114–133. https://doi.org/10.1107/S2052252521011891

Moore, P. B. (2009). On the relationship between diffraction patterns and motions in macromolecular crystals. *Structure, 17*, 1307–1315.

Moreau, D. W. (2020). *Ice formation and solvent nanoconfinement in protein crystallography* [Ph. D. thesis, Cornell University].

Moriarty, N. W., Janowski, P. A., Swails, J. M., Nguyen, H., Richardson, J. S., Case, D. A., & Adams, P. D. (2020). Improved chemistry restraints for crystallographic refinement by integrating the Amber force field into Phenix. *Acta Crystallographica. Section D, 76*, 51–62. https://doi.org/10.1107/S2059798319015134

Peck, A., Poitevin, F., & Lane, T. J. (2018). Intermolecular correlations are necessary to explain diffuse scattering from protein crystals. *IUCrJ, 5*, 211–222.

Pei, X., Bhatt, N., Wang, H., Ando, N., & Meisburger, S. P. (2023). Introduction to diffuse scattering and data collection. submitted.

Perez, J., Faure, P., & Benoit, J. (1996). Molecular rigid-body displacements in a tetragonal lysozyme crystal confirmed by X-ray diffuse scattering. *Acta Crystallographica. Section D, 52*, 722–729. https://doi.org/10.1107/S0907444996002594

Ploscariu, N., Burnley, T., Gros, P., & Pearce, N. M. (2021). Improving sampling of crystallographic disorder in ensemble refinement. *Acta Crystallographica. Section D, 77*, 1357–1364. https://doi.org/10.1107/S2059798321010044

Polikanov, Y. S., & Moore, P. B. (2015). Acoustic vibrations contribute to the diffuse scatter produced by ribosome crystals. *Acta Crystallographica. Section D, 71*, 2021–2031.

Roe, D. R., & Cheatham, T. E. (2013). PTRAJ and CPPTRAJ: Software for processing and analysis of molecular dynamics trajectory data. *Journal of Chemical Theory and Computation, 9*, 3084–3095.

Ross, G. A., Russell, E., Deng, Y., Lu, C., Harder, E. D., Abel, R., & Wang, L. (2020). Enhancing water sampling in free energy calculations with grand canonical Monte Carlo. *Journal of Chemical Theory and Computation, 16*, 6061–6076.

Rupp, B. (2010). *Biomolecular crystallography: Princples, practice, and application of structural biology*. New York: Garland Science.

Samways, M. L., BruceMacdonald, H. E., & Essex, J. W. (2020). grand: A Python module for grand canonical water sampling in OpenMM. *Journal of Chemical Information and Modeling, 60*, 4436–4441.

Schanda, P., & Ernst, M. (2016). Studying dynamics by magic-angle spinning solid-state NMR spectroscopy: Principles and applications to biomolecules. *Progress in Nuclear Magnetic Resonance Spectroscopy, 96*, 1–46.

Schnieders, M. J., Fenn, T. D., Pande, V. S., & Brunger, A. T. (2009). Polarizable atomic multipole X-ray refinement: Application to peptide crystals. *Acta Crystallographica. Section D, 65*, 952–965.

Stachowski, T. R., Vanarotti, M., Seetharaman, J., Lopez, K., & Fischer, M. (2022). Water networks repopulate protein-ligand interfaces with temperature. *Angewandte Chemie International Edition, 61*, e202112919.

Van Benschoten, A. H., Liu, L., Gonzalez, A., Brewster, A. S., Sauter, N. K., Fraser, J. S., & Wall, M. E. (2016). Measuring and modeling diffuse scattering in protein X-ray crystallography. *Proceedings of the National Academy of Sciences of the United States of America, 113*, 4069–4074.

van Zundert, G. C., Hudson, B. M., de Oliveira, S. H. P., Keedy, D. A., Fonseca, R., Heliou, A., ... van den Bedem, H. (2018). qFit-ligand reveals widespread conformational heterogeneity of drug-like molecules in X-ray electron density maps. *Journal of Medicinal Chemistry, 61*, 11183–11198.

Wall, M. E. (2018). Internal protein motions in molecular-dynamics simulations of Bragg and diffuse X-ray scattering. *IUCrJ, 5*, 172–181.

Wall, M. E., Wolff, A. M., & Fraser, J. S. (2018). Bringing diffuse X-ray scattering into focus. *Current Opinion in Structural Biology, 50*, 109–116.

Welberry, T. R. (2004). *Diffuse X-ray scattering and models of disorder.* Oxford: Oxford University Press.

Winn, M. D., Ballard, C. C., Cowtan, K. D., Dodson, E. J., Emsley, P., Evans, P. R., ... Wilson, K. S. (2011). Overview of the CCP4 suite and current developments. *Acta Crystallographica. Section D, 67*, 235–242. https://doi.org/10.1107/S0907444910045749

Woldeyes, R. A., Sivak, D. A., & Fraser, J. S. (2014). E pluribus unum, no more: From one crystal, many conformations. *Current Opinion in Structural Biology, 28*, 56–62.

Wych, D. C., & Wall, M. E. (2023a). Molecular-dynamics simulations of macromolecular diffraction, part I: Preparation of protein crystal simulations. submitted.

Wych, D. C., & Wall, M. E. (2023b). Molecular-dynamics simulations of macromolecular diffraction, part II: Analysis of protein crystal simulations. submitted.

Wych, W. C., Fraser, J. S., Mobely, D. L., & Wall, M. E. (2019). Liquid-like and rigid-body motions in molecular-dynamics simulations of a crystalline protein. *Structural Dynamics, 6*, 064704.

Xue, Y., & Skrynnikov, N. R. (2014). Ensemble MD simulations restrained via crystallographic data: Accurate structure leads to accurate dynamics. *Protein Science, 23*, 488–507.

Yabukarski, F., Doukov, T., Mokhtari, D. A., Du, S., & Herschlag, D. (2022). Evaluating the impact of X-ray damage on conformational heterogeneity in room-temperature (277 K) and cryo-cooled protein crystals. *Acta Crystallographica. Section D, 78*, 945–963.

CHAPTER SIX

Modeling diffuse scattering with simple, physically interpretable models

Ariana Peck[a,*], Thomas J. Lane[b], and Frédéric Poitevin[a]
[a]Linac Coherent Light Source, SLAC National Accelerator Laboratory, Menlo Park, CA, United States
[b]CHARM Therapeutics, London, United Kingdom
*Corresponding author. e-mail address: apeck@slac.stanford.edu

Contents

Abstract

Diffuse scattering has long been proposed to probe protein dynamics relevant for biological function, and more recently, as a tool to aid structure determination. Despite recent advances in measuring and modeling this signal, the field has not been able to routinely use experimental diffuse scattering for either application. A persistent challenge has been to devise models that are sophisticated enough to robustly reproduce experimental diffuse features but remain readily interpretable from the standpoint of structural biology. This chapter presents *eryx*, a suite of computational tools to evaluate the primary models of disorder that have been used to analyze protein diffuse scattering. By facilitating comparative modeling, *eryx* aims to provide insights into the physical origins of this signal and help identify the sources of disorder that are critical for reproducing experimental features. This framework also lays the groundwork for the development of more advanced models that integrate different types of disorder without loss of interpretability.

Methods in Enzymology, Volume 688
ISSN 0076-6879, https://doi.org/10.1016/bs.mie.2023.06.022

1. Introduction

Disorder in crystals redistributes the scattered intensity to produce a diffuse background signal between and beneath the Bragg peaks (Welberry & Weber, 2016). Whereas Bragg diffraction reports on mean atomic positions, diffuse scattering is sensitive to the covariation in atomic displacements (Meisburger et al., 2017). In protein crystals, this may include the concerted motions of residues and domains that are critical for their biological function (Meisburger et al., 2017). As a result, there has long been interest in using diffuse scattering to probe protein dynamics, which remain challenging to study experimentally. However, debate about the physical origins of protein diffuse scattering continues (de Klijn et al., 2019; Meisburger et al., 2020; Peck et al., 2018; Wall et al., 2018; Wall, 2018). Studies in recent years have pointed to the contribution of correlations between neighboring protein molecules that arise as a consequence of crystal packing (Meisburger et al., 2020; Peck et al., 2018). For diffuse scattering analysis to become a useful tool in structural biology, readily interpretable models that can account for both internal and lattice dynamics are needed.

One of the most pressing challenges faced by the field lies in robustly modeling this signal and determining whether, once the signal is adequately reproduced, information of interest to structural biologists can be extracted. Of particular interest are the concerted motions of protein domains and residues that underlie processes like catalysis, allostery, and ligand binding (Meisburger et al., 2017). In certain circumstances, diffuse scattering has also been suggested to provide information about the static protein structure when the resolution of the Bragg diffraction is limited by crystalline disorder (Ayyer et al., 2016). Ideally we would approach the challenge of modeling the diffuse signal by inferring which atoms undergo collective motions directly from the data. If these displacements are small and Gaussian, mathematically this could be achieved by solving for a covariance matrix whose entries describe the magnitude and direction of correlated displacements between pairs of atoms (Meisburger et al., 2017). An advantage of representing the disorder with a covariance matrix is that it provides a readily interpretable dynamics model that can be used to directly, if not uniquely, predict the diffuse scattering. However, even assuming a threshold distance beyond which correlated motions do not extend, the number of entries in this matrix far exceeds the number of unique data points in experimental diffuse scattering maps (Mazumder &

Ayyer, 2021; Peck et al., 2018). Since the data to parameter regime is insufficient to solve the inverse problem, forward modeling is needed instead.

Here we present *eryx*, a unified framework for testing the principal models of disorder that have been proposed to explain diffuse scattering from protein crystals. The exception is molecular dynamics (MD) simulations (Wall, 2018; Wych et al., 2019, 2023), which are discussed in other chapters of this issue (Case, 2013; Wych & Wall, 2023a, 2023b). Though the models implemented in *eryx* are less sophisticated than MD simulations, the lower complexity aids in their interpretability. These models often have only a few free parameters that are directly related to the magnitude, directionality, or spatial extent of the correlated disorder involved. As a result, comparative modeling can offer insights into the physical origins of diffuse scattering even if individual models do not fully reproduce the experimental signal (Peck et al., 2018). Further, we anticipate that collecting these models in a single, flexible framework will lend itself to the development of more advanced models that combine multiple modes of disorder without sacrificing interpretability.

2. Set up and resources

Eryx is a Python-based package that can be downloaded from Github at https://github.com/apeck12/eryx. We recommend creating a specific *conda* environment for the repository and then using *pip* to install the package's dependencies. In addition to common scientific computing libraries, *eryx* leverages *gemmi* for parsing PDB files (Wojdyr, 2021), *plotly* for interactive visualizations, and the *multiprocess* library for parallel computing.

Eryx was designed to prioritize interactive visualization rather than speed. Given the complexity of experimental diffuse scattering, visualization remain a powerful tool to gain intuition about the physical origins of this signal and assess model quality. Because of the reciprocal nature between real and Fourier space, features that span a small volume of the diffuse map indicate the presence of long-range correlations in the crystal and vice versa. To facilitate interactive visualization, we recommend running the code described in the following sections in JupyterLab, a browser-based interactive coding environment. Exact run times will vary according to the size of the map and atomic model. When possible, the code leverages the crystal's symmetry to reduce the run time by restricting the time-consuming structure factor calculation to the asymmetric unit and expanding the intensity predictions to the rest of reciprocal space using symmetry operations. We have found that for maps with fewer than 2 million symmetry-unique grid points and atomic models with roughly 1000 non-hydrogen atoms, most disorder models can be assessed within a few minutes on a standard laptop. In the sections that follow, the diffuse scattering predicted by different disorder models is shown for a triclinic model system containing ~1000 non-hydrogen atoms on a finely sampled map spanning ~120 million unique grid points; in this case, half an hour to a couple of hours was required to assess most disorder models. Currently *eryx* is designed to provide a platform for initial exploratory analysis and modeling, but we intend to leverage GPU-based acceleration to improve performance for finely-sampled maps and large protein systems in a future release of the code.

As an alternative to JupyterLab, *eryx* also provides a command-line approach for evaluating different models of disorder. This approach may be especially useful for larger maps with longer run times, as the code can be submitted to a high-performance computing cluster. The simulated diffuse scattering maps will be saved to disk, and the storage requirements will vary depending on the dimensions of the map. In this approach, parameters for each model are defined in a text-editable configuration file (config.yaml). Each disorder model is then assessed as follows:

```
python [path_to_eryx]/eryx/scripts/main.py -c config.yaml -t [DisorderModel].
```

A sample configuration file that lists the available disorder models and recommended initial parameters can be found in the examples folder. Results will be saved to the path specified in the `root_dir` line in the configuration file.

3. Modeling disorder

The starting point for all forward models considered to date is an atomic model, ideally determined from the Bragg diffraction data measured alongside the experimental diffuse scattering. The Bragg and diffuse signals originate from the same scattering process so should be self-consistent (Meisburger et al., 2017). In *eryx*, the atomic model should be supplied as a single PDB file of the asymmetric unit. By default, solvent molecules are stripped from the atomic coordinates, with the expectation that these are less likely to move collectively with the protein atoms and thus contribute less significantly to the diffuse signal. Hydrogen atoms are also ignored. Internally, the asymmetric unit is expanded to the unit cell or a supercell depending on the disorder model. For cases that contain noncrystallographic symmetry, an external program like Chimera (Pettersen et al., 2004) should be used to expand the asymmetric unit to the unit cell, and the `expand_p1` flag should be set to False when setting up the disorder model.

All disorder models implemented in *eryx* can be set up in a similar fashion (Fig. 1). In addition to supplying a PDB file (`pdb_path`), the positions in reciprocal space at which the diffuse scattering will be evaluated must be defined. We constrain these positions to lie on a regular grid in hkl space whose nodes are determined by three tuples (`hsampling`, `ksampling`, and `lsampling`). Each of these tuples lists the minimum, maximum, and sampling frequency values along the h, k, and l axes. The minimum and maximum values should be chosen such that the defined volume at minimum encompasses the asymmetric unit in reciprocal space. The sampling frequency is the oversampling rate relative to integral Miller indices and does not need to be the same along each reciprocal axis. While the minimum and maximum values control the resolution limits of the map, the sampling frequency determines the length scale across which correlated displacements can be resolved. A sampling frequency of three, for instance, results in a map that is sensitive to correlations that span up to three unit cell lengths. Finally, the high resolution limit of the map (`res_limit`) can be set so that the diffuse scattering is only evaluated at grid points within this boundary. For comparison to experiment, the grid defined in setting up the disorder model should match the grid on which the experimental map was reconstructed.

Two additional arguments common to all disorder models may require adjustment depending on the size of the map and model and the compute resources available. The structure factor calculation used by all models performs direct summation using the four Gaussian atomic form factor parameterization from the International Tables of Crystallography (Prince, 2004). Multiprocessing is used to speed up this calculation, so `n_processes` should be set to the maximum number of processes. This calculation is also performed on batches of q-vectors, the size of which can be reduced by changing the `batch_size` argument to avoid memory errors. Where relevant, arguments specific to the different disorder models are described below.

The models implemented in *eryx* are summarized in Table 1. While all of these models have few refinable parameters, they differ in their basis

A.

```
from eryx.models import LiquidLikeMotions

# define the map's grid in hkl space as tuples corresponding to the
# (min, max, oversampling frequency) along the three reciprocal axes
hsampling = (-22.461538, 22.46153, 13)
ksampling = (-27.454545, 27.454545, 11)
lsampling = (-30.454545, 30.454545, 11)

# set up class
model = LiquidLikeMotions("6o2h.pdb",
                          hsampling,
                          ksampling,
                          lsampling,
                          res_limit=1.25, # high-resolution limit in Angstrom
                          expand_p1=True, # PDB contains the asymmetric unit
                          n_processes=8, # number of CPU processes available
                          batch_size=10000 # number of q-vectors to process per batch
                          )

# compute diffuse scattering map for specified parameters
Id = model.apply_disorder(sigmas=0.24, gammas=10)

# optimize model parameters, specifying bounds of search space
target = np.load("experimental_full.npy")
model.optimize(target, sigmas_min=0.24, sigmas_max=0.44,
               gammas_min=20, gammas_max=36, ns_search=6, ng_search=5)

# visualize the best fit map, stored in the opt_map class variable
visualize_central_slices(model.opt_map, vmax_scale=10)
```

B. `Optimal sigma: 0.36, optimal gamma: 28.0, with correlation coefficient 0.7425`

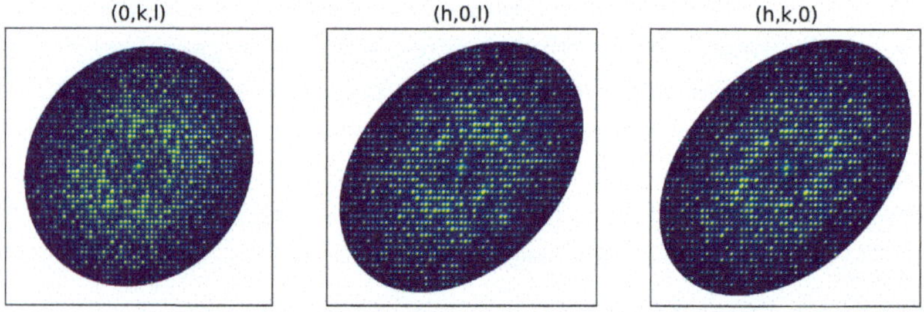

Fig. 1 Common interface for disorder models. (A) All disorder models are instantiated following the same pattern. In addition to providing the path to a PDB file, the grid in hkl-space that the map occupies is defined by the `hsampling`, `ksampling`, and `lsampling` tuples that dictate the extents and sampling frequency along the reciprocal lattice vectors. In the case of the `LiquidLikeMotions` model, the grid must be symmetric about the origin in reciprocal space. For other models, it suffices to define a grid that encompasses the reciprocal asymmetric unit. If a resolution limit is provided (`res_limit`), grid voxels outside this threshold will be set to a value of NaN. The values shown in this example match a publicly available experimental diffuse scattering map of triclinic lysozyme (CXIDB ID: 128). The `n_processes` and `batch_size` arguments can be adjusted depending on the availability of compute resources and memory. All disorder models have an `apply_disorder` routine that can be used to predict a diffuse scattering map; whether model parameters should be supplied to this function or when first calling the class varies by model. The `RigidBodyRotations`, `RigidBodyTranslations`, and `LiquidLikeMotions` classes also have an `optimize` routine that fits the model parameters to a target map. (B) The output from running the code in (A) is shown.

Table 1 Overview of disorder models available in *eryx*.

Model	Parameters	Mathematical basis	Long-range disorder modeled
`NonInteractingDeformableMolecules`	`enm_cutoff`, `gamma_intra`[a]	Molecular dynamical matrix from a GNM in the one-phonon approximation	No
`Ensemble`	`weights`	Variance of the structure factors from a discrete ensemble	No
`RigidBodyRotations`	`sigma`	Blurred molecular transform[b]	No
`RigidBodyTranslations`	`sigma`	Scaled molecular transform	No
`LiquidLikeMotions`	`sigma`, `gamma`, `asu_confined`	Convolution between a disorder kernel and the crystal/molecular transform	Yes/No
`OnePhonon`	`enm_cutoff`, `gamma_intra`, `gamma_inter`	Crystal dynamical matrix from a GNM in the one-phonon approximation	Yes

[a]Though technically a refinable parameter of this class, adjusting this spring constant only changes the overall scale of the predicted intensities but otherwise does not alter the diffuse features. By default the predicted diffuse scattering is subsequently rescaled based on the atomic displacement parameters of the refined structural model.
[b]Though the diffuse scattering predicted by this model is calculated from an ensemble of rotated molecules using Guinier's equation, the signal resembles a blurred copy of the molecular transform.

function, application, and whether they allow for long-range correlations. Two of these—the `NonInteractingDeformableMolecules` and `Ensemble` models—predict motions that can be interpreted in terms of biological function, while the rigid body and liquid-like motions models suggest that the diffuse scattering contains useful information for structure determination instead. By contrast, the `OnePhonon` model describes lattice dynamics, a type of disorder without a readily apparent biophysical application. In the following sections, we demonstrate the use of these models by comparison to a publicly available experimental diffuse scattering map of triclinic lysozyme (Meisburger et al., 2020) (CXIDB ID: 128). This map is oversampled relative to integral Miller indices by factors of 13, 11, and 11 along the $a^\star$, $b^\star$, and $c^\star$ reciprocal lattice vectors. The diffuse scattering intensities correspond to the "variational" component of the total scattering, i.e. the diffuse scattering with the solvent-dominated isotropic background removed (Meisburger et al., 2020). The refined structure determined from that experiment was used as the starting point for forward modeling (PDB ID: 6O2H).

3.1 Models of biological dynamics

Many proteins undergo large-scale collective motions that arise from fluctuations centered on a ground-state structure (Gerstein & Krebs, 1998). In the harmonic approximation, the diffuse scattering predicted by these dynamics can be computed from a covariance matrix that describes the magnitude and direction of correlated displacements between pairs of atoms (Meisburger et al., 2017). As noted in the Introduction, the high dimensionality of this matrix prevents it from being inferred directly from experimental diffuse scattering maps. However, various approaches exist to construct the covariance matrix. The approach implemented in the `NonInteractingDeformableMolecules` class relies on an elastic network model, in which internal dynamics emerge from modeling pairs of atoms in the molecule as connected by harmonic springs with a uniform spring constant (Bray et al., 2011). In the current implementation, the elastic network model is a Gaussian Network Model (GNM) (Haliloglu et al., 1997), which models pairwise interactions between atoms as a function of interatomic distance, in contrast to the Anisotropic Network Model (Eyal et al., 2006), which introduces a directional force. The Hessian matrix of the GNM can be inverted to estimate the isotropic covariances of interatomic displacements, which are normalized to put the diagonal elements on the same scale as the atomic displacement parameters from the refined structural model (Riccardi et al., 2010). This model has two free parameters: the maximum interatomic distance, `enm_cutoff`, for modeling pairs of atoms as connected by harmonic springs with a spring constant of `gamma_intra`. However, adjusting `gamma_intra` introduces a global scale factor without otherwise altering the features and consequently will not affect the predicted map, whose final scale will be determined from normalizing the isotropic covariance matrix to match the atomic displacement parameters. This disorder model can be set up using the approach shown in Fig. 1A, and the predicted diffuse scattering can be calculated using the `apply_disorder` function. Fig. 2 visualizes a central slice from the map computed by this approach for triclinic lysozyme. This elastic network model approach could be extended in various ways, for instance by treating the spring constants as a function of interatomic distance (Van Benschoten et al., 2016) or extending the pairwise interactions to span neighboring protein molecules in the crystal (Meisburger et al., 2020). The latter approach is discussed in Section 3.3.

However, many protein motions of biochemical interest cannot be adequately described as harmonic fluctuations around a mean conformation. For example, residues tend to populate discrete rotamers rather than a continuum between these positions (Lang et al., 2010), and proteins often adopt distinct conformations in their active and inactive states (Nussinov, 2016). Electron density maps solved from Bragg diffraction data frequently provide evidence for such conformational disorder, resulting in efforts to refine multi-conformer ensemble models against the maps (Fraser et al., 2011; Keedy et al., 2015b; Lang et al., 2010). However, while Bragg

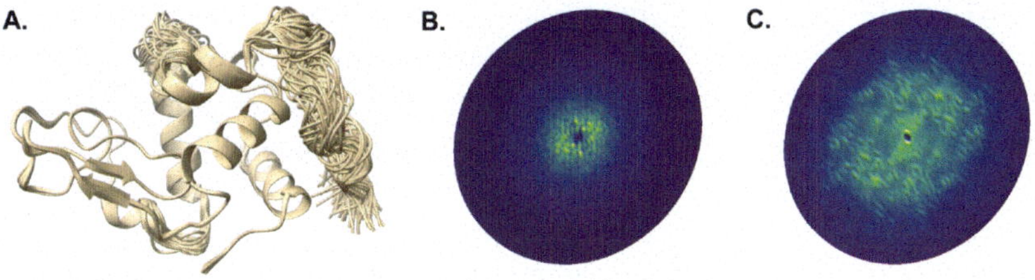

Fig. 2 Models of internal dynamics relevant to protein function. (A) To describe anharmonic motions, diffuse scattering can be predicted from an ensemble of molecules using Guinier's equation. A representative conformational ensemble of lysozyme was generated using the *phenix.ensemble_refinement* program, which refines the ensemble model against the Bragg diffraction data. The backbone atoms of the 102 discrete conformations of this ensemble are overlaid. (B) The h=0 slices from diffuse scattering maps predicted by the (B) `Ensemble` and (C) `NonInteractingDeformableModels` classes are shown for triclinic lysozyme.

diffraction data cannot distinguish between ensembles that predict distinct correlated dynamics but the same mean electron density, diffuse scattering can (Meisburger et al., 2017). As a result, there have been proposals to use diffuse scattering to discriminate between ensemble models that are refined against but indistinguishable based on the Bragg diffraction data (Moore, 2009; Van Benschoten et al., 2015; Wall et al., 2014).

There are numerous options available to generate ensembles from electron density maps, such as qFit (Riley et al., 2021), Ringer (Fraser et al., 2011), and *phenix.ensemble_refinement* (Burnley et al., 2012). In *eryx*, the `Ensemble` class can be used to predict the diffuse scattering from an ensemble using Guinier's equation (Guinier, 1963):

$$I_{\text{diffuse}}(\mathbf{q}) \propto \sum_{\text{m}=1}^{M} \langle |F_{\text{asu}}(R_m \mathbf{q})|^2 \rangle_\Omega - \langle |F_{\text{asu}}(R_m \mathbf{q})| \rangle_\Omega^2 \tag{1}$$

where F_{asu} and R_m respectively correspond to the complex structure factors of the asymmetric unit and the rotational component of the crystal's mth symmetry operator, and $\langle \cdots \rangle_\Omega$ denotes averaging over an ensemble Ω of protein conformations. The diffuse scattering is thus computed as the variance of the structure factors of the asymmetric unit across the ensemble, and the intensity contributions from all asymmetric units are summed incoherently by assuming no coupling of conformational disorder between asymmetric units. In *eryx*, Debye-Waller factors are computed from the B-factors reported in the PDB file and applied to the structure factors to account for atomic thermal vibrations, which are assumed to be independent of the ensemble disorder. The diffuse scattering from this disorder model can be computed using the `Ensemble` class, where the atomic model input is a single PDB file in which each frame specifies a distinct conformation. An example map is shown in Fig. 2, in which the conformational ensemble was generated by *phenix.ensemble_refinement* (Burnley et al., 2012). This ensemble refinement program rescales the B-factors to fit the Bragg diffraction data, but in this particular case the resulting atomic displacement parameters concentrate high-intensity features at low resolution to a greater extent than observed experimentally (c.f. Fig. 3, lower right). Although currently *eryx* does not provide an option to optimize the B-factors, population weights for each state can be specified when calling the class's `apply_disorder` function; otherwise, uniform state weights are applied. In principle, these weights could be optimized by fitting an experimental diffuse scattering map. In practice, however, ensemble models generated by refinement against the Bragg

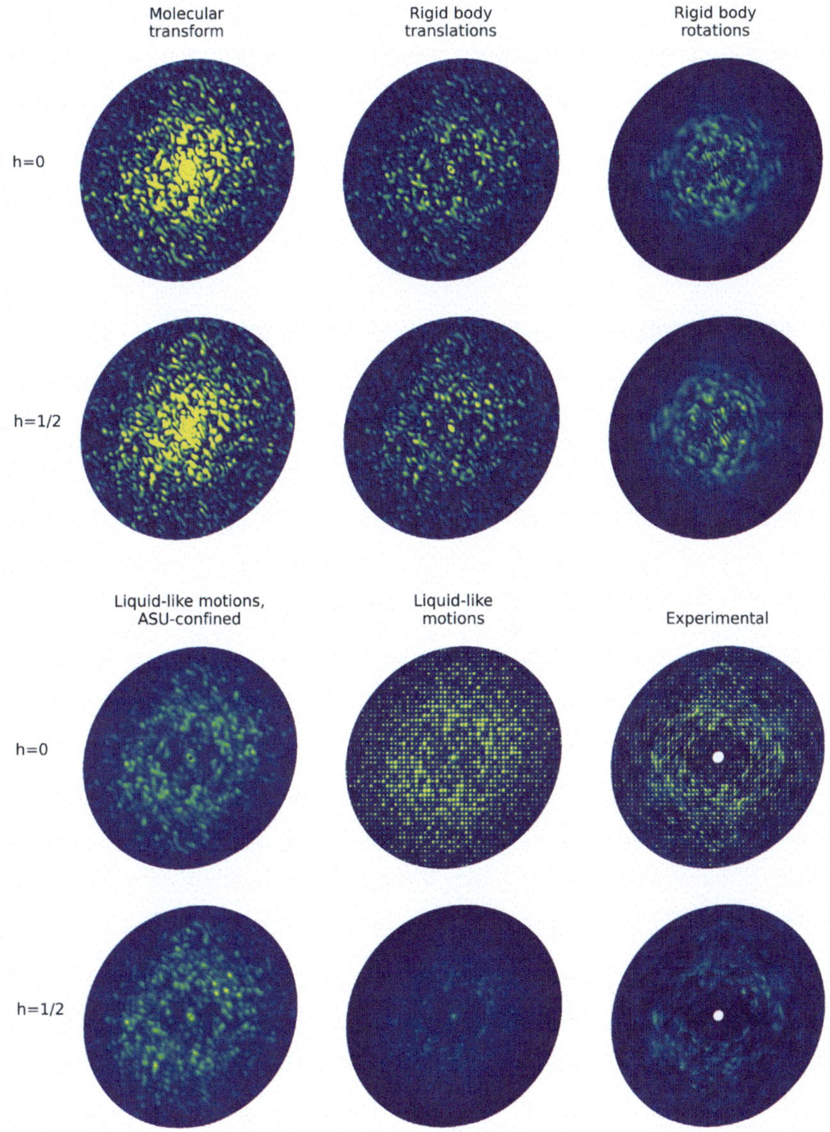

Fig. 3 Models that share the molecular or crystal transform as their basis. The h=0 and h=1/2 planes are shown for the molecular transform, experimental map of triclinic lysozyme, and the indicated models of disorder. Of these models, only the classic LLM predicts the halos around Bragg peaks observed at reciprocal lattice points (e.g. the h=0 plane) in the experimental map. The other models assume that correlated disorder is confined to the asymmetric unit, so predict features whose size does not depend on proximity to reciprocal lattice points. The overall correlation coefficient of the predicted and experimental diffuse scattering maps ranged from 0.34 to 0.39 for the rigid body motions and asymmetric unit-confined LLM models and was 0.74 for the classic LLM model. For triclinic space groups, the crystal transform has the same intensity at integral Miller indices as the molecular transform but is zero at fractional Miller indices.

data–for example, from translation-libration-screw or multiconformer ensembles—have historically done a poor job of reproducing the experimental signal, so it seems unlikely that fitting the population weights would be successful (Peck et al., 2018; Van Benschoten et al., 2015). Given that the Bragg data provide evidence for the presence of conformational ensembles, this type of disorder must contribute the diffuse scattering as well. However, it does not appear to do so in a dominant fashion (Peck et al., 2018; Van Benschoten et al., 2015).

3.2 Models relevant to structure determination

The disorder models considered in this section share the molecular or crystal transform as their basis. While the molecular transform is the incoherent sum of the structure factors of each asymmetric unit, the crystal transform corresponds to the coherent sum over all scatterers in the unit cell and thus contains the same information as the Bragg peaks. Models that fall into this category are described in more detail below.

In rigid body rotational disorder, the contents of each asymmetric unit (which often coincides with a single protein) are assumed to rotate as a rigid body. In *eryx*, the simplest form of this model is implemented, in which there is no preferred axis of rotation and isotropic rotations are independently drawn from a normal distribution with standard deviation σ. The diffuse signal is then computed by applying Guinier's equation (Eq. 1) to an ensemble of rotated molecules (Guinier, 1963). This approach requires a sufficiently large ensemble to ensure that the predicted features are reliable, with minimal error from modeling a continuous distribution of rotations with a discrete representation. The required ensemble size depends on how finely and to what resolution the map is sampled, in addition to the magnitude of σ. For the triclinic lysozyme map described above, we found that an ensemble size of 200 yielded replicate maps with a correlation coefficient of > 0.95 for $\sigma \leq 2°$, including at higher resolution where features are expected to converge more slowly. The ensemble size can be set by changing the `num_rot` argument when calling the class's `optimize` function (Fig. 1A). An alternative approach to Guinier's equation would be to predict the diffuse signal using the Translation-Libration-Screw (TLS) formalism to represent rigid body displacements, as the libration component corresponds to rotations (Meisburger et al., 2020). Though technically only applicable to purely translational motion, this approach can approximately model sufficiently small rotations (Urzhumtsev et al., 2013). It has been estimated that this approximation breaks down for rotations exceeding 8° (Cruickshank, 1956), a threshold that exceeds the expected amplitude for proteins constrained in a crystal lattice. The TLS formalism is related to the `OnePhonon` model described in Section 3.3, but in that case the entries of the TLS matrix are derived by building an elastic network model that accounts for correlated displacements—including translations and screw motions—between neighboring protein molecules in the crystal.

In rigid body translational disorder, each asymmetric unit behaves as a rigid body displaced from its ideal position in the crystal lattice. These displacements are assumed to be normally distributed and uncorrelated between asymmetric units. The scale of disorder is determined by the parameter σ, which is a scalar or vector for isotropic and anisotropic displacements, respectively, and related to the Debye-Waller factor. The diffuse scattering predicted by this disorder model at the wavector $\mathbf{q}$ is (Ayyer et al., 2016; Moore, 2009):

$$I_{\text{diffuse}}(\mathbf{q}) \propto \left[1 - e^{-q^2\sigma^2}\right] I_{\text{m}}(\mathbf{q}) \tag{2}$$

where $q = |\mathbf{q}| = 2\pi/d = 4\pi \sin\theta/\lambda$ is the amplitude of the wavevector $\mathbf{q}$ and related as noted to the Bragg spacing d, the scattering angle 2θ, and the wavelength λ. The disorder is captured in the Debye-Waller term, so the molecular transform, I_m, is computed with the temperature factors set to zero. It has been suggested that this model may be most applicable for membrane proteins, as the detergent used for crystallization may result in fewer crystal contacts than are typical for soluble proteins (Birch et al., 2018) and consequently reduce the correlated disorder between protein neighbors (communication with K. Ayyer).

In the liquid like motions (LLM) model, proteins are treated as objects that can be internally deformed rather than as rigid structural units (Caspar et al., 1988; Clarage et al., 1992; Wall et al., 1997). The covariance of atomic displacements scales with the displacement parameter σ and decays exponentially with interatomic distance according to a characteristic correlation length, γ. The expression for the predicted diffuse scattering is:

$$I_{\text{diffuse}}(\mathbf{q}) \propto q^2\sigma^2\, e^{-q^2\sigma^2} \left[I_{\text{c}}(\mathbf{q}) \star \tilde{\Gamma}(\mathbf{q})\right]. \tag{3}$$

where $\tilde{\Gamma}(\mathbf{q})$ is a correlation kernel describing the distance-dependent decay in covariances, $I_{\text{c}}(\mathbf{q})$ is the crystal transform, and $\star$ indicates a convolution. Temperature factors are set to zero when calculating the crystal transform, as this model assumes a global displacement parameter that applies to all atoms (Caspar et al., 1988; Van Benschoten et al., 2016). When the map is oversampled relative to the Bragg peaks, the convolution results in the appearance of halos around reciprocal lattice points, as the scattered intensity is redistributed from integral Miller indices where the crystal transform is nonzero to surrounding regions in reciprocal space. In *eryx*, a modified version of this model in which correlations are restricted to the asymmetric unit can also be evaluated by passing the flag `asu_confined=True` when setting up the class (Fig. 1A). In this variant of the LLM, the diffuse scattering is the convolution between the correlation kernel and the molecular rather than the crystal transform. Because the molecular transform is defined throughout reciprocal space, this model does not predict the halos characteristic of experimental diffuse scattering maps and underperforms the classic LLM in reproducing the observed signal (Peck et al., 2018). However, we include it as an option in *eryx* as this variant provides a useful bridge between the rigid body translational disorder and classic LLM models. Comparison with the former can offer insight into the effect of treating proteins as deformable rather than perfectly rigid objects. In the limit that the correlation length exceeds the dimensions of the protein, the asymmetric-unit confined LLM and rigid body translations models predict diffuse scattering patterns that bear a strong resemblance to one another and mathematically differ only by a Debye-Waller scale factor and higher-order terms from a Taylor expansion (Peck et al., 2018). On the other hand, comparison to the classic LLM provides an indication of the contribution of intermolecular correlations—i.e., those extending between proteins in neighboring asymmetric units—to the measured diffuse scattering (Peck et al., 2018).

The diffuse scattering predicted by these models can be simulated with the `RigidBodyRotations`, `RigidBodyTranslations`, and `LiquidLikeMotions` classes, which can be run as shown in Fig. 1A. Because the LLM relies on convolution, the map's dimensions should encompass a full sphere in reciprocal space centered about the origin rather than a unique wedge to avoid artifacts from boundary conditions. For the LLM, correlated disorder will only be restricted to the asymmetric unit if `asu_confined=True` is supplied as an argument when setting up the model. For each model, an `optimize` function can be called to fit the (an)isotropic displacement parameter or standard deviation of rotations (σ) and, in the case of the LLM, the correlation length (γ). This function takes as input an experimental map, in addition to the lower and upper bounds for the parameter search (Fig. 1A). Optimization is then performed by scanning between these thresholds to identify the parameter value(s) that maximize(s) the overall correlation coefficient between the predicted and experimental maps. In general we have found that the best fit isotropic displacement parameter σ is within three-fold of the Wilson B factor of the Bragg diffraction data for both the rigid body translations and LLM models, while the value of the correlation length γ has ranged from a few to tens of Ångstroms across multiple systems (Clarage et al., 1992; Peck et al., 2018; Van Benschoten et al., 2016; Wall et al., 1997). To avoid memory errors for finely-sampled maps, the optimize function can be called iteratively to search different regions of parameter space. The results from all calls will be stored inside the class and can be visualized using the class's `plot_scan` function.

The diffuse scattering maps predicted by these models are visualized and compared to the experimental map of triclinic lysozyme in Fig. 3. The model parameters were fit to maximize the overall correlation coefficient with the experimental signal. Comparison of the h=0 and h=1/2 planes is instructive in identifying the classic LLM as the only model that can reproduce the observed halos that are centered at integral Miller indices and decay toward the edge of the Brillouin zone (the boundary between adjacent reflections). By contrast, the other models do not predict correlations that extend between neighboring asymmetric units. Because of the reciprocal nature between real and Fourier space, the shorter length-scale correlations inherent to the rigid body motions and asymmetric unit-confined LLM models yield larger features in reciprocal space whose size does not change depending on their proximity to integral Miller indices. In addition to qualitatively reproducing the halo features, the classic LLM substantially improves the quantitative agreement with the experimental signal as well, with an overall correlation coefficient of 0.74 compared to 0.3 to 0.4 for the other models presented in this section.

The models considered in this section do not report on biologically-relevant protein dynamics. The rigid body motions experienced by proteins confined to a crystal lattice will not resemble their movement under physiological conditions. Similarly, liquid-like motions are a phenomenological description that cannot capture the idiosyncratic dynamics that underlie protein function. However, if these models adequately reproduce experimental diffuse scattering, the signal could potentially aid structure determination instead. The models of rigid body translational disorder and liquid-like motions share the molecular or crystal transform as their basis and predict that the diffuse scattering is a copy of this transform adjusted by a scale factor or convolved by some kernel whose mathematical form is known. [While visually the diffuse signal from rigid body rotational disorder also resembles a blurred version of the molecular transform, the mathematical relationship is not as straightforward (Moore, 2009)]. In principle, it should therefore be possible to extract the molecular or crystal transform from an experimental map by performing the inverse of the specific scaling or convolution operation, assuming that rigid body translations or liquid-like motions give rise to the observed signal.

An estimate of the molecular transform would be particularly useful for structure determination, as iterative phase retrieval algorithms developed for single-particle imaging could be used to solve the protein structure (Miao et al., 1998). By contrast, the crystal transform is not oversampled relative to integral Miller indices, so it does not contain sufficient information for *ab initio* phasing. However, in cases where the diffuse scattering extends beyond the Bragg diffraction, an estimate of the crystal transform would enable extending the structural model solved from the Bragg data to the resolution limit of the measurable diffuse data. Although the possibility of leveraging diffuse scattering to aid structure determination was recognized decades ago (Miao et al., 1998; Sayre, 1952), there has only been one report of the signal being used in this fashion (Ayyer et al., 2016). Thus, significant questions remain about how general and robust this potential application of diffuse scattering is. In most cases, it appears that the classic LLM model better reproduces experimental diffuse scattering than rigid body motions models (Peck et al., 2018), suggesting that the diffuse signal could provide information about the crystal rather than the molecular transform. However, the ability to reliably deconvolve the crystal transform from experimental maps is an open question. Further, how accurate the extracted transform needs to be to permit resolution extension remains unknown.

3.3 Models of lattice dynamics

Diffuse features that occupy small regions in reciprocal space, including halos centered on reciprocal lattice points and streaks along rows of reflections, indicate long-range correlations that span multiple unit cells in real space (Glover et al., 1991; Meinhold et al., 2007; Peck et al., 2018; Polikanov & Moore, 2015). A source of disorder that could account for these features is acoustic phonons, which result from thermally-excited lattice vibrations. As elastic waves propagate through the crystal, atoms across different unit cells will oscillate in a correlated fashion. In reciprocal space, this gives rise to sharp features that peak at reciprocal lattice points and decay with an inverse squared dependence on the distance from these nodes (Meisburger et al., 2017).

One approach to model lattice vibrations is to construct an elastic network model that extends throughout the crystal. In this model, pairs of atoms behave as though connected by harmonic springs. All atom pairs within each asymmetric unit are treated this way, while pairs of atoms in different asymmetric units are only connected if their interatomic distance is below a threshold value. This behavior is then extended across the crystal by applying periodic boundary conditions. Following the approach described by Meisburger et al. (2020), the dynamical matrix of the elastic network representation is projected on six degrees of freedom for each asymmetric unit, effectively treating them as rigid bodies. This coarse-graining reduces the full phonon spectrum from 3N-3 optical modes (where N is the number of atoms in the unit cell) to 3, in addition to 3 acoustic modes. Collectively, these modes describe the covariance of instantaneous displacements experienced by each asymmetric unit, which emerge from the combination of internal and intermolecular interactions represented in the elastic network model. In the one-phonon approximation, the diffuse scattering can be computed by integrating the contributions from the phonon spectrum (Meisburger et al., 2017, 2022).

Diffuse scattering from this model of lattice dynamics can be computed using the `OnePhonon` class. The free parameters of this model are the intramolecular and intermolecular spring constants, in addition to the cutoff distance below which pairs of atoms are treated as connected by springs. These parameters can be defined by respectively setting the `gamma_intra`, `gamma_inter`, and `enm_cutoff` arguments when creating an instance of the class. Slices through the h=0 and h=1/2 reciprocal planes from a simulated map are displayed in Fig. 4A. Similar to the LLM, halos are observed around reciprocal lattice nodes as a consequence of accounting for intermolecular disorder. Examining the phonon modes shows that the intermolecular contribution dominates the covariance (Fig. 4B), and decomposing the phonon spectrum into acoustic and optical modes shows that these respectively give rise to halos and cloudier diffuse features (Fig. 4C). In contrast to the halos, which decay with an inverse squared dependence on the distance from the nearest reciprocal lattice point, the spatial extent of the cloudier features that report on shorter-range correlations is independent of position in reciprocal space.

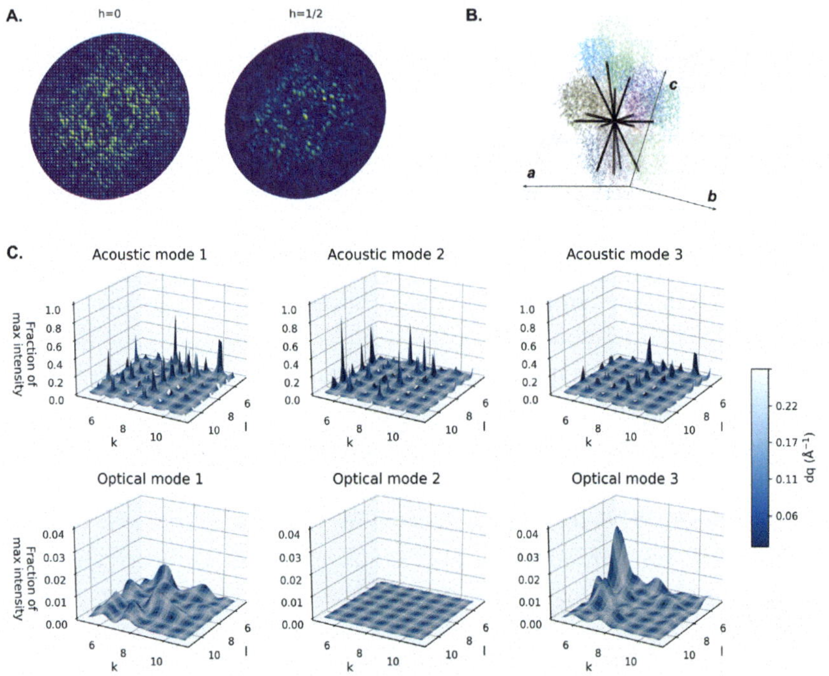

Fig. 4 Lattice dynamics from a crystalline elastic network model. (A) The diffuse scattering predicted in the one-phonon approximation for triclinic lysozyme is shown for the indicated slices in reciprocal space. The values of the inter- and intramolecular spring constants were set to 2 and 1, respectively, with a cut-off distance of 4 Å to treat intermolecular atom pairs as connected. (B) The `VisualizeCrystal` class can be used to display the supercell used to construct the elastic network model. Individual protein molecules are colored differently, and the dominant entries in the covariance matrix are depicted with black rods. Their extent and direction indicate that intermolecular correlations between neighboring proteins account for most of the covariance in this model. (C) The phonon spectrum was decomposed into acoustic and optical modes. The contribution from each mode for a small region of the h=0 slice is shown, with the intensities normalized based on the maximum intensity value of the summed modes. The surface is colored by the distance to the nearest reciprocal lattice node.

Lattice vibration models have outperformed other models in reproducing experimental diffuse scattering (Meisburger et al., 2020, 2022). However, questions remain about whether acoustic phonons specifically are responsible for the observed features. Acoustic phonons predict that the strength and shape of diffuse scattering will reflect the bulk mechanical properties of the crystal rather than the specific internal structure, and further, will be a function of temperature (correspondence with P. Moore). There have been no systematic studies examining these trends, but the characteristic features predicted by this model have been observed both in room temperature and cryogenic datasets (Glover et al., 1991; Meinhold et al., 2007; Meisburger et al., 2020; Peck et al., 2018; Polikanov & Moore, 2015). Anecdotally, we have also observed very similar diffuse features across a temperature-series of the same crystal system (data not shown). While it is possible that dynamic disorder at room temperature is converted to static disorder when crystals are flash-frozen, studies have suggested that cryo-cooling can alter the distribution of states (and presumably the atomic covariances arising from lattice vibrations) (Keedy et al., 2015a; Polikanov & Moore, 2015). Systematic studies investigating how the diffuse signal and especially the halos are modulated by temperature could provide more definitive answers about whether acoustic phonons are responsible for this signal. For now, we note that models of phonon-induced dynamic disorder do not uniquely predict these features; certain models of static disorder could do so as well.

Regardless, these lattice dynamics models seem a promising approach to develop further given their ability to simultaneously reproduce both the sharp and cloudy features characteristic of protein diffuse scattering. A limitation of these models is that they assume the harmonic approximation so cannot capture discrete conformational disorder (described in Section 3.1). However, these models could be extended in various ways to adjust the underlying disorder and predicted signal. For example, the inclusion of a directional component to the interatomic springs has enabled modeling halo anisotropy observed in experimental maps (Meisburger et al., 2020). Other modifications could more specifically target the cloudier features arising from short-range collective motions. For instance, instead of treating the contents of each asymmetric unit as a rigid body, individual domains or residues could be treated as rigid groups instead. This finer decomposition of the protein would extend the phonon spectrum with additional optical modes that correspond to more detailed internal protein dynamics in real space. Though related to the LLM (Meisburger et al., 2022), these lattice dynamics models are more adaptable because the covariance is not pre-defined by a specific kernel, but instead computed from interatomic interactions that can be represented in different ways.

4. Model evaluation

Eryx provides various tools to assess model performance (Fig. 5). The primary quantitative metric used in the field is the Pearson correlation

```
# specify atomic model and map grid
pdb_path = '6o2h.pdb'
hsampling = (-22.461538, 22.46153, 13)
ksampling = (-27.454545, 27.454545, 11)
lsampling = (-0.454545, 30.454545, 11)
res_limit = 1.25

# compute a resolution map and mask in which voxels outside the
# resolution limit or reciprocal asymmetric unit are masked
model = AtomicModel(pdb_path, expand_p1=True, frame=-1)
hkl_grid, map_shape = generate_grid(model.A_inv,
                                    hsampling,
                                    ksampling,
                                    lsampling,
                                    return_hkl=True)
mask, res_map = get_resolution_mask(model.cell, hkl_grid, res_limit)
asu_mask = get_asu_mask(model.space_group, hkl_grid)
mask *= asu_mask

# compute the cross-correlation by resolution shell
# Id is a dict of the experimental and simulated maps
cc_vals = {}
for model in prefices:
    if model != 'exp':
        bins, cc_vals[model] = compute_cc_by_shell(Id['exp'].flatten(),
                                                   Id[model].flatten(),
                                                   res_map=res_map,
                                                   mask=mask)

# visualize the region centered on the (1,2,5) reflection
visualize_brillouin_zone(Id['llm'], (1,2,5), hsampling, ksampling,
                         lsampling, hkl_grid=hkl_grid, ylabel='LiquidLikeMotions')
```

Fig. 5 Tools to assess model performance. The sample code demonstrates how to compute the CC by resolution bin between experimental and simulated diffuse scattering maps. The mask provided to the `compute_cc_by_shell` function causes voxels that lie outside resolution limit or reciprocal asymmetric unit to be ignored. The returns of this function can be plotted as in Fig. 6A. The final line shows how to examine the features in the immediate vicinity of a reciprocal lattice point; the resulting plot is shown in Fig. 6B.

coefficient (CC) between experimental and simulated maps. This metric can be computed for the full unique wedge in reciprocal space, by resolution shell, or as a function of $|d\mathbf{q}|$, the distance between grid nodes and the nearest reciprocal lattice point (Fig. 6A). In addition to these quantitative metrics, qualitative agreement can be assessed by different methods. Given hkl planes in reciprocal space can be visualized using the `visual_reciprocal_plane` function; as discussed in Section 3.3, comparing the spatial extent of features on reciprocal lattice planes and midway between them often provides insight into the length scale of disorder in real space. The fall-off in the intensity from reciprocal lattice

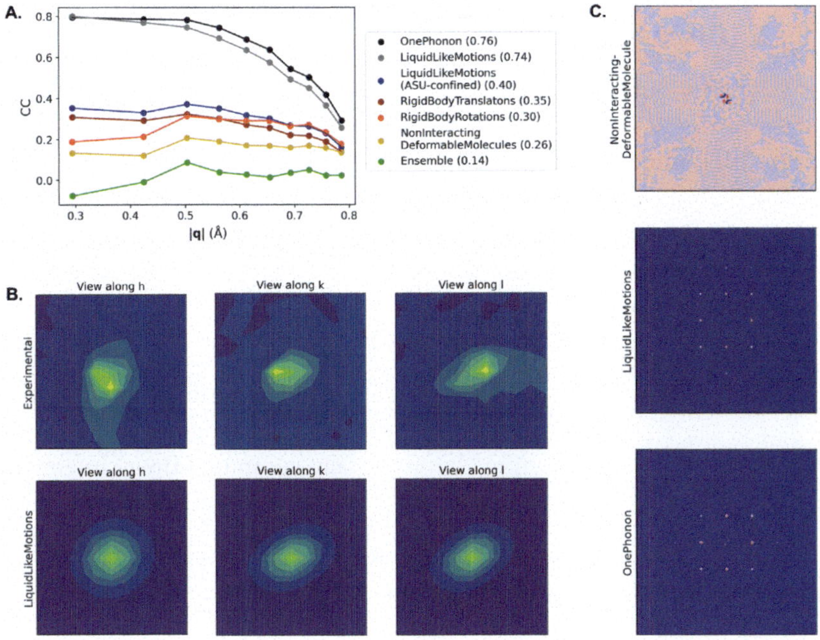

Fig. 6 Evaluating model performance. (A) The CC between the experimental diffuse scattering and indicated simulated map was computed using the `compute_cc_by_shell` function and is plotted by resolution shell. The overall CC between each model and the experimental map is noted in parenthesis. (B) The `visualize_brillouin_zone` function can be used to examine the region in the immediate vicinity of reciprocal lattice nodes where halos are observed in experimental maps. (C) 3D-ΔPDFs can be computed using the `DeltaPDF` class. The central b-c plane of the 3D-ΔPDF for the indicated model is shown. Periodic peaks are observed in the 3D-ΔPDFs of models that predict intermolecular correlations (the LLM and phonon models), where the inter-peak vectors correspond to the reciprocal lattice vectors. The central region reports on short-range correlations in real space.

points can be visualized using the `visualize_brillouin_zone` function, which displays the region surrounding a given Miller index (Fig. 6B). Finally, the 3D-ΔPDF, or diffuse Patterson, of any map can be calculated using the `DeltaPDF` class as shown in Fig. 6C. The 3D-ΔPDF is computed by subtracting the mean intensity in each resolution shell from the diffuse scattering map, followed by taking the Fourier transform of the resulting signal. The advantage of the 3D-ΔPDF is its ability to separate the contributing correlations based on their length scale in real space (Meisburger et al., 2020). The periodic peaks in these plots correspond to long-range disorder across multiple unit cells, while the central region reports on short-range correlations in the crystal.

5. Discussion

Recently our ability to model diffuse scattering has seen significant improvements in terms of the overall agreement with the experimental signal (Meisburger et al., 2020,2022). However, further advances in modeling are needed before the field can routinely interpret diffuse scattering in terms of internal protein dynamics relevant for function or use this signal to aid structure determination. *Eryx* provides a unified framework for evaluating many of the models previously proposed to explain this signal. Some limitations of these models are known; for instance, the observation of halos around Bragg peaks indicates the presence of intermolecular correlations that conflicts with the assumptions of rigid body motions models (Peck et al., 2018). However, these models do not need to fully reproduce the observed signal in order to inform on the nature of the disorder involved. Comparative modeling can offer useful insights into the ability of different types of disorder to reproduce specific features in experimental maps. One of the aims of *eryx* is to facilitate such exploratory analysis.

An alternative approach to forward modeling is MD simulations. Early studies recognized that the sampling problem inherent to this technique is particularly acute for diffuse scattering, as far more sampling is required for convergence of the covariation in atomic displacements than the mean atomic positions (Clarage et al., 1995). However, advances in computational resources over the past decade have enabled much larger simulations. The extension of these simulations both temporally, from picoseconds to microseconds, and spatially, from a single asymmetric unit to tens or hundreds of unit cells, has significantly improved the ability of these models to predict experimental diffuse scattering (Meisburger et al., 2020; Wall, 2018; Wych et al., 2019, 2023). However, the improved agreement comes at the cost of interpretability. Dimensionality reduction techniques required to analyze such large-scale MD simulations can introduce approximation error and bias (Suarez et al., 2021). Further, although by construction this technique integrates multiple modes of disorder, their contributions cannot be easily tuned to assess their relative importance in reproducing the experimental signal.

Models that combine different sources of disorder in a readily adjustable manner would deepen our understanding of the physical origins of diffuse scattering. However, constructing a combined disorder model that is mathematically and physically justified is challenging. The simplest approach of optimizing a linear combination of models commonly used to explain this signal is unlikely to succeed. There is no guarantee that the sources of disorder predicted by these models are fully uncoupled, a condition that is required for their contributions to be additive. Further, reweighting their relative contributions would only adjust the degree to which each model is represented but not vary the scale of disorder (e.g. the standard deviation of rotations or correlation length of liquid-like motions) predicted by each. As a result, such a linear combination would not be easy to interpret. A more promising direction may be to expand on efforts to find an alternative basis set that describes protein dynamics in some reduced dimensional space (Mazumder & Ayyer, 2021). Ideally, this basis set would permit decoupling the contribution of lattice dynamics, which are biologically uninteresting. If the crystal transform can be deconvolved from this component of the signal, resolution extension could be pursued. Once this portion of the signal is accounted for, it may then be possible to study the residual in terms of internal protein dynamics that may be relevant for biological function (Meisburger et al., 2020).

For now, forward modeling remains a necessary tool for studying protein diffuse scattering since the underlying correlations cannot be directly inferred from the data. As the field advances, however, it may be possible to make simplifying assumptions about the nature of the disorder that gives rise to this signal. Mathematically this may permit imposing constraints on the covariance matrix that describes the correlated displacements between all pairs of atoms in the crystal. The application of constraints based on known chemical bond distances and angles has proven very effective in the analysis of Bragg diffraction data (Prince et al., 2006). In the case of diffuse scattering, constraints on dynamics rather than structure would be needed but could permit reducing the space of all possible disorder models, which is otherwise intractably large. If the number of free parameters can be decreased sufficiently relative to the number of unique data points in experimental maps, solving the inverse problem may be brought within reach.

Acknowledgments

We thank Peter Moore and Kartik Ayyer for discussions about the origins of diffuse scattering. A.P and F.P are supported by the Exascale Computing Project (17-SC-20-SC), a collaborative effort of the U.S. Department of Energy Office of Science and the National Nuclear Security Administration. Use of the Linac Coherent Light Source (LCLS), SLAC National Accelerator Laboratory, is supported by the U.S. Department of Energy, Office of Science, Office of Basic Energy Sciences under Contract No. DE-AC02-76SF00515.

Code availability

The *eryx* source code and additional examples are available at https://github.com/apeck12/eryx.

References

Ayyer, K., Yefanov, O. M., Oberthur, D., Roy-Chowdhury, S., Galli, L., Mariani, V., Basu, S., Coe, J., Conrad, C. E., Fromme, R., Schaffer, A., Dorner, K., James, D., Kupitz, C., Metz, M., Nelson, G., Xavier, P. L., Beyerlein, K. R., Schmidt, M., Sarrou, I., Spence, J. C., Weierstall, U., White, T. A., Yang, J. H., Zhao, Y., Liang, M., Aquila, A., Hunter, M. S., Robinson, J. S., Koglin, J. E., Boutet, S., Fromme, P., Barty, A., & Chapman, H. N. (2016). Macromolecular diffractive imaging using imperfect crystals. *Nature, 530*(7589), 202–206.

Birch, J., Axford, D., Foadi, J., Meyer, A., Eckhardt, A., Thielmann, Y., & Moraes, I. (2018). The fine art of integral membrane protein crystallisation. *Methods, 147*, 150–162.

Bray, J. K., Weiss, D. R., & Levitt, M. (2011). Optimized torsion-angle normal modes reproduce conformational changes more accurately than cartesian modes. *Biophys. J. 101*(12), 2966–2969.

Burnley, B. T., Afonine, P. V., Adams, P. D., & Gros, P. (2012). Modelling dynamics in protein crystal structures by ensemble refinement. *Elife, 1*, e00311.

Case, D. (2013). Simulations of macromolecular crystals: Implications for the analysis of Bragg and diffuse scattering. *Methods in Enzymology*.

Caspar, D. L., Clarage, J., Salunke, D. M., & Clarage, M. (1988). Liquid-like movements in crystalline insulin. *Nature, 332*(6165), 659–662.

Clarage, J. B., Clarage, M. S., Phillips, W. C., Sweet, R. M., & Caspar, D. L. (1992). Correlations of atomic movements in lysozyme crystals. *Proteins, 12*(2), 145–157.

Clarage, J. B., Romo, T., Andrews, B. K., Pettitt, B. M., & Phillips, G. N. (1995). A sampling problem in molecular dynamics simulations of macromolecules. *Proceedings of the National Academy of Sciences of the United States of America, 92*(8), 3288–3292.

Cruickshank, D. W. J. (1956). Errors in bond lengths due to rotational oscillations of molecules. *Acta Crystallographica, 9*(9), 757–758. https://doi.org/10.1107/S0365110×56002059

de Klijn, T., Schreurs, A. M. M., & Kroon-Batenburg, L. M. J. (2019). Rigid-body motion is the main source of diffuse scattering in protein crystallography. *IUCrJ, 6*(Pt 2), 277–289.

Eyal, E., Yang, L.-W., & Bahar, I. (2006). Anisotropic network model: Systematic evaluation and a new web interface. *Bioinformatics, 22*(21), 2619–2627. https://doi.org/10.1093/bioinformatics/btl448

Fraser, J. S., van den Bedem, H., Samelson, A. J., Lang, P. T., Holton, J. M., Echols, N., & Alber, T. (2011). Accessing protein conformational ensembles using room-temperature X-ray crystallography. *Proceedings of the National Academy of Sciences of the United States of America, 108*(39), 16247–16252.

Gerstein, M., & Krebs, W. (1998). A database of macromolecular motions. *Nucleic Acids Research, 26*(18), 4280–4290.

Glover, I. D., Harris, G. W., Helliwell, J. R., & Moss, D. S. (1991). The variety of X-ray diffuse scattering from macromolecular crystals and its respective components. *Acta Crystallographica Section B, 47*(6), 960–968. https://doi.org/10.1107/S0108768191004585

Guinier, A. (1963). *X-ray diffraction in crystals, imperfect crystals, and amorphous bodies*. San Francisco, CA: W. H. Freeman.

Haliloglu, T., Bahar, I., & Erman, B. (1997). Gaussian dynamics of folded proteins. *Physical Review Letters, 79*(16), 3090–3093. https://doi.org/10.1103/physrevlett.79.3090

Keedy, D. A., Kenner, L. R., Warkentin, M., Woldeyes, R. A., Hopkins, J. B., Thompson, M. C., Brewster, A. S., Van Benschoten, A. H., Baxter, E. L., Uervirojnangkoorn, M., McPhillips, S. E., Song, J., Alonso-Mori, R., Holton, J. M., Weis, W. I., Brunger, A.

T., Soltis, S. M., Lemke, H., Gonzalez, A., Sauter, N. K., Cohen, A. E., van den Bedem, H., Thorne, R. E., & Fraser, J. S. (2015a). Mapping the conformational landscape of a dynamic enzyme by multitemperature and XFEL crystallography. *Elife, 4.*

Keedy, D. A., Fraser, J. S., & van den Bedem, H. (2015b). Exposing hidden alternative backbone conformations in X-ray crystallography using qFit. *PLoS Computational Biology, 11*(10), e1004507.

Lang, P. T., Ng, H. L., Fraser, J. S., Corn, J. E., Echols, N., Sales, M., Holton, J. M., & Alber, T. (2010). Automated electron-density sampling reveals widespread conformational polymorphism in proteins. *Protein Science, 19*(7), 1420–1431.

Mazumder, P., & Ayyer, K. (2021). ***Understanding conformational dynamics from macromolecular crystal diffuse scattering***. bioRxiv arXiv: ⟨https://www.biorxiv.org/content/early/2021/03/04/2021.02.11.429988.full.pdf⟩, doi: 10.1101/2021.02.11.429988.

Meinhold, L., Merzel, F., & Smith, J. C. (2007). Lattice dynamics of a protein crystal. *Physical Review Letters, 99*(13), 138101.

Meisburger, S. P., Thomas, W. C., Watkins, M. B., & Ando, N. (2017). X-ray scattering studies of protein structural dynamics. *Chemical Reviews,* 7615–7672.

Meisburger, S. P., Case, D. A., & Ando, N. (2020). Diffuse X-ray scattering from correlated motions in a protein crystal. *Nature Communications, 11*(1), 1271.

Meisburger, S. P., Case, D. A., & Ando, N. (2022). ***Robust, total x-ray scattering workflow to study correlated motion of proteins in crystals***. bioRxiv arXiv: ⟨https://www.biorxiv.org/content/early/2023/01/02/2022.08.22.504832.full.pdf⟩, doi: 10.1101/2022.08.22.504832.

Miao, J., Sayre, D., & Chapman, H. N. (1998). Phase retrieval from the magnitude of the fourier transforms of nonperiodic objects. *Journal of the Optical Society of America A, 15*(6), 1662–1669.

Moore, P. B. (2009). On the relationship between diffraction patterns and motions in macromolecular crystals. *Structure, 17*(10), 1307–1315.

Nussinov, R. (2016). Introduction to Protein Ensembles and Allostery. *Chemical Reviews, 116*(11), 6263–6266.

Peck, A., Poitevin, F., & Lane, T. J. (2018). Intermolecular correlations are necessary to explain diffuse scattering from protein crystals. *IUCrJ, 5*(Pt 2), 211–222.

Pettersen, E. F., Goddard, T. D., Huang, C. C., Couch, G. S., Greenblatt, D. M., Meng, E. C., & Ferrin, T. E. (2004). UCSF Chimera–a visualization system for exploratory research and analysis. *Journal of Computational Chemistry, 25*(13), 1605–1612.

Polikanov, Y. S., & Moore, P. B. (2015). Acoustic vibrations contribute to the diffuse scatter produced by ribosome crystals. *Acta Crystallographica. Section D, Biological Crystallography, 71*(Pt 10), 2021–2031.

Prince, E., Finger, L. W., & Konnert, J. H. (2006). *Constraints and restraints in refinement.* John Wiley & Sons, Ltd,694–701. https://doi.org/10.1107/97809553602060000611 Ch. 8.3.

Prince, E. (2004). *International tables for crystallography. Volume C, Mathematical, physical and chemical tables* (3rd ed.). Dordrecht; Boston; London: Published for the International Union of Crystallography by Kluwer Academic Publishers, 2004. previous edition: edited by A.J.C. Wilson and E. Prince. 1999.; Includes bibliographical references and indexes ⟨https://search.library.wisc.edu/catalog/999703926102121⟩.

Riccardi, D., Cui, Q., & Phillips, G. N. (2010). Evaluating elastic network models of crystalline biological molecules with temperature factors, correlated motions, and diffuse x-ray scattering. *Biophysical Journal, 99*(8), 2616–2625.

Riley, B. T., Wankowicz, S. A., de Oliveira, S. H. P., van Zundert, G. C. P., Hogan, D. W., Fraser, J. S., Keedy, D. A., & van den Bedem, H. (2021). qFit 3: Protein and ligand multiconformer modeling for X-ray crystallographic and single-particle cryo-EM density maps. *Protein Science, 30*(1), 270–285.

Sayre, D. (1952). Some implications of a theorem due to Shannon. *Acta Crystallographica, 5*(6), 843. https://doi.org/10.1107/S0365110×52002276
Suarez, E., Wiewiora, R. P., Wehmeyer, C., é, F., Chodera, J. D., & Zuckerman, D. M. (2021). What Markov State Models Can and Cannot Do: Correlation versus Path-Based Observables in Protein-Folding Models. *Journal of Chemical Theory and Computation, 17*(5), 3119–3133.
Urzhumtsev, A., Afonine, P. V., & Adams, P. D. (2013). TLS from fundamentals to practice. *Crystallography Reviews, 19*(4), 230–270.
Van Benschoten, A. H., Afonine, P. V., Terwilliger, T. C., Wall, M. E., Jackson, C. J., Sauter, N. K., Adams, P. D., Urzhumtsev, A., & Fraser, J. S. (2015). Predicting X-ray diffuse scattering from translation-libration-screw structural ensembles. *Acta Crystallographica. Section D, Biological Crystallography, 71*(Pt 8), 1657–1667.
Van Benschoten, A. H., Liu, L., Gonzalez, A., Brewster, A. S., Sauter, N. K., Fraser, J. S., & Wall, M. E. (2016). Measuring and modeling diffuse scattering in protein X-ray crystallography. *Proceedings of the National Academy of Sciences of the United States of America, 113*(15), 4069–4074.
Wall, M. E., Clarage, J. B., & Phillips, G. N. (1997). Motions of calmodulin characterized using both Bragg and diffuse X-ray scattering. *Structure, 5*(12), 1599–1612.
Wall, M. E., Adams, P. D., Fraser, J. S., & Sauter, N. K. (2014). Diffuse X-ray scattering to model protein motions. *Structure, 22*(2), 182–184.
Wall, M. E., Wolff, A. M., & Fraser, J. S. (2018). Bringing diffuse X-ray scattering into focus. *Current Opinion in Structural Biology, 50*, 109–116.
Wall, M. E. (2018). Internal protein motions in molecular-dynamics simulations of Bragg and diffuse X-ray scattering. *IUCrJ, 5*(Pt 2), 172–181.
Welberry, T., & Weber, T. (2016). One hundred years of diffuse scattering. *Crystallography Reviews, 22*(1), 2–78. https://doi.org/10.1080/0889311X.2015.1046853
Wojdyr, M. (2021). ***Gemmi – library for structural biology*** [software]. ⟨https://github.com/project-gemmi/gemmi⟩.
Wych, D., & Wall, M. (2023a). Molecular-dynamics simulations of macromolecular diffraction, part I: Preparation of protein crystal simulations. *Methods in Enzymology*.
Wych, D., & Wall, M. (2023b). Molecular-dynamics simulations of macromolecular diffraction, part II: Analysis of protein crystal simulations. *Methods in Enzymology*.
Wych, D. C., Fraser, J. S., Mobley, D. L., & Wall, M. E. (2019). Liquid-like and rigid-body motions in molecular-dynamics simulations of a crystalline protein. *Structural Dynamics, 6*(6), 064704.
Wych, D. C., Aoto, P. C., Vu, L., Wolff, A. M., Mobley, D. L., Fraser, J. S., Taylor, S. S., & Wall, M. E. (2023). Molecular-dynamics simulation methods for macromolecular crystallography. *Acta Crystallographica Section D: Structural Biology, 79*(Pt 1), 50–65.

CHAPTER SEVEN

Interpreting macromolecular diffraction through simulation

Iris D. Young[a], Derek Mendez[a,b], Billy K. Poon[a], Johannes P. Blaschke[c], Felix Wittwer[c], Michael E. Wall[d], and Nicholas K. Sauter[a,*]

[a]Molecular Biophysics and Integrated Bioimaging Division, Lawrence Berkeley National Laboratory, Berkeley, CA, United States
[b]Stanford Synchrotron Radiation Lightsource, SLAC National Accelerator Laboratory, Menlo Park, CA, United States
[c]National Energy Research Scientific Computing Center, Lawrence Berkeley National Laboratory, Berkeley, CA, United States
[d]Computer, Computational and Statistical Sciences Division, Los Alamos, NM, United States
*Corresponding author. e-mail address: nksauter@lbl.gov

Contents

Methods in Enzymology, Volume 688
ISSN 0076-6879, https://doi.org/10.1016/bs.mie.2023.06.011

Abstract

This chapter discusses the use of diffraction simulators to improve experimental outcomes in macromolecular crystallography, in particular for future experiments aimed at diffuse scattering. Consequential decisions for upcoming data collection include the selection of either a synchrotron or free electron laser X-ray source, rotation geometry or serial crystallography, and fiber-coupled area detector technology vs. pixel-array detectors. The hope is that simulators will provide insights to make these choices with greater confidence. Simulation software, especially those packages focused on physics-based calculation of the diffraction, can help to predict the location, size, shape, and profile of Bragg spots and diffuse patterns in terms of an underlying physical model, including assumptions about the crystal's mosaic structure, and therefore can point to potential issues with data analysis in the early planning stages. Also, once the data are collected, simulation may offer a pathway to improve the measurement of diffraction, especially with weak data, and might help to treat problematic cases such as overlapping patterns.

1. Introduction

Most macromolecular crystallographers are familiar with one or more software packages for data reduction, such as *xds* (Kabsch, 2010b) or *dials* (Winter et al., 2018), which take the raw X-ray diffraction patterns, and convert the Bragg spot intensities to real-valued amplitudes of structure factors, which are the Fourier coefficients for calculating the electron density map. Diffraction simulation is the opposite concept—it is the idea of having a program that would accurately and quantitatively predict every pixel on the diffraction pattern, including the location, size, shape, and intensity profile of every Bragg spot, given a set of underlying conditions. There is a long history of software packages geared toward such simulation. Recent authors have used them for instructive purposes; for example, Diederichs illustrated the effects of mosaicity and spectral dispersion on the diffraction pattern (Diederichs, 2009), while Holton & Frankel calculated the minimum crystal size needed for a complete diffraction data set (Holton & Frankel, 2010). Given the current progress in diffuse scattering, where it seems that the analysis of correlated motion may be within reach (Wall, Wolff, & Fraser, 2018), it is appropriate to ask whether simulation programs can and should be extended to cover the diffuse pattern.

Let us first ask whether the purpose of simulation is solely educational, or does it play a critical role in data analysis? After all, if we have a powerful data reduction package, can we not simply interpret the data by direct measurement? For example, to process Bragg spots, we could run a spot-finding program, and then integrate the signal by summing up all the photons on the Bragg-spot pixels. The reality, however, is that Bragg spot measurement is never direct. Rather, quantitative modeling has been an important part of data reduction for decades, both for locating the Bragg spots on the image, and for accurately finding their intensities. The critical thing to appreciate is that the diffraction pattern is a combination of Bragg reflections and background diffraction (which includes diffuse scattering). Measuring a given Bragg spot always means quantifying the total signal and then subtracting the background, all the while accounting for experimental noise, which itself consists of several parts (counting uncertainty, instrument readout, and systematic effects) and potential overlap of neighboring spots. It is easily shown (Bourenkov & Popov, 2006) that the Bragg signal is often weaker than the background, especially at higher resolutions. While many strong spots may be noticed on a diffraction pattern, the weaker ones are not visually apparent by simply looking at an image, and would not be detected by a naïve spotfinding algorithm. Therefore, in real data reduction, the strong spots are used to deduce the lattice (the unit cell and crystal orientation), then all the spots are predicted (both strong and weak), and Bragg intensity is summed on the pixels considered to be at the predicted positions (Leslie, 2006).

Therefore it is clear that quantitative modeling has an established role in processing the data *post*-experiment, but what about for planning purposes, *pre*-experiment? Here it needs to be acknowledged that the required modeling tools are so familiar that they are used routinely without discussion. For example, synchrotron beamline user interfaces like *Blu-Ice* allow the scientist to decide whether the detector position will allow for high-angle Bragg spots, or if instead the detector needs to be repositioned closer to the sample (McPhillips et al., 2002). However, with today's expanded focus on collecting data to reveal diffuse scattering, it is prudent to re-examine the toolset to make sure we are asking the right questions. Since the diffuse signal lies between Bragg spots, it may be sensible to focus on whether the unit cell parameters and detector distance permit sufficient pixels between the Bragg orders to distinguish the low-intensity diffuse signal from the Bragg spikes. Also, we may ask if the Bragg peaks might be broadened by other factors such as beam divergence or dispersion, or the crystal mosaic structure. If the Bragg peaks are too wide, neighboring spots may partially overlap and preclude the observation of diffuse signal. The software described herein is capable of illustrating these effects for a specific crystal form identified by the program user, in the form of a PDB-format structure file. It thus serves as a mechanism for general education, with the added potential of driving the successful design and execution of future experiments.

To frame the discussion that follows, it is useful to draw a distinction between *empirical* and *physical* modeling approaches. In many cases, data reduction workflows are empirical in the sense of not necessarily breaking down the underlying physics of diffraction. A key example is the spot analysis performed by the program *xds* (Kabsch, 2010a). The popularity of that software is due in part to its approach to modeling of the size and shape of the Bragg spot, employing a clever transformation to a coordinate system in which all Bragg reflections share a common intensity profile described by two empirical parameters. Under these conditions, the structure factor (Fourier coefficient) intensity can be modeled by fitting a single scale factor to the data.

In contrast to the empirical approach, we will take a physical approach in view of our pedagogical purpose, to show how subtle properties of the diffraction pattern arise based on more fundamental building blocks. Familiar elements of the experiment are shown in Fig. 1, including the X-ray beam, detector, crystal, and rotating goniometer, along with parameters used to express the experimental geometry and sample properties. While the list is not comprehensive, it is at least a minimum

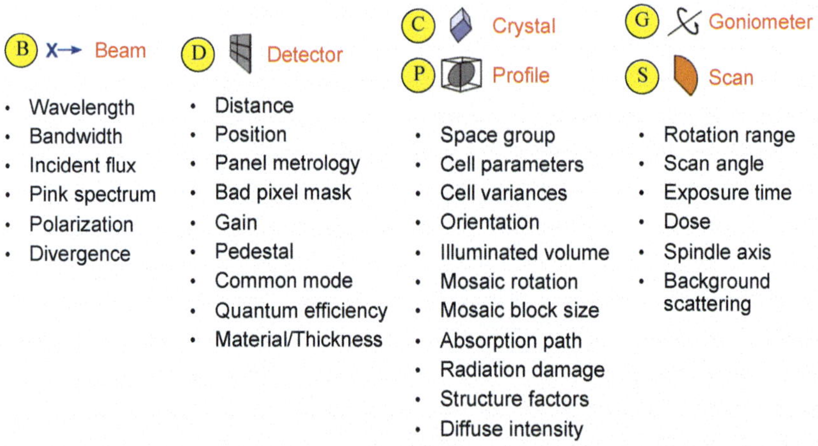

Fig. 1 A general list of experimental elements for physical modeling. The first three columns list parameters relevant to all experiments, while the fourth column applies only to rotation experiments, and not to serial crystallographic data, where every shot is static with no known rotational relationship to any other shot.

starting point. We are now poised to construct a diffraction model from this set of parameters. There are indeed data reduction workflows that adhere to physical modeling principles, in particular *eval15* (Schreurs, Xian, & Kroon-Batenburg, 2009). Here the physics is represented by four statistical distributions, representing respectively the crystal volume, mosaic texture, divergent beam, and wavelength spread. Random drawings are performed from the four distributions, with the diffraction pattern built up by thousands of repeated drawings. Parameters describing the width of these distributions are fit interactively through a graphical interface. Ultimately the information carried forward to structure solution is derived from the simulation rather than directly measured from the raw data. While primarily meant for small molecules, *eval15* has also been applied to protein crystallography (Kroon-Batenburg, Schreurs, Ravelli, & Gros, 2015). Separately, James Holton described a similar ray-tracing approach in great mathematical detail (Holton, Classen, Frankel, & Tainer, 2014), and together we have collaborated to include his program *nanoBragg* (Sauter, Kern, Yano, & Holton, 2020) in the Computational Crystallography Toolbox (*cctbx*). We subsequently published the code *diffBragg* (Mendez et al., 2020), to calculate first derivatives of the *nanoBragg* simulation, for future use in parameter optimization calculations that will ultimately form a new data reduction pipeline. Other current authors are also using simulation to address issues in data reduction (Brehm, White, & Chapman, 2023).

This chapter is an attempt to highlight our physical modeling approach to a more general audience, while its eventual applications to data reduction (and particularly to diffuse scattering data) are still under development. We describe *simtbx.sim_view*, which supports the simulation functionality of *diffBragg* within an interactive image viewer, and which was used as a teaching tool at the 2022 School of Crystallography in Erice, Italy. Section 2 discusses current best practices for the experimental measurement of protein diffuse scattering, Section 3 presents the general principles of pixelwise image simulation, while Section 4 considers how to extend these pixel models to at least some aspects of diffuse scattering. Section 5 explains how *simtbx* models the complexity of real-world experiments, while Section 6 enumerates the model parameters that may be adjusted by the software user, and some that await implementation. Section 7 speculates on future directions for data interpretation, while Section 8 gives instructions for downloading and using the package.

2. Present status and future prospects for diffuse scattering

The above discussion of macromolecular diffuse scattering emphasizes the importance of purposeful planning, with careful optimization of the experimental geometry so that the diffuse pattern is well-measured and clearly distinguishable from the Bragg pattern. It is therefore useful to examine recent experience, noting how experimental choices can lead to successful data collection. We focus the discussion on MCA, the 2020 paper of Meisburger, Case, and Ando (2020). Not only does this paper present an exciting result dealing with correlated motion, it also has a noteworthy methods section, serving as a road map for future experiments by identifying critical best practices for measurement of diffuse scattering data. We examine these practices here, especially considering their relation to diffraction simulation.

2.1 Pixel array detectors

The MCA authors' first experimental choice was to use a Pilatus imaging detector, which uses pixel array detection (PAD), meaning the direct detection of X-ray photons in a silicon crystal via the production of photoelectrons. Critically, a knife-edge test (Fig. 2B) shows that PAD technology reads out the photon signal in essentially the same pixel where the photon is absorbed, thus producing a very sharp Bragg spot (Fig. 2D) absent other effects [in this test, the edge of a knife held against the detector at a slight angle is used to generate a sequence of pixels with smoothly varying X-ray exposure (Koerner, Philipp, Hromalik, Tate, & Gruner, 2009; Philipp, Koerner, Hromalik, Tate, & Gruner, 2010)]. This is in contrast to fiber-optic coupled detectors (Fig. 2A), where X-ray photons are first converted to visible photons within a thin phosphor sheet, which are in turn coupled through a fiber optic taper to a charge-coupled device (CCD) detector. Here, the secondary visible photons generated by the phosphor spread out (within the phosphor) in all directions (Fig. 2C), by one or two pixels (Holton, Nielsen, & Frankel, 2012). Mathematically this is described as a point-spread function (PSF) that is applied to the incident diffraction pattern. While fiber-optic coupled detectors are very popular for general structure solution, the PSF makes them less than ideal for diffuse scattering, as it leaves less room to record the diffuse scattering pattern in between Bragg spots. In addition, commercially available CCD detectors exhibit artifacts that interfere with measurement of diffuse scattering when

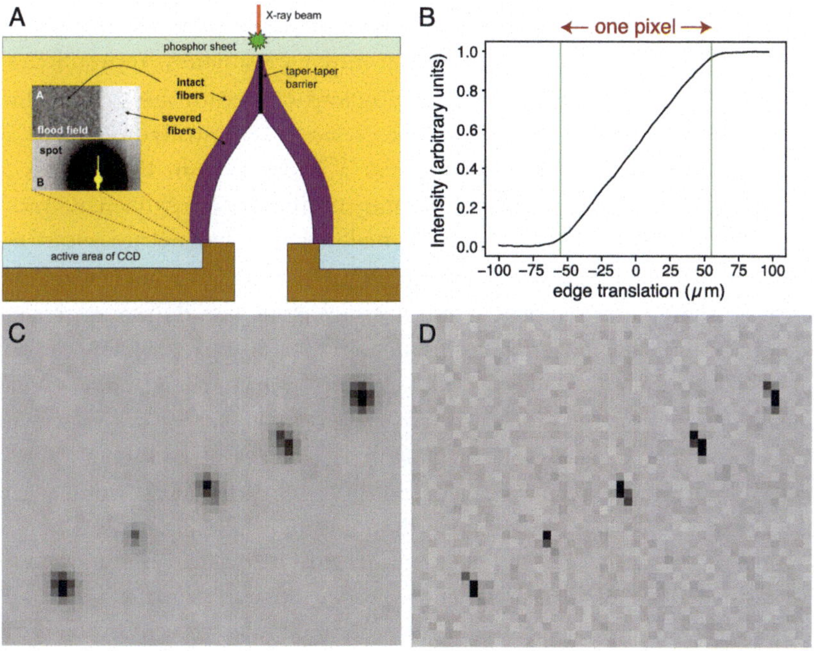

Fig. 2 A comparison of imaging detector types. (A) In a fiber-coupled area detector, X-rays are converted to visible photons in the phosphor layer (green), then coupled through a fiber-optic taper (orange) to a charge-coupled device (blue). (B) In a pixel array detector (PAD), the intensity response of one pixel as a knife edge is translated across the detector surface, exposed to an X-ray flood field. Photons are directly converted to detected charge in silicon, with most of the response confined to one pixel. (C) A simulated image models the point-spread function of a fiber-coupled detector, while (D) the corresponding simulation of a PAD shows much sharper Bragg spots. *Panel A: Reproduced from Holton et al. (2012), with permission from IUCr Journals. Panel B: Adapted from Koerner et al. (2009), with permission from IOP Publishing.*

the number of X-rays recorded on a pixel becomes too large, due to the spread of charge to neighboring pixels after saturation. While it is possible to simulate the PSF with *nanoBragg*, and therefore possible in theory to deconvolute its effects, we will only consider PAD technology for the discussion of simulation.

2.2 Space group and unit cell

Studying lysozyme was a best-case scenario, as it is a well-known, relatively small protein. Larger proteins to be studied in the future will have larger unit cells, and by the nature of reciprocal space the resulting Bragg spots will be closer together in image space. Crowded patterns will make it harder to detect the diffuse signal in between Bragg spots, but this might be mitigated by simple measures such as moving the detector further from the sample. For this paper it was a judicious choice to use a crystalline lysozyme form in space group *P*1, since lysozyme normally crystallizes in space group $P4_32_12$, and thus produces a Bragg pattern eight times as crowded as the one studied here. Higher-symmetry crystals provide more multiplicity, which is advantageous for increasing signal-to-noise and improving characterization of data quality. In contrast, lower-symmetry crystals provide more independent measurements for diffuse scattering, which can be advantageous in modeling. These examples support the argument that there is a potential role for simulation in the planning of difficult future experiments. One can read a PDB file to set the symmetry and unit cell, set the detector type and geometric parameters, and use a physical model to predict the size and shape of spots. One can then inspect for overlap, either from spot to spot, or between the spots and the diffuse halo.

2.3 Ambient temperature

It has become clear over the last decade that proteins exist in a conformational ensemble that can be observed crystallographically (Woldeyes, Sivak, & Fraser, 2014), with the likelihood that enzyme function and regulation are determined by contact networks that can be modulated in response to the local environment (van den Bedem, Bhabha, Yang, Wright, & Fraser, 2013). Furthermore, cryopreservation techniques that have long guarded against radiation damage are best avoided, in order to sample conformation space under operating conditions (Keedy et al., 2015). Therefore the MCA authors were particularly concerned with keeping the radiation absorbed dose below 65 kGy, thus protecting against both general damage that reduces resolution, and specific damage that accrues to relevant chemical species like disulfide bonds and metallocenters in enzymes. Monitoring the electron density maps in real time during rotational data collection, they translated the crystal to a fresh position in the crystal at the first sign of damage. This was possible because large crystals were available in the 100–200 μm range.

If large crystals are not available in future studies, the best path forward may lie with serial crystallographic methods. Here the experiment is performed at ambient temperature, but only one shot is taken per crystal to avoid radiation damage. Thus there is a need to accumulate the full pattern by merging thousands of patterns from small crystals in random orientations. The serial approach is available at either synchrotron sources (with millisecond up to second time scales), or at X-ray free electron lasers (XFELs, where pulses, tens of femtoseconds long, enable pump-probe experiments in the femtosecond to microsecond regimes). In both cases, there is a capability for performing time-domain studies under ambient conditions. However, serial crystallography also presents challenges such as the partiality of Bragg spots, as the data are collected with still shots rather than with goniometer rotation. XFEL X-ray pulses also have significant X-ray dispersion. Simulation programs offer tools for understanding (and therefore mitigating) these problems, as will be discussed.

3. Image simulation methods

Before mentioning the details of image simulation, we will outline the mathematical framework, which is the same framework used to explain crystallography in a graduate-level class. The crystal is thought of as existing in *direct space*, and the diffraction pattern in *reciprocal space*, with the two adjunct spaces co-rotating, and related to each other by Fourier transformation (Scheme 1).

A copy of the molecule is found at each position in the crystal lattice. Mathematically this is the operation of convolution (⊗), and its Fourier transform is a multiplication. Specifically, in the diffraction pattern, the molecular transform is sampled at each reciprocal lattice point. Experimentally, the Bragg spot intensity gives us the structure factor (squared) for the corresponding Miller index. Furthermore, the Ewald sphere construction (Fig. 3) reminds us that we only observe Bragg spots that are exactly in the diffraction condition.

We need a few proportionality factors to make sure our simulated image is calculated with sensible units (photons/pixel) and is quantitative on an absolute scale. Various formulae go back at least a hundred years (Darwin, 1914) but the modern equivalent may be summarized as (Eq. (1)):

	Fourier Transform	
Crystal	⟶	Diffraction Pattern
Molecule ⊗ Crystal Lattice	⟶	Molecular Transform × Reciprocal Lattice

Scheme 1 Initial framework for simulation.

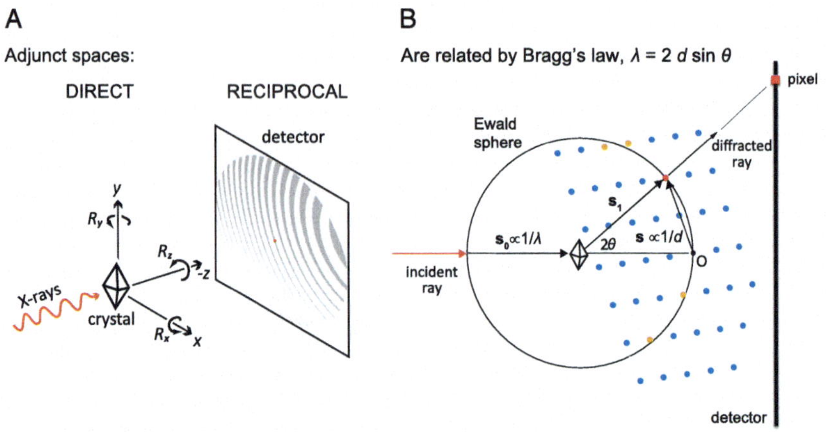

Fig. 3 Fundamental framework for image simulation. As implied by Scheme 1, (A) the adjunct spaces representing crystal (direct) and diffraction pattern (reciprocal) are (B) related by Bragg's law, relating the wavelength λ, the scattering angle 2θ, and the spatial resolution d. The red reciprocal lattice point (located a reciprocal distance $1/d$ from the reciprocal origin O) satisfies the diffraction condition and therefore produces a scattered ray, reaching the indicated detector pixel. Although the adjunct spaces co-rotate, their rotation origins differ in the Ewald sphere construction (B), with the crystal rotating around the sphere center, and the reciprocal lattice rotating around O. *Adapted from Sauter et al. (2014).*

$$I_{\text{pixel}} = J_0(\lambda) r_e^2 P \Delta\Omega_{\text{pixel}} N_{\text{cells}} \,|\, \mathbf{F}_{\text{cell}}(\mathbf{h_0}) \,|^2 \, \mathrm{F}_{\text{latt}}^2(\mathbf{s})$$

where $J_0(\lambda)$ is the number of photons incident on the crystal, r_e is the radius of the electron, P is a polarization factor (Kahn et al., 1982), $\Delta\Omega_{\text{pixel}}$ is the solid angle subtended by the pixel, and N_{cells} is the number of illuminated unit cells. The factor $|\, \mathrm{F}_{\text{cell}}(\mathbf{h_0}) \,|^2$ is just the familiar structure factor intensity at a particular Miller index $\mathbf{h_0}$. Finally for our initial discussion, $\mathrm{F}_{\text{latt}}(\mathbf{s})$ is a simple delta function, telling us if the reciprocal lattice point is on the Ewald sphere (in diffracting condition), given our unit cell and crystal orientation that determines the scattering vector **s**.

From here we can add nuanced detail to our mathematical treatment. The statement that $\mathbf{F}_{\text{latt}}(\mathbf{s})$ is just a delta function must be an over-simplification, since that would imply seeing only one Bragg spot at a time as the crystal rotates on the goniometer, rotating successive reciprocal lattice points through the Ewald sphere. In reality we know that diffraction patterns are crowded with Bragg spots. Let's re-examine the math implied by Scheme 1, and introduce further complexity, Scheme 2:

	Fourier Transform	
Crystal	$\longrightarrow$	Diffraction Pattern
Molecule ⊗ Crystal Lattice	$\longrightarrow$	Molecular Transform × Reciprocal Lattice
(Molecule ⊗ Lattice) × Block mask	$\longrightarrow$	(Molecular Transform × Reciprocal Lattice) ⊗ RLP volume

Scheme 2 Revised simulation framework taking into account the finite size of the mosaic block.

On reconsidering, the reciprocal lattice point is a delta function only if the crystal lattice is infinite in extent (due to the Fourier transform of an infinite sine wave being a single spatial frequency). We may think of a real crystal being constructed, mathematically, as an infinite crystal multiplied by a block mask that describes the volume of matter that diffracts coherently, i.e., where the scattering from unit cells constructively interferes to produce Bragg spots. As Scheme 2 shows, this has the effect of placing a reciprocal lattice point volume (RLP volume) at every reciprocal lattice position, again the operation of a convolution. This reciprocal object, the Fourier transform of the coherently-diffracting mosaic block, is finite in size and thus gives rise to the crowded pattern of Bragg spots.

Let's consider a possible mathematical form for the RLP volume. Suppose that the block mask (the coherently scattering mosaic domain) is a rectangular parallelepiped of $N^3 = 7 \times 7 \times 7$ unit cells. We get the familiar Laue interference pattern (Eq. (2)),

$$F_{\mathrm{latt}}(\mathbf{s}) \propto \frac{\sin\pi N h}{\sin\pi h} \times \frac{\sin\pi N k}{\sin\pi k} \times \frac{\sin\pi N l}{\sin\pi l}$$

where h, k, and l are the reciprocal space coordinates (Miller indices), and specifically for $N = 7$, this gives rise to five rectangular interference fringes between every pair of Bragg spots (Fig. 4A). While it is difficult to observe this pattern in practice, there is one famous XFEL experiment where diffraction was collected from single mosaic blocks, giving rise to precisely this type of pattern (Chapman et al., 2011). In a more typical experiment, the crystal might consist of numerous mosaic blocks all of slightly different sizes and shapes, with the total diffraction being the incoherent sum of the coherent diffraction from each block (Fig. 4B). Therefore it no longer makes sense to speak of a specific value of N, but rather a Gaussian distribution, such that the overall RLP shape might reflect the average block size $\langle N \rangle$ while taking a Gaussian form (Eq. (3)),

$$F_{\mathrm{latt}}(\mathbf{s}) \propto e^{-(\langle N \rangle \Delta \mathbf{h})^2}$$

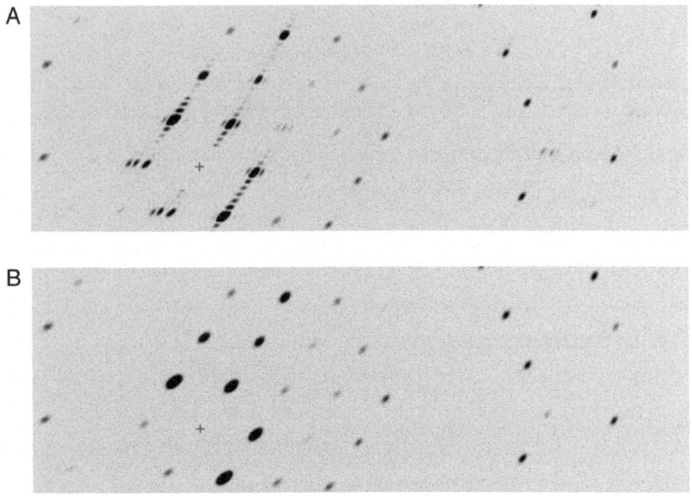

Fig. 4 Size and shape of the mosaic block affects the Bragg spot geometry. Each reciprocal lattice point may be thought of as a Fourier transform of the average coherently scattering block within the crystal (Scheme 2). If the block is a rectangular parallelepiped, seven unit cells on an edge (Eq. (2)), the simulated pattern is as shown in (A), with five fringes appearing between every Bragg spot center. In contrast, if the block geometry draws from a 3D Gaussian distribution with a mean width of $\langle N \rangle = 7$ cells (Eq. (3)), the generated pattern is as shown in (B). Crosshairs indicate the direct beam.

Importantly, this implies a Gaussian falloff of pixel intensity when the RLP is a reciprocal space distance $\Delta\mathbf{h}$ away from the Ewald sphere. The behavior may be summarized as follows: when the average block size $\langle N \rangle$ is small, many Bragg reflections are visible because RLPs are in diffracting condition relatively far from the Ewald sphere; and vice versa, with few reflections being visible if $\langle N \rangle$ is large (Sauter et al., 2014).

This gives us some fairly sophisticated grounding for understanding the finer points of diffraction (Pflugrath, 1999; Sauter & Adams, 2017). As the RLP volume passes through the Ewald sphere (due to a goniometer rotation), the intensity of a Bragg spot first increases, then reaches a maximum, and then decreases, eventually to zero (Fig. 5B). This is the famous "rocking curve". The rocking curve might be narrower (in rotation) than one X-ray frame, or it might extend over several consecutive frames in the rotation series. To correctly measure the structure factor, we must integrate the number of photons over the full rocking curve, and such a Bragg spot would be considered to be "fully measured". Interestingly, as the RLP passes through the Ewald sphere (Fig. 5A), the $\mathbf{s}_1$ vector representing the scattering direction (and which points from the Ewald sphere

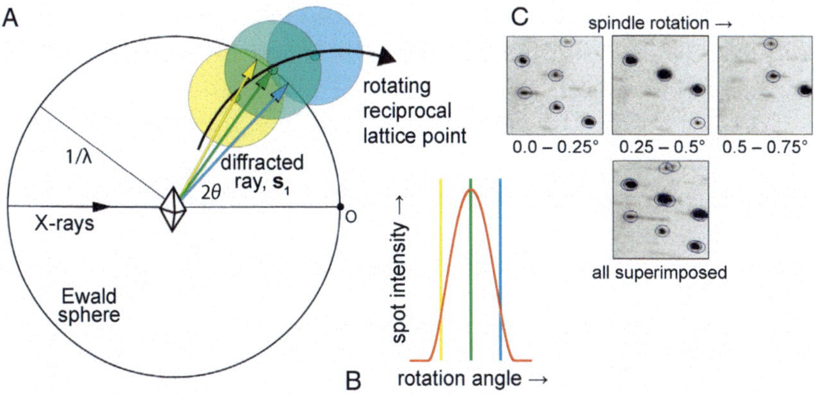

Fig. 5 Bragg spot position changes slightly with crystal rotation. (A) As the crystal rotates on the goniometer spindle, a reciprocal lattice point (RLP) co-rotates thru the diffraction condition (the Ewald sphere). However, the $\mathbf{s}_1$ vector, indicating the direction of the corresponding Bragg spot, makes a slightly larger scattering angle 2θ when the RLP is inside the sphere, than when the RLP is outside. (B) The rocking curve indicates that the diffracted spot intensity reaches a maximum when the RLP intersects the Ewald sphere surface. (C) Experimental diffraction patterns collected at increasing spindle rotations reflect these effects. *Panels A, B: Reproduced from Sauter and Adams (2017) with permission from the Royal Society of Chemistry. Panel C: Adapted from Pflugrath (1999).*

center to the RLP center) changes slightly, just a fraction of a degree. The visible manifestation is that Bragg spots change their position very slightly (Fig. 5C), as the crystal rotates on an axis perpendicular to the beam. It is a small effect but it should be accurately reproduced by the simulator.

What happens with the rocking curve during serial crystallography? Here there is no goniometer rotation; the rocking curve is sampled at a single orientation only. Without the context of a model, we don't know if we are near the rocking curve peak or near the tails. In short, we are in the uncomfortable position that the thing we measure (the Bragg spot intensity) is not at all proportional to the thing we are seeking (the integrated area of the rocking curve). The initial approach in serial crystallography was to simply improve the outcome with overwhelming multiplicity of measurement. A more sophisticated approach is to use a model to determine a separate correction factor for each Bragg intensity, so our partial measurements can be converted to the full spot equivalent. Physics-based modeling is especially applicable here, especially if iterative parameter estimation (*postrefinement*) can be used to derive best estimates of the crystal, detector, and beam parameters, along with the mosaic block size $\langle N \rangle$. Many software authors addressed this problem in the last decade, leading to a dramatic improvement in the electron density maps (Ginn et al., 2015; Kabsch, 2014; Kroon-Batenburg et al., 2015; Sauter, 2015; Uervirojnangkoorn et al., 2015; White, 2014).

4. Extending simulators to diffuse scattering

We are now in a position to include some limited information about diffuse scattering in the simulator. First let's give a plausibility argument. We know that there are several contributing factors to the diffuse scattering (Fig. 6) among which the inelastic Compton scattering is the least interesting for molecular function, and will not be addressed here. The isotropic diffuse signal contains interesting information about internal motions, but varies on a reciprocal length scale longer than we are concerned with here. Of the remaining variational diffuse features, there is a cloudy pattern observed between lattice layers that informs us about short-range correlations in the molecular motion; this will also not be discussed here. Finally there is a pattern of intense halos that surround each Bragg spot, which are visible at the lattice positions even after Bragg spots have been removed (Fig. 6A). These halos reflect the long-range correlations on scales of a few unit cells. An alternative representation is shown in Fig. 6B taken from the MCA paper, showing the scattering intensity around one Bragg spot as a function of goniometer rotation. The largest signal (blue) is just the rocking curve for the Bragg spot. In this particular experiment, where the spindle was rotated in steps of 0.1°, the full width of the Bragg spot was typically found to be on a single rotational step. The intense halo also has a characteristic footprint (black), with a width extending about 1 or 2° on either side of the central peak. The fact that we can clearly see the intense halo as a separate phenomenon, which mimics the well-understood rocking curve producing the $F_{\text{latt}}(\mathbf{s})$ factor in Eq. (1), suggests that we could simply add another term to our simulation that would represent the reciprocal space form of the intense halos (Eq. (4)):

$$I_{\text{pixel}} = J_0(\lambda)\, r_e^2\, P\, \Delta\Omega_{\text{pixel}}\, N_{\text{cells}}\, |\, F_{\text{cell}}(\mathbf{h_0})\, |^2 \times \left(F_{\text{latt}}^2(\mathbf{s}) + F_{\text{diffuse halo}}^2(\mathbf{s})\right)$$

Possible functional forms for the $\mathbf{F}_{\text{diffuse halo}}^2(\mathbf{s})$ term for protein crystals have been discussed in the literature for at least 30 years. Caspar et al. expressed the idea of an autocorrelation function to express an exponentially

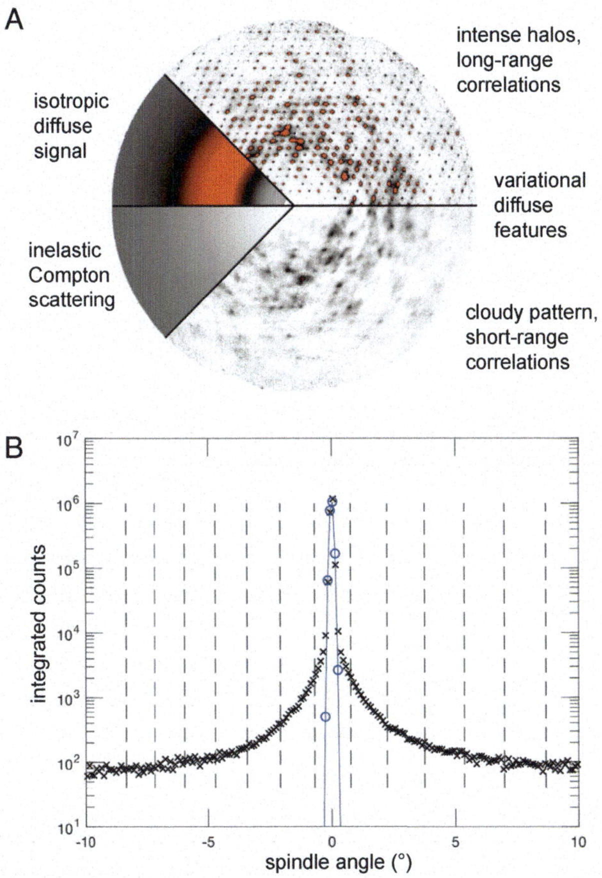

Fig. 6 Summary of the diffuse scattering pattern from macromolecular crystallography. (A) A section through reciprocal space, highlighting various contributions to the diffuse isotropic scattering (left) and diffuse anisotropic, or variational scattering (right). (B) Detail of the scattering focusing on one Bragg spot, as the goniometer spindle rotation brings the reciprocal point through the Ewald sphere. The blue curve is the rocking curve proper, with its sharp peak at the central position (note the logarithmic scale of intensity), while the black data originate from the long-range correlations, forming an intense halo around the tails of the Bragg peak. *Adapted from Meisburger et al. (2020).*

decreasing effect with correlation length γ (Caspar, Clarage, Salunke, & Clarage, 1988). The Fourier transform of this function gives an expression for the diffuse halo density in reciprocal space, which models the black curve of Fig. 6B. Since we are discussing long-range correlations between unit cells, a first estimate of the length γ could be 50 Å. Anisotropic correlation can also be represented, in which case the single parameter γ would be replaced with a tensor Λ. Finally, an additional parameter σ gives the vibrational amplitude of the motion, typically of magnitude <1 Å. We cannot go into mathematical detail here, but the reader may refer to Michael Wall's calmodulin paper (Wall, Clarage, & Phillips, 1997) where Wall Eq. (8) essentially gives the $\mathbf{F}^2_{\text{diffuse halo}}(\mathbf{s})$ term, with Wall Eq. (9) used for the autocorrelation function. The reader is also referred to Meisburger Eqs. (138) to (141) in a recent review (Meisburger, Thomas, Watkins, & Ando, 2017).

Eq. (4), with its diffuse halo term, is incorporated into the *simtbx.sim_view* simulator associated with this chapter (Fig. 7).

5. Complexity of the simulation

It should be emphasized that the Eq. (4) model for pixel intensity is valid only for a still shot. If the experiment includes goniometer rotation, we take a simple summation over intermediate goniometer rotational positions (Eq. (5)),

$$I_{\text{pixel}} = J_0(\lambda) r_e^2 P \Delta\Omega_{\text{pixel}} N_{\text{cells}} \mid \mathrm{F}_{\text{cell}}(\mathbf{h}_0) \mid^2 \times \sum_{\substack{\text{spindle} \\ \text{rotation}}} \left(\mathrm{F}^2_{\text{latt}}(\mathbf{s}) + \mathrm{F}^2_{\text{diffuse halo}}(\mathbf{s})\right).$$

For example, we might represent a 0.1° rotation as a sum over 10 still shots spaced 0.01° apart. The changing rotational position does not appear explicitly but is implicit in Eq. (5). The exact math is beyond the scope of this chapter, but we can show the basic idea (and the reader is invited to skip this paragraph if only interested in the broad framework). In simplest terms (Fig. 3B), let's define the incident beam vector as $\mathbf{s}_0$, and the scattered ray vector as $\mathbf{s}_1$, both with a length of inverse wavelength ($1/\lambda$), and $\mathbf{s} = \mathbf{s}_1 - \mathbf{s}_0$. Then the fractional Miller index $\mathbf{h}$ implicitly used for $\mathrm{F}^2_{\text{latt}}(\mathbf{s}) + \mathrm{F}^2_{\text{diffuse halo}}(\mathbf{s})$ is (Eq. (6)),

$$\mathbf{h} = \mathbf{R}_\phi \mathbf{A}\mathbf{s}$$

where $\mathbf{R}_\varphi$ is the rotation matrix corresponding to a spindle rotation of φ, and $\mathbf{A}$ is the direct space orientation matrix (Eq. (7)),

$$\mathbf{A} = \begin{pmatrix} a_x & a_y & a_z \\ b_x & b_y & b_z \\ c_x & c_y & c_z \end{pmatrix}$$

Fig. 7 Snapshot of the *simtbx.sim_view* image viewer, simulating the diffuse halo pattern from calmodulin crystals (PDB entry 1CM1), which reflects a long-range correlation between unit cells, according to the model of Wall et al. (1997), Eq. (8). Parameters for the correlation length γ and vibrational amplitude-squared σ^2 may be specified by the user to empirically fit experimental data. In this case, streaked diffuse features were produced by multiplying γ by an anisotropy factor along the a–b and c unit cell directions in direct space (leaving a+b alone), and multiplying σ^2 by the same anisotropy factor along the a–b direction (leaving a+b and c alone). The Laue symmetry was then applied, leading to streaks along both the $a^* - b^*$ and $a^* + b^*$ directions in reciprocal space. In the present case, this type of anisotropy was purely an *ad hoc* decision allowing us to generate a simulation that resembles the data in the 1997 paper, but this exercise suggests that correct models may require similar anisotropic features in the general case.

consisting of the x, y, and z components of the unit cell basis vectors **a**, **b**, and **c**.

Another source of computational complexity is the phenomenon of mosaic rotation. In Eq. (5) we are treating all the mosaic blocks of the crystal as having the same exact orientation represented by matrix **A**. This may be a reasonable approximation for work at ambient temperature.

However, from several decades experience with cryopreserved crystals, we know that the quick cooling process essentially shatters the crystal into many mosaic blocks of slightly different orientations, such that the Bragg spots can actually appear as tiny arcs (Fig. 8D) concentric on the direct beam (Nave, 2014). We handle this in our simulation by adding another large sum over individual mosaic rotations (Eq. (8)),

$$I_{\text{pixel}} = J_0(\lambda)\, r_e^2\, P\, \Delta\Omega_{\text{pixel}} N_{\text{cells}}\, |\, F_{\text{cell}}(\mathbf{h_0})\, |^2 \times \sum_{\substack{\text{spindle}\\ \text{rotation}}} \sum_{\substack{\text{mosaic}\\ \text{rotations}}} \left(F^2_{\text{latt}}(\mathbf{s}) + F^2_{\text{diffuse halo}}(\mathbf{s})\right)$$

where the rotations represent a Gaussian distribution over small angles with randomly distributed axes of rotation.

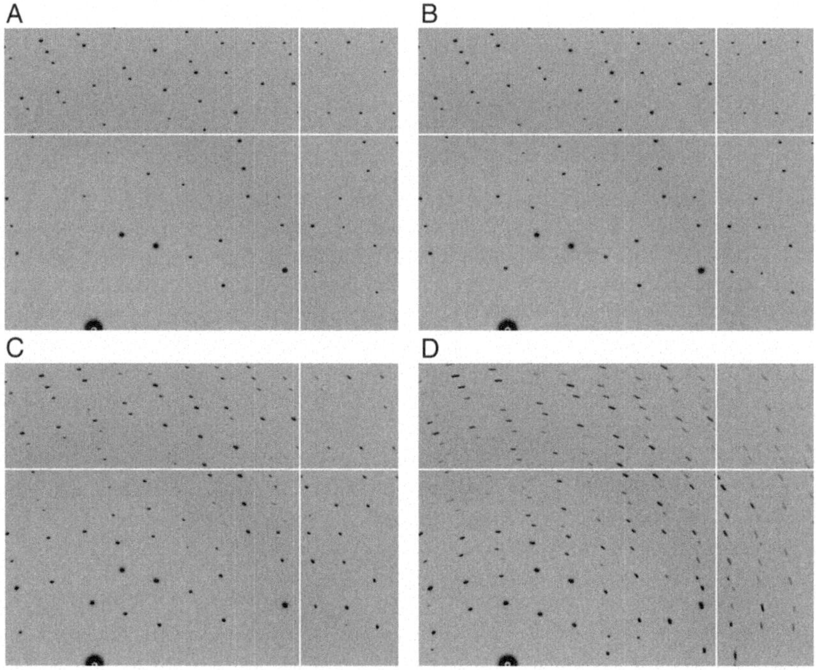

Fig. 8 Simulations that differ in the crystal's mosaic rotational disorder but that are otherwise identical. (A) Perfectly ordered mosaic blocks (rotational full width $\eta = 0°$). (B) Small mosaicity typical of ambient temperature crystals ($\eta = 0.1°$). (C) Medium-high mosaicity found with cryopreserved samples ($\eta = 0.6°$). (D) Large mosaicity, with arclets indicating high disorder ($\eta = 1.5°$). Courtesy of James Holton.

A final source of complexity is the beam dispersion. While synchrotron beamlines are often monochromatic (with a dispersion of ~0.01% in energy), XFEL sources have a broader dispersion of ~0.3%, with a unique mean energy for each shot, and a spiky spectrum with a stochastic shape. We incorporated 100 experimental XFEL spectra into the *simtbx.sim_view* simulator so that the user can sample the slight differences in the diffraction. These are accessed by the "New XFEL pulse" button in the GUI. Computationally we incorporate spectral shape into the calculation with (Eq. (9)):

$$I_{\text{pixel}} = \sum_{\text{wavelength } \lambda} J_0(\lambda)\, r_e^2\, P\, \Delta\Omega_{\text{pixel}} N_{\text{cells}} \,|\, F_{\text{cell}}(\mathbf{h_0}) \,|^2 \times \sum_{\substack{\text{spindle}\\ \text{rotation}}} \sum_{\substack{\text{mosaic}\\ \text{rotations}}} \left(F_{\text{latt}}^2(\mathbf{s}) + F_{\text{diffuse halo}}^2(\mathbf{s}) \right)$$

where the factor $J_0(\lambda)$ refers to the number of incident photons within each wavelength interval. The effect of spectral dispersion is to transform Bragg spots into radial streaks emanating from the direct beam position (Fig. 9).

Considering the triple summation, Eq. (9) may not be the most computationally efficient design for image simulation, and alternate ideas are welcome. To help compensate for the combinatorial footprint, we have implemented a GPU-accelerated version, presently available for users with Linux machines with Nvidia graphics cards.

6. Parameter space and implications for simulation

Here we offer further discussion of each of the parameters thus far identified as relevant to simulating the diffraction experiment, with a brief overview of how we anticipate the user may interact with them via the *simtbx.sim_view* display.

6.1 Incident beam

The X-ray beam can be described in terms of its wavelength, flux, divergence, polarization, shape, and bandpass or spectrum. We ignore shape as the area exposed to the beam is bundled into illuminated volume, and the shape itself has no effect. We are also implicitly ignoring the pulsed nature of an XFEL beam as well as flux, which again do not matter if we are not addressing radiation damage or any other changes over the course of the exposure. Wavelength is clearly important—this determines diffracting conditions—and therefore bandpass also concerns us. Here we

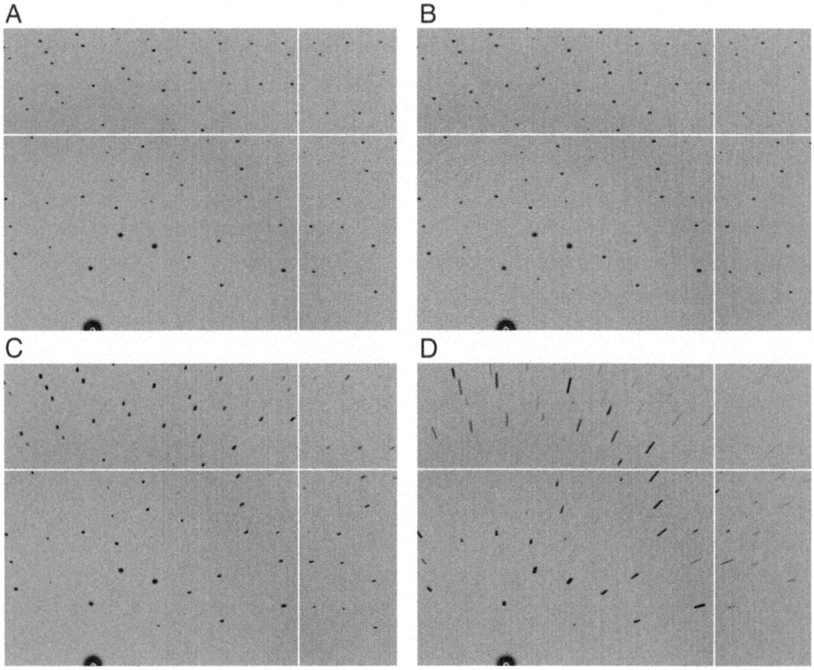

Fig. 9 Simulations that differ in the dispersion of the incident beam but that are otherwise identical. (A) Low dispersion ($\Delta E/E = 0.014\%$) typically produced by a synchrotron monochromator equipped with a Si 111 crystal. (B) Medium dispersion ($\Delta E/E = 0.25\%$) found in an XFEL pulse. (C) Pink beam, $\Delta E/E = 1.6\%$ and (D) Laue, $\Delta E/E = 5.1\%$. Courtesy of James Holton.

must consider the possibilities case wise. If we are working with a monochromatic beam that we are content to model as perfectly monochromatic, we use a single wavelength. If we choose to model bandpass, our model currently approximates this as a histogram of contributions from different energy channels and sums these contributions. This approach can handle a pink beam or an XFEL spectrum equally well. We have three spectrum shapes available in *simtbx.sim_view*: monochromatic, Gaussian, or SASE (self-amplified spontaneous emission, applicable to XFEL experiments). As just mentioned, we include 100 measured SASE spectra from LCLS in our simulator, and the binning of this histogram was selected to preserve the different shapes of these spectra. The binning is not a parameter exposed to the user at present, but it would be trivial to make this change if the need arises. The effect of a noticeable bandpass or spectrum is a radial streaking of each Bragg spot (Fig. 9D). It is more noticeable when a large mosaic rotation also spreads peaks over many pixels, and under the right conditions, the user can observe realistic shot-to-shot differences when simulating the exposure of the same crystal to different XFEL pulses. Finally, divergence and polarization are properties that are important to model correctly, but typically are fixed at a particular beamline, and are known quantities. We therefore incorporate these into the model but do not expose them as variables to adjust or refine.

6.2 Unit cell and symmetry

The reciprocal lattice itself scales with the unit cell of the crystal, producing a global expansion or contraction of the diffraction pattern on the detector. We include each unit cell parameter as an independently adjustable variable in the simulation, as we have often found ourselves in the position of observing a slightly different unit cell than expected during an experiment. Where applicable, crystal symmetry is enforced so that symmetry-related axes scale together. The user is encouraged to explore multiple orientations when adjusting unit cell parameters, since the effects of scaling each axis will be more or less visible in different orientations—when the beam is aligned with one of the unit cell axes, changes in that axis length will not alter the spot-to-spot distance. The "randomize orientation" button can be useful for exploring many orientations.

6.3 Crystal orientation

The orientation of the crystal with respect to the beam determines which Bragg reflections will be observed (Fig. 3B). An initial crystal orientation is acquired during indexing, but in practice, especially for serial crystallography, small errors may remain. We measure this misorientation as rotational offsets on the x-, y-, and z-axes, and denote these rot_x, rot_y and rot_z, respectively. Rot_z is easy to detect and correct, since this angle rotates spots about the origin in the plane of the detector. Rot_x and rot_y are more difficult. These latter two have almost no effect on the position of spots on the detector, but do impact which reflections meet the diffracting conditions. In rotation crystallography we have the opportunity to integrate over the entire rocking curve as the reciprocal lattice point rotates through the Ewald sphere, so a slight misorientation is tolerable. In contrast, recall that in serial crystallography we accumulate only a large number of slices on individual still shots. This renders us extremely sensitive to any misestimation of a reflection's position relative to the Ewald sphere: Although each point in the reciprocal lattice is smeared out into a small (reciprocal) volume by crystal imperfections, allowing us to record a partial spot even when we are not perfectly in diffracting conditions, the resulting intensity changes dramatically with slight rotations of the crystal. All three (mis) orientation angles may be adjusted in the *simtbx* simulation. To aid the user in viewing the different effects of rotations in x, y and z, we also simulate a "reference" crystal, which begins as identical to the primary simulated crystal but does not change as the simulation parameters are updated. (It can be brought back into register with the "Update pinned image" button.) The user toggles whether the reference is displayed in red; the primary simulation is always displayed in cyan, and the two combine to produce white where pixel intensities match perfectly. A spectrum emerges at positions where intensities are only slightly different.

6.4 Mosaic character

Mosaicity describes variation among individual coherently scattering blocks within a crystal, and there are multiple approaches to parameterizing this effect (Nave et al., 2016). In *simtbx.sim_view* we focus on the distribution of mosaic block orientations. We presume this to be a three-dimensional normal distribution (over rot_x, rot_y, and rot_z) and thus sample a representative set of orientations (Eq. (8)). Interference between the mosaic blocks is ignored: the coherent diffraction from each block is computed separately and summed to produce the simulated diffraction pattern. These simplifications appear to be reasonable for both cryocooled microcrystals at synchrotrons and crystals at ambient temperature at XFELs, although when mosaic domains are individually, explicitly modeled, there may be further improvement in the agreement between measured and simulated images. Additional parameters of mosaicity not modeled here include variation in unit cell dimensions and variation in the mosaic block size, N. The former would produce uniform radial streaking of Bragg peaks and the latter would produce nonuniform distortion of the peaks. As we are primarily interested in applying these simulations to cases where other effects on spot shape are much stronger, we have omitted these variables for now. For simplicity we model the mosaicity as isotropic. It is possible to model anisotropic rotations and block sizes in *nanoBragg* and *diffBragg*, and this functionality could be extended to *simtbx.sim_view* in the future.

6.5 Diffuse scattering

As discussed earlier in this chapter, several types of diffuse scattering arise from different physical phenomena, and only the diffuse halos around Bragg peaks produced by long-range correlations on the order of several unit cells are modeled in our simulation. By default diffuse scattering is disabled in *simtbx.sim_view* since it is an expensive calculation, but it can be enabled with a toggle switch. (We recommend selecting a monochromatic beam or turning off diffuse scattering while exploring other parameters.) Halos are parameterized by a correlation length γ in Ångstroms, and the amplitude σ of the correlated vibrations also in Ångstroms (Fig. 7). These also govern the strength of the diffuse signal relative to the Bragg signal. Diffuse scattering may be either isotropic or anisotropic, and we have incorporated a unitless anisotropy term in our simulation to reflect this, which at present is assigned a specific directionality. An increasingly thorough treatment of diffuse scattering is a high priority for ongoing work (Peck, Poitevin, & Lane, 2018; Wall et al., 2018).

6.6 Oscillation width and rotation

Our simulator was designed with serial crystallography as the target use case, but we later added a rotation mode in order to explore rotation experiments as well. Note that this mode is much slower because each oscillation image is produced as the sum of a series of still shots, so we generally do not recommend adjusting other variables with this mode enabled, and we especially caution against enabling it at the same time as diffuse scattering. Rotation experiments are presumed to be carried out at synchrotron sources, so we fix the spectrum shape to monochromatic in this mode. The subset of parameters displayed in the GUI is adjusted accordingly when switching between modes.

6.7 Background and noise

The *simtbx.sim_view* viewer still awaits features that usefully depict background and noise. Air and water scattering are well-understood and in principle easily modeled. There may be other known sources of scattering or absorption as well, but these two typically dominate, producing familiar isotropic patterns. Several types of noise arise from the experiment. Shot noise is a consequence of the statistical likelihood of observing a particular photon given a ground-truth photon rate, and it follows a Poissonian distribution. The shorter the exposures (and fewer photons collected on each image), the greater the fractional contribution of shot noise to each image. There is also readout noise, a result of dark current, which is the signal registered by the detector in the absence of any photons. This is because at the stage where photons impinging on the detector are converted to electrical signal, the same process of charge separation into electrons and holes very occasionally occurs spontaneously. The likelihood of these spontaneous events is governed by the Boltzmann distribution and increases with temperature. Readout noise, or simply read noise, follows a Gaussian distribution. Detectors that do not convert X-ray photons directly to electron–hole pairs—indirect detectors, as opposed to direct detectors—are also subject to a point-spread function (PSF; Fig. 2A). X-ray detectors also do not have perfect linear responses, and diffraction requires a very large dynamic range compared with direct imaging. Ideally each additional photon hitting the detector would register as one additional photon event, but at very high photon counts, this response drops off. This last effect can be surprisingly difficult to identify from a diffraction pattern directly, as it does not alter the expected Bragg peak shapes. Detectors usually have some variation in responsiveness from pixel to pixel even after calibration, manifesting as slightly different gains, and typically have a few "hot" and "cold" pixels that deviate more dramatically. Cosmic radiation occasionally interacts with the detector, producing a strong, sharp signal at a random location; this can be distinguished from a hot pixel by the fact that it appears on only one image. In *simtbx.sim_view*, we presume an ideal detector and avoid modeling most of these phenomena. *nanoBragg* and *diffBragg* do, however, have the functionality necessary to handle them. The relative contributions of various sources of noise are well worth exploring in a detailed simulation.

7. Conclusions

With the expectation that image simulation can serve as both an informative teaching reference and an aid for planning future experiments, this chapter provides a first implementation of the graphical program *simtbx.sim_view*. However, given its focus on pre-experiment modeling, the present software represents only a starting point. Ultimately we wish to include a mechanism for post-experiment data interpretation. At minimum, this would require the development of new code to simultaneously display the computational model alongside the experimental data. With this type of interactive display, the software user could then adjust the model parameters so that the model matches the data as closely as possible. This is essentially the data processing paradigm of *eval15* (Schreurs et al., 2009), except that the parameter set would be expanded to include aspects of the diffuse scattering such as correlation length, vibrational amplitude and anisotropy.

We certainly expect that most parameter optimization will be performed automatically, just as is currently done with conventional diffraction experiments. Indeed, the program *diffBragg*, which provides the foundation for *simtbx.sim_view,* is capable of computing the first-derivative of the simulated diffraction pattern with respect to each parameter. Therefore it is envisioned that it can be used for iterative inverse modeling, using gradient-descent methods to minimize the variance between model and experiment (Mendez et al., 2020). However, we caution that this prospect of automated optimization does not obviate the need for developing an intuitive understanding of the underlying physics. We can easily imagine a situation where numerous physical parameters (such as the unit cell and mosaic structure) are refined within an automatic data reduction pipeline, only to find that the model has radically diverged. Interactive display of model and data is the final safeguard, to verify that the numerical metrics of variance are telling the truth, and that published model is comfortably within reason.

The Meisburger study of diffuse scattering from lysozyme (Meisburger et al., 2020) benefited from an uncrowded diffraction pattern where the separation between Bragg and diffuse features was readily accomplished. In future experiments, a combination of factors may contribute to more crowded and complex patterns; these factors include larger unit cells, smaller mosaic block sizes, and the use of disperse X-ray beams at XFEL sources, used in the pursuit of time-resolved data that are free from radiation damage. In this event, image simulation provides a potential avenue to deconvolute overlapping signals: either overlapping Bragg spots, or Bragg spots overlapping with diffuse halos. Deconvolution has been described in the literature (Bourgeois, 1999; Schreurs et al., 2009), but has not been common practice. However, in the intervening years computational hardware has advanced such that GPU acceleration is widely available. A fresh look at simulators is thus warranted, both to facilitate deconvolution and simply to provide the best physics-based profiles for data reduction to most effectively account for experimental uncertainty.

8. Software availability

Instructions and files for running *simtbx.sim_view* may be found at https://github.com/cctbx-xfel/erice_2022. Optimal performance requires a Linux machine accelerated with an Nvidia GPU; currently both Volta and Ampere architectures are supported. Instructions for installing and running the software, plus a short program description, will be kept up to date and accessible in the aforementioned Github repository. Github makes provision for users to report bugs, request features, and contribute code.

Acknowledgements

Thanks to James Holton for initially implementing the *nanoBragg* simulator within the *cctbx* toolbox. This led to the *diffBragg* extension to facilitate parameter fitting, in turn laying the groundwork for the graphical interface *simtbx.sim_view*. Aaron Brewster maintained the *cctbx* testing suite, and David Wych helped implement the diffuse scattering equations. We acknowledge funding from U. S. National Institutes of Health grants R01-GM117126 and P30-GM133894, and from the Exascale Computing Project (17-SC-20-SC), a collaborative effort of the U.S. Department of Energy (DOE) Office of Science and the DOE National Nuclear Security Administration. This work used resources of the National Energy Research Scientific Computing Center (NERSC), a DOE User Facility located at Lawrence Berkeley National Laboratory, operated under Contract No. DE-AC02–05CH11231 using NERSC award ASCR-ERCAP0024135. D.M. is supported by the DOE Office of Science, Biological and Environmental Research program.

References

Bourenkov, G. P., & Popov, A. N. (2006). A quantitative approach to data-collection strategies [Review]. *Acta Crystallographica. Section D, Biological Crystallography, 62*(1), 58–64. https://doi.org/10.1107/S0907444905033998

Bourgeois, D. (1999). New processing tools for weak and/or spatially overlapped macromolecular diffraction patterns. *Acta Crystallographica. Section D, Biological Crystallography, 55*(10), 1733–1741. https://doi.org/10.1107/s0907444999008355

Brehm, W., White, T., & Chapman, H. N. (2023). Crystal diffraction prediction and partiality estimation using Gaussian basis functions. *Acta Crystallographica Section A: Foundations and Advances, 79*(2), 145–162. https://doi.org/10.1107/S2053273323000682

Caspar, D. L. D., Clarage, J., Salunke, D. M., & Clarage, M. (1988). Liquid-like movements in crystalline insulin. *Nature, 332*, 659–662. https://doi.org/10.1038/332659a0

Chapman, H. N., Fromme, P., Barty, A., White, T. A., Kirian, R. A., Aquila, A., ... Spence, J. C. (2011). Femtosecond X-ray protein nanocrystallography. *Nature, 470*(7332), 73–77. https://doi.org/10.1038/nature09750

Darwin, C. G. (1914). XXXIV. The theory of X-ray reflexion. *The London, Edinburgh, and Dublin Philosophical Magazine and Journal of Science, 27*(158), 315–333. https://doi.org/10.1080/14786440208635093

Diederichs, K. (2009). Simulation of X-ray frames from macromolecular crystals using a ray-tracing approach. *Acta Crystallographica. Section D, Biological Crystallography, 65*(6), 535–542. https://doi.org/10.1107/S0907444909010282

Ginn, H. M., Brewster, A. S., Hattne, J., Evans, G., Wagner, A., Grimes, J. M., ... Stuart, D. I. (2015). A revised partiality model and post-refinement algorithm for X-ray free-electron laser data. *Acta Crystallographica. Section D, Biological Crystallography, 71*(6), 1400–1410. https://doi.org/10.1107/s1399004715006902

Holton, J. M., Classen, S., Frankel, K. A., & Tainer, J. A. (2014). The R-factor gap in macromolecular crystallography: An untapped potential for insights on accurate structures. *The FEBS Journal, 281*(18), 4046–4060. https://doi.org/10.1111/febs.12922

Holton, J. M., & Frankel, K. A. (2010). The minimum crystal size needed for a complete diffraction data set. *Acta Crystallographica. Section D, Biological Crystallography, 66*(4), 393–408. https://doi.org/10.1107/S0907444910007262

Holton, J. M., Nielsen, C., & Frankel, K. A. (2012). The point-spread function of fiber-coupled area detectors. *Journal of Synchrotron Radiation, 19*(6), 1006–1011. https://doi.org/10.1107/S0909049512035571

Kabsch, W. (2010a). Integration, scaling, space-group assignment and post-refinement. *Acta Crystallographica. Section D, Biological Crystallography, 66*(2), 133–144. https://doi.org/10.1107/S0907444909047374

Kabsch, W. (2010b). XDS. *Acta Crystallographica. Section D, Biological Crystallography, 66*(2), 125–132. https://doi.org/10.1107/S0907444909047337

Kabsch, W. (2014). Processing of X-ray snapshots from crystals in random orientations. *Acta Crystallographica. Section D, Biological Crystallography, 70*(8), 2204–2216. https://doi.org/10.1107/S1399004714013534

Kahn, R., Fourme, R., Gadet, A., Janin, J., Dumas, C., & André, D. (1982). Macromolecular crystallography with synchrotron radiation: Photographic data collection and polarization correction. *Journal of Applied Crystallography, 15*(3), 330–337. https://doi.org/10.1107/s0021889882012060

Keedy, D. A., Kenner, L. R., Warkentin, M., Woldeyes, R. A., Hopkins, J. B., Thompson, M. C., ... Fraser, J. S. (2015). Mapping the conformational landscape of a dynamic enzyme by multitemperature and XFEL crystallography. *Elife, 4*, e07574. https://doi.org/10.7554/eLife.07574

Koerner, L. J., Philipp, H. T., Hromalik, M. S., Tate, M. W., & Gruner, S. M. (2009). X-ray tests of a pixel array detector for coherent x-ray imaging at the Linac Coherent Light Source. *Journal of Instrumentation, 4*(03), P03001. https://doi.org/10.1088/1748-0221/4/03/p03001

Kroon-Batenburg, L. M., Schreurs, A. M., Ravelli, R. B., & Gros, P. (2015). Accounting for partiality in serial crystallography using ray-tracing principles. *Acta Crystallographica. Section D, Biological Crystallography, 71*(9), 1799–1811. https://doi.org/10.1107/S1399004715011803

Leslie, A. G. W. (2006). The integration of macromolecular diffraction data. *Acta Crystallographica. Section D, Biological Crystallography, 62*(1), 48–57. https://doi.org/10.1107/S0907444905039107

McPhillips, T. M., McPhillips, S. E., Chiu, H. J., Cohen, A. E., Deacon, A. M., Ellis, P. J., ... Kuhn, P. (2002). Blu-Ice and the distributed control system: Software for data acquisition and instrument control at macromolecular crystallography beamlines. *Journal of Synchrotron Radiation, 9*(6), 401–406. https://doi.org/10.1107/s0909049502015170

Meisburger, S. P., Case, D. A., & Ando, N. (2020). Diffuse X-ray scattering from correlated motions in a protein crystal. *Nature Communications, 11*(1), 1271. https://doi.org/10.1038/s41467-020-14933-6

Meisburger, S. P., Thomas, W. C., Watkins, M. B., & Ando, N. (2017). X-ray scattering studies of protein structural dynamics. *Chemical Reviews, 117*(12), 7615–7672. https://doi.org/10.1021/acs.chemrev.6b00790

Mendez, D., Bolotovsky, R., Bhowmick, A., Brewster, A. S., Kern, J., Yano, J., ... Sauter, N. K. (2020). Beyond integration: Modeling every pixel to obtain better structure factors from stills. *IUCrJ, 7*(6), 1151–1167. https://doi.org/10.1107/S2052252520013007

Nave, C. (2014). Matching X-ray beam and detector properties to protein crystals of different perfection. *Journal of Synchrotron Radiation, 21*(3), 537–546. https://doi.org/10.1107/s1600577514003609

Nave, C., Sutton, G., Evans, G., Owen, R., Rau, C., Robinson, I., & Stuart, D. I. (2016). Imperfection and radiation damage in protein crystals studied with coherent radiation. *Journal of Synchrotron Radiation, 23*(1), 228–237. https://doi.org/10.1107/S1600577515019700

Peck, A., Poitevin, F., & Lane, T. J. (2018). Intermolecular correlations are necessary to explain diffuse scattering from protein crystals. *IUCrJ, 5*(2), 211–222. https://doi.org/10.1107/s2052252518001124

Pflugrath, J. W. (1999). The finer things in X-ray diffraction data collection. *Acta Crystallographica. Section D, Biological Crystallography, 55*(10), 1718–1725. https://doi.org/10.1107/s090744499900935x

Philipp, H. T., Koerner, L. J., Hromalik, M. S., Tate, M. W., & Gruner, S. M. (2010). Femtosecond radiation experiment detector for X-ray free-electron laser (XFEL) coherent X-ray imaging. *IEEE Transactions on Nuclear Science, 57*(6), 3795–3799. https://doi.org/10.1109/TNS.2010.2085445

Sauter, N. K. (2015). XFEL diffraction: Developing processing methods to optimize data quality. *Journal of Synchrotron Radiation, 22*(2), 239–248. https://doi.org/10.1107/S1600577514028203

Sauter, N. K., & Adams, P. D. (2017). Overcoming data processing challenges for breakthrough crystallography. In U. Bergmann, V. K. Yachandra, & J. Yano (Eds.). *X-Ray free electron lasers: Applications in materials, chemistry and biology* (pp. 70–87). Royal Society of Chemistry. ⟨https://doi.org/10.1039/9781782624097⟩.

Sauter, N. K., Hattne, J., Brewster, A. S., Echols, N., Zwart, P. H., & Adams, P. D. (2014). Improved crystal orientation and physical properties from single-shot XFEL stills. *Acta Crystallographica. Section D, Biological Crystallography, 70*(12), 3299–3309. https://doi.org/10.1107/s1399004714024134

Sauter, N. K., Kern, J., Yano, J., & Holton, J. M. (2020). Towards the spatial resolution of metalloprotein charge states by detailed modeling of XFEL crystallographic diffraction. *Acta Crystallographica. Section D, Structural Biology, 76*(2), 176–192. https://doi.org/10.1107/S2059798320000418

Schreurs, A. M. M., Xian, X., & Kroon-Batenburg, L. M. J. (2009). EVAL15: A diffraction data integration method based on *ab initio* predicted profiles. *Journal of Applied Crystallography, 43*(1), 70–82. https://doi.org/10.1107/s0021889809043234

Uervirojnangkoorn, M., Zeldin, O. B., Lyubimov, A. Y., Hattne, J., Brewster, A. S., Sauter, N. K., ... Weis, W. I. (2015). Enabling X-ray free electron laser crystallography for challenging biological systems from a limited number of crystals. *Elife, 4*, e05421. https://doi.org/10.7554/eLife.05421

van den Bedem, H., Bhabha, G., Yang, K., Wright, P. E., & Fraser, J. S. (2013). Automated identification of functional dynamic contact networks from X-ray crystallography. *Nature Methods, 10*(9), 896–902. https://doi.org/10.1038/nmeth.2592

Wall, M. E., Clarage, J. B., & Phillips, G. N. J. (1997). Motions of calmodulin characterized using both Bragg and diffuse X-ray scattering. *Structure (London, England: 1993), 5*(12), 1599–1612. https://doi.org/10.1016/s0969-2126(97)00308-0

Wall, M. E., Wolff, A. M., & Fraser, J. S. (2018). Bringing diffuse X-ray scattering into focus. *Current Opinion in Structural Biology, 50*, 109–116. https://doi.org/10.1016/j.sbi.2018.01.009

White, T. A. (2014). Post-refinement method for snapshot serial crystallography. *Philosophical Transactions of the Royal Society of London. Series B, Biological Sciences, 369*, 20130330. https://doi.org/10.1098/rstb.2013.0330

Winter, G., Waterman, D. G., Parkhurst, J. M., Brewster, A. S., Gildea, R. J., Gerstel, M., ... Evans, G. (2018). DIALS: Implementation and evaluation of a new integration package. *Acta Crystallographica. Section D, Structural Biology, 74*(2), 85–97. https://doi.org/10.1107/S2059798317017235

Woldeyes, R. A., Sivak, D. A., & Fraser, J. S. (2014). *E pluribus unum*, no more: From one crystal, many conformations [Review]. *Current Opinion in Structural Biology, 28C*, 56–62. https://doi.org/10.1016/j.sbi.2014.07.005

CHAPTER EIGHT

Refinement of multiconformer ensemble models from multi-temperature X-ray diffraction data

Siyuan Du[a,b], Stephanie A. Wankowicz[c], Filip Yabukarski[a,d], Tzanko Doukov[e], Daniel Herschlag[a,f,g], and James S. Fraser[c,h,*]
[a]Department of Biochemistry, Stanford University, Stanford, CA, United States
[b]Department of Chemistry, Stanford University, Stanford, CA, United States
[c]Department of Bioengineering and Therapeutic Sciences, University of California, San Francisco, CA, United States
[d]Bristol-Myers Squibb, San Diego, CA, United States
[e]Stanford Synchrotron Radiation Lightsource, SLAC National Accelerator Laboratory, Menlo Park, CA, United States
[f]Department of Chemical Engineering, Stanford University, Stanford, CA, United States
[g]Stanford ChEM-H, Stanford University, Stanford, CA, United States
[h]Quantitative Biosciences Institute, University of California, San Francisco, CA, United States
*Corresponding author. e-mail address: jfraser@fraserlab.com

Contents

Methods in Enzymology, Volume 688
ISSN 0076-6879, https://doi.org/10.1016/bs.mie.2023.06.009

Abstract

Conformational ensembles underlie all protein functions. Thus, acquiring atomic-level ensemble models that accurately represent conformational heterogeneity is vital to deepen our understanding of how proteins work. Modeling ensemble information from X-ray diffraction data has been challenging, as traditional cryo-crystallography restricts conformational variability while minimizing radiation damage. Recent advances have enabled the collection of high quality diffraction data at ambient temperatures, revealing innate conformational heterogeneity and temperature-driven changes. Here, we used diffraction datasets for Proteinase K collected at temperatures ranging from 313 to 363 K to provide a tutorial for the refinement of multiconformer ensemble models. Integrating automated sampling and refinement tools with manual adjustments, we obtained multiconformer models that describe alternative backbone and sidechain conformations, their relative occupancies, and interconnections between conformers. Our models revealed extensive and diverse conformational changes across temperature, including increased bound peptide ligand occupancies, different Ca^{2+} binding site configurations and altered rotameric distributions. These insights emphasize the value and need for multiconformer model refinement to extract ensemble information from diffraction data and to understand ensemble–function relationships.

1. Introduction

All molecular processes are defined by energy landscapes, which are in turn manifested by an ensemble of interconverting conformational states (Austin, Beeson, Eisenstein, Frauenfelder, & Gunsalus, 1975; Benkovic & Hammes-Schiffer, 2003; Benkovic, Hammes, & Hammes-Schiffer, 2008; Frauenfelder, Parak, & Young, 1988; Frauenfelder, Sligar, & Wolynes, 1991; Hammes, Benkovic, & Hammes-Schiffer, 2011). For example, ligand binding affinity is defined by the relative population of the bound to the unbound state(s), and enzymatic rates by the possibility of crossing to the transition state from the ground state. Therefore, understanding protein functions requires obtaining and comparing conformational ensembles in different ligand-bound states under physiologically-relevant conditions. Because conformational ensembles reveal probabilities of states and therefore their underlying energetics, they provide the possibility to relate structural features to thermodynamic quantities for molecular processes. This is a goal unattainable using single conformer structural models and an essential step towards a quantitative and predictive understanding of protein functions.

The need for conformational ensembles to decipher protein functions has long been recognized, yet experimental approaches to obtain ensemble information are limited by their resolution or by technological challenges.

For example, nuclear magnetic resonance (NMR) methods allow us to determine the degree of motion of protein groups and the rate of interconversions between sub-states but do not reveal atomic-level details of these sub-states (Ishima & Torchia, 2000; Kempf & Loria, 2003; Kleckner & Foster, 2011; Kovermann, Rogne, & Wolf-Watz, 2016; Mittermaier & Kay, 2006). Similarly, Förster resonance energy transfer (FRET) experiments are used to study protein conformational dynamics, but only reveal large conformational changes reported by the distance between two groups (the donor and the acceptor) (Mazal & Haran, 2019; Okamoto & Sako, 2017; Schuler & Eaton, 2008).

In contrast, X-ray crystallography provides atomic-level information about protein three-dimensional structures. The ability to model individual atom positions from diffraction data has allowed us to relate the shape of a protein to its function (Indiani & O'Donnell, 2006; Kato, Miyakawa, & Tanokura, 2018), identify specific residues involved in biological processes, and propose models for how they function (Robertus, Kraut, Alden, & Birktoft, 1972; Tsukada & Blow, 1985). In principle, X-ray diffraction data represent an ensemble average from multiple conformational states (DePristo, de Bakker, & Blundell, 2004; Rejto & Freer, 1996; Smith, Hendrickson, Honzatko, & Sheriff, 1986), but obtaining and modeling ensembles from X-ray data have been challenging for two practical reasons. First, the majority of structures deposited in the Protein Data Bank (PDB) are obtained under cryogenic conditions (~100 K) (Garman, 2003). While useful in reducing radiation damage, cryo-cooling alters the conformational landscape of a protein because the low temperature strongly favors low enthalpy states and quenches many degrees of freedom (Frauenfelder, Petsko, & Tsernoglou, 1979; Weik & Colletier, 2010). As shown in multiple studies, protein dynamics typically undergo a significant change (termed "glass transition") at ~180–200 K, suggesting that structural features from models obtained under this temperature range may reflect cryo-artifacts instead of physiologically-relevant protein features (Fraser et al., 2011; Halle, 2004; Keedy et al., 2014; Rasmussen, Stock, Ringe, & Petsko, 1992; Tilton, Dewan, & Petsko, 1992). Indeed, crystallographic data obtained at ambient temperatures reveal conformational states that are hidden or different from cryo structures (Fraser et al. 2009, 2011; Keedy, Kenner et al., 2015; Yabukarski et al., 2022).

Second, most of the structures deposited in the PDB are modeled as single conformers, which in many cases do not explain the full density data (Furnham, Blundell, DePristo, & Terwilliger, 2006; Gutermuth, Sieg, Stohn, & Rarey, 2023; Smith et al., 1986). Single conformer models typically use isotropic or anisotropic B-factors to represent variability of atomic positions, but these parameters can only account for harmonic deviations from the average positions, with the assumption that atoms fluctuate within a single local minimum (Kuzmanic, Pannu, & Zagrovic, 2014). Nevertheless, a large fraction of residues may be able to occupy multiple local minima of similar energies, resulting in anharmonic electron density distributions (Kuriyan, Petsko, Levy, & Karplus, 1986). More recently, modeling techniques have emerged to model anharmonic displacements from the underlying diffraction data to reveal the alternative conformations that the protein can adopt (Burnley & Gros, 2013; Burnley, Afonine, Adams, & Gros, 2012; Forneris, Burnley, & Gros, 2014; Fraser et al., 2011; Ginn, 2021; Keedy, Fraser, & van den Bedem, 2015; Riley et al., 2021; van Zundert et al., 2018). However, unlike methods to obtain single conformer models which have become standardized and widely applied, methods to efficiently search for and model alternative conformations require specialized software and techniques that are only used by a relatively small community.

To address these challenges in obtaining ensemble models via X-ray crystallography, we recently described an improved data-collection pipeline to minimize radiation damage at ambient temperatures (up to 363 K) that can be broadly implemented for different proteins and at other beamlines (Doukov, Herschlag, & Yabukarski, 2020). Here, we focus on the refinement of X-ray diffraction data obtained at ambient temperatures to generate multiconformer ensemble models of high quality and interpretability. Using diffraction datasets of Proteinase K collected at a series of temperatures above the glass-transition range (313–363 K), we provide a practical roadmap to guide multiconformer model refinement and discuss refinement choices and their advantages and limitations. In addition, in these datasets across temperature, we observed changes in the binding positions of a Ca^{2+} ion that are required for catalysis, and we describe the modeling and refinement of alternative Ca^{2+} binding configurations and coupled conformational preferences of Ca^{2+}-coordinating residues. Finally, we show the profound impact of temperature on the Proteinase K conformational ensemble revealed by our models, including changes in conformational heterogeneity (such as altered rotamer distributions) and compositional heterogeneity (such as increased peptide-bound states at higher temperatures), emphasizing the need for ambient- and multi-temperature X-ray crystallography to probe protein conformational landscapes and reveal hidden conformational features.

2. Collection of multi-temperature X-ray diffraction data

We followed the approaches described in Doukov et al. (2020) to collect diffraction datasets at and above room temperature.

2.1 Obtaining crystals for X-ray diffraction

Tritirachium album proteinase K (Sigma, catalog # P2308) was dissolved at 30 mg/mL in a 50 mM TRIS pH 7.5 (Sigma, T1699) buffer. The protein was crystallized using a hanging drop setup on a 24 well VDX plate with sealant (Hampton Research, HR3-171) and 22 mm thick siliconized circle cover slides (Hampton Research, HR3-247) by mixing 2 μL of protein solution with 2 μL 1.2 M ammonium sulfate (AS; Sigma, A4915) on the coverslip, which was placed over 1 mL 1.2 M AS in the VDX plate well. Prior to data collection, the aqueous layer around the crystals was exchanged to an inert Paratone-N oil (Hampton Research; # HR2-643). Paratone-N oil layer significantly reduces evaporation (Hope, 1990; Pflugrath, 2015; Weik et al., 2005). Oil-exchanged crystals were mounted on Dual-Thickness MicroLoops LD™ (Mitegen, SKU:M2-L18SP-200) and MicroGrippers™ loops (Mitegen, SKU:M7-L18SP-300). Excess oil was removed, and pins were manually mounted on the BL14-1 goniometer at Stanford Synchrotron Radiation Lightsource (SSRL) for data collection (Doukov et al., 2020). Additional information on the crystallization protocol can be found at https://www.moleculardimensions.com/products/ready-to-grow-crystallization-kit.

2.2 Achieving high-temperature capabilities and temperature control

An Oxford Cryosystems Cryostream 800 model N_2 cooler/heater (https://www.oxcryo.com/single-crystal-diffraction/cryostream-800) with a temperature range of 80–400 K was installed to collect high temperature data at the SSRL beamline 14-1. Because the physical properties of protein crystals deteriorate over time when exposed to high temperatures, we adapted the standard nozzle-closing crystal annealer operation to control the crystal exposure to the heated N_2 stream and minimize time at high temperature as follows. First, the outer layer of the crystal's aqueous mother liquor was exchanged with an inert oil (paratone-N) to eliminate the risk of potential irreproducible crystal dehydration by exposing the crystal to air (Doukov et al., 2020). After the N_2 gas is heated to the desired (high) temperature, the annealer paddle is placed in the "closed" position to prevent the gas flow from reaching the sample and heating it during the experimental setup [i.e., crystal mounting and centering, closing the experimental hutch, entering the experimental parameters into the *Blu-Ice* control software (McPhillips et al., 2002)]. Control kinetic measurements showed that a J thermocouple placed from room temperature (~293 K) to a 363 K N_2 stream (the highest temperature used in this work) was within 0.5% of the desired temperature in ≤10 s (not shown) and we used this equilibration time prior to data collection (see below). For data collection, the annealer paddle is moved to the "open" position via the beamline control software *Blu-Ice* and data collection is initiated after a ≤10 s temperature equilibration delay (Doukov et al., 2020).

2.3 Diffraction data collection

Proteinase K crystals with dimensions 0.3–0.4 mm on each side were used for data collection. To approach resolutions similar to cryogenic temperatures, larger crystals are required for the collection of X-ray diffraction data at and above room temperature, because higher temperature can lead to more radiation damage (Garman & Owen, 2006; Garman & Weik, 2017; Nave & Garman, 2005; Roedig et al., 2016; Southworth-Davies, Medina, Carmichael, & Garman, 2007; Warkentin & Thorne, 2010; Warkentin, Badeau, Hopkins, & Thorne, 2011). To maximize diffraction intensity while minimizing the number of absorbed photons per unit cell, the beam and crystal size are matched as closely as possible. We routinely used the highest beam size of 250 μm (horizontal) by 80 μm (vertical). At least 100 degrees of rotation data were collected as quickly as possible for each crystal to avoid dehydration and any macroscale defects in the crystal that can happen alongside microscopic radiation damage. Usually, each degree frame was collected for 0.04–0.2 s with the detector distance and energy adjusted to achieve highest resolution and high quality dataset (see Table 1, Table S1 from Doukov et al., 2020).

2.4 Data processing

Diffraction data recorded on Eiger 16M PAD detector (Casanas et al., 2016) was processed with the *XDS* package (Kabsch, 2010) and the programs *Pointless* (Evans, 2006) and *Aimless* (Evans & Murshudov, 2013), as implemented in the *autoxds* in-house processing script at SSRL (https://smb.slac.stanford.edu/facilities/software/xds/). Absorbed doses were calculated using *RADDOSE-3D* (Bury, Brooks-Bartlett, Walsh, & Garman, 2018; Zeldin, Gerstel, & Garman, 2013).

Table 1 Data collection and refinement statistics. Statistics for the highest-resolution shell are shown in parentheses.

	313 K	**333 K**	**343 K**	**353 K**	**363 K**
PDB id	8SOG	8SQV	8SPL	8SOV	8SOU
Wavelength	0.95369	1.03316	1.03316	1.03316	1.12709
Resolution range	35.37–1.13 (1.17–1.13)	35.21–1.22 (1.264–1.22)	35.43–1.21 (1.254–1.21)	32.47–1.291 (1.337–1.291)	34.23–1.542 (1.597–1.542)
Space group	P 43 21 2	P 43 21 2	P 43 21 2	P 43 21 2	P 43 21 2
Unit cell	68.40 68.40 103.7 90.00 90.00 90.00	68.07 68.07 103.3 90.00 90.00 90.00	68.40 68.40 104.1 90.00 90.00 90.00	68.36 68.36 104.0 90.00 90.00 90.00	68.46 68.46 105.0 90.00 90.00 90.00
Total reflections	783,043 (73,918)	512,064 (47,881)	500,061 (44,549)	441,430 (38,293)	74,611 (7362)
Unique reflections	92,190 (8318)	72,388 (5444)	74,345 (7072)	62,442 (5564)	37,357 (3000)
Multiplicity	8.5 (8.1)	7.1 (6.7)	6.7 (6.0)	7.1 (6.3)	2.0 (2.0)
Completeness (%)	93.84 (91.27)	94.21 (76.10)	97.43 (94.91)	98.85 (90.59)	96.51 (81.34)
Mean I/sigma(I)	11.17 (1.13)	11.29 (0.76)	12.72 (0.65)	13.98 (0.87)	6.54 (0.79)
Wilson B-factor	11.22	12.04	11.76	12.89	17.95
R_{merge}	0.096 (1.4)	0.092 (2.3)	0.079 (2.5)	0.078 (2.1)	0.047 (1.1)
R_{meas}	0.1027 (1.503)	0.09991 (2.489)	0.0859 (2.694)	0.08437 (2.289)	0.06705 (1.522)
R_{pim}	0.03461 (0.5238)	0.03705 (0.9605)	0.03182 (1.071)	0.03117 (0.8963)	0.04741 (1.076)

(continued)

Table 1 Data collection and refinement statistics. Statistics for the highest-resolution shell are shown in parentheses. (*cont'd*)

	313 K	**333 K**	**343 K**	**353 K**	**363 K**
$CC_{1/2}$	0.999 (0.519)	0.999 (0.305)	1 (0.339)	0.999 (0.333)	0.999 (0.383)
CC^*	1 (0.827)	1 (0.683)	1 (0.711)	1 (0.707)	1 (0.744)
Reflections used in refinement	86,766 (8318)	68,444 (5443)	73,822 (7072)	61,857 (5563)	36,237 (3000)
Reflections used for R-free	2000 (191)	2000 (159)	1998 (191)	1999 (180)	1999 (165)
R_{work}	0.1194 (0.2494)	0.1315 (0.2531)	0.1742 (0.3797)	0.1306 (0.3122)	0.1721 (0.3369)
R_{free}	0.1515 (0.2877)	0.1688 (0.3030)	0.2091 (0.4308)	0.1651 (0.3540)	0.2226 (0.4128)
CC_{work}	0.986 (0.768)	0.982 (0.780)	0.979 (0.608)	0.982 (0.671)	0.977 (0.675)
CC_{free}	0.985 (0.752)	0.972 (0.722)	0.973 (0.527)	0.977 (0.653)	0.973 (0.563)
Total number (N) of non-hydrogen atoms	8190	7926	7649	6441	5326
N, macromolecules	7919	7725	7486	6250	5204
N, ligands	21	26	21	21	23
N, solvent	250	175	142	170	99
Protein residues	279	279	285	285	285
RMS (bonds) (Å)	0.007	0.007	0.004	0.004	0.002

RMS (angles) (°)	0.98	1.04	0.66	0.70	0.51
Ramachandran favored (%)	96.57	95.98	95.33	96.92	97.26
Ramachandran allowed (%)	3.43	4.02	4.67	3.08	2.05
Ramachandran outliers (%)	0.00	0.00	0.00	0.00	0.68
Rotamer outliers (%)	1.42	1.33	1.26	0.60	0.73
Clashscore	2.90	3.57	2.99	2.61	1.37
Average B-factor ($Å^2$)	12.7	14.1	13.7	15.1	21.5
Average B-factor, macromolecules ($Å^2$)	11.8	13.5	13.2	14.3	20.9
Average B-factor, ligands ($Å^2$)	65.1	65.2	70.6	82.6	82.7
Average B-factor, solvent ($Å^2$)	38.2	35.2	33.0	38.1	40.2

3. Single conformer model refinement

Fig. 1 summarizes all refinement steps from the processed reflection data obtained above to the final multiconformer model. The first part of this process involves obtaining single conformer models via standard molecular replacement methods and iterative improvement of the model, which we briefly describe here.

3.1 Molecular replacement

Multiconformer modeling requires high quality data that is free of pathologies. These can be assessed using tools such as *phenix.xtriage* that can reveal the presence of twinning and translational noncrystallographic symmetry (tNCS). There are no pathologies in these high resolution Proteinase K datasets. Noting this, we proceeded to molecular replacement to obtain the initial phases (PDB: 3q5g; 100% sequence identity to wild type Proteinase K from *Parengyodontium album*). This search model was

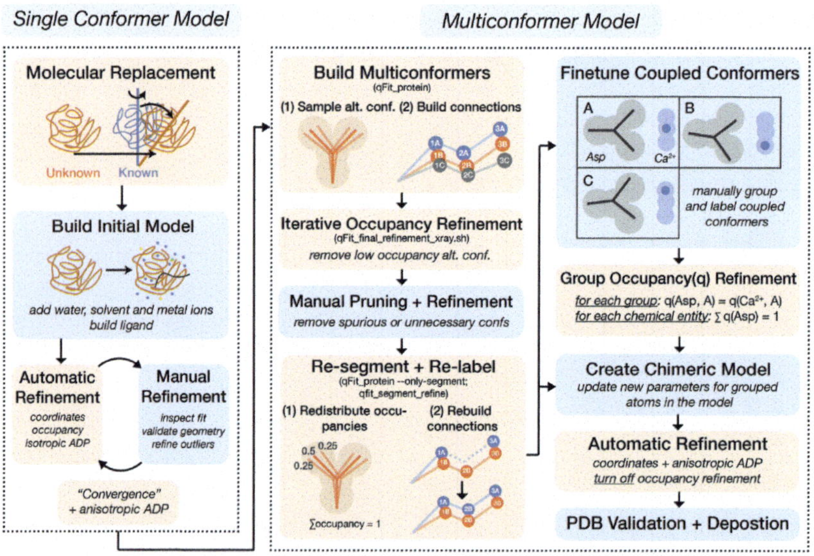

Fig. 1 Flowchart for the refinement of a multiconformer model from diffraction data. Orange boxes indicate steps that are automated by refinement softwares such as *refmac* or *phenix*, or by *qFit* features; Blue boxes indicate steps that need manual interventions (e.g. in *Coot*). Abbreviations: ADP (atomic displacement parameter), alt. conf. (alternative conformations), q (occupancy).

chosen because its crystallization was done in the same solvent as in our experiment. Molecular replacement (MR) was performed using the program *Phaser* after adding R_{free} labels to the reflection data.

3.2 Initial model building

We used the program *Coot* to examine the MR-generated model (.pdb) along with the density maps (.mtz), and manually complete an initial model. First, the C-terminal carboxylate group was added to the model (using the *Add OXT at C terminus* tool) and the N- and C-terminus were refined (using *Real Space Refine Zone* and *Regularize Zone*). Next, we checked for the presence of any cis peptide bonds, as they are highly unfavorable (unless involving a proline residue) and may indicate model errors. One proline cis peptide bond was found for Proteinase K (as was present in the molecular replacement model) and was determined to be real as the model agrees with the $2F_o - F_c$ density map. Prior to refinement and after MR, we deleted all alternative conformers to obtain a single conformer model that is needed for later steps. We then cleaned up inorganic molecules (SO_4, glycerol) from the search model that were not present in our datasets. Next, we added water molecules with using the *Find Waters* tool, which identifies cluster of unmodeled density with more than 1.4 angstroms (Å) from any other heavy atom in the model and a volume of under 4.2 $Å^3$.

Before the refinement cycles, several simple validation metrics available in *Coot* were examined, including (1) Ramachandran plot, (2) geometry analysis, and (3) rotamer analysis. Any outliers where atoms do not fit the densities well were refined using *Real Space Refine Zone* and *Regularize Zone*. Water molecules were examined using *Check/Delete Waters* where problematic water models were identified. In many cases, water molecules were too close to each other (<2.4 Å), suggesting partial occupancies. These water pairs were edited so that they are alternative conformations of the same water molecule, and their occupancies were adjusted so that their combined occupancies do not exceed 1. Lastly, inorganic and water molecules were renumbered such that residue numbers are continuous within each chain.

3.3 Iterative model refinement

To improve the model and phases calculated from the model, it is necessary to perform multiple rounds of automatic refinement followed by manual adjustments until the convergence of a final single conformer model.

In each initial round of automatic refinement (using the programs *refmac* and *phenix.refine*), 5–30 cycles of maximum likelihood refinement were performed for atomic coordinates, isotropic B-factors[a] and occupancies; for final rounds of initial refinement, given the high resolution of the data, anisotropic B-factors were refined instead of isotropic. After each round of automatic refinement was completed, we manually inspected the $F_o - F_c$ and $2F_o - F_c$ maps and the model in *Coot*. Difference ($F_o - F_c$) map peaks above 5σ were examined in addition to the validation metrics mentioned above; any regions where the1 model did not match the $2F_o - F_c$ map were adjusted. Some of these peaks appeared to result from unmodeled alternative conformations, as indicated by $F_o - F_c$ peaks resembling the shape of the sidechain, and were expected to resolve after multiconformer modeling.

For the Proteinase K datasets, 5–6 iterations were performed until "convergence". Here, we note that "convergence" is assessed remembering the adage that "refinement is never finished, but can be abandoned". While one can continue the iterative refinement cycles infinitely, further improvements of model quality and agreement with experimental data will become lower in magnitude. Practically, we need to navigate these diminishing returns to determine whether we have arrived at a "final" model. We considered three aspects: (1) whether the models gave reasonable chemical representations of molecules, judged by the presence of outliers in torsion angles and geometry; (2) qualitatively, whether the model explains the density map well, judged mainly by the presence of interpretable $F_o - F_c$ map peaks (above 4–5σ) and how well the $2F_o - F_c$ map contours around the model; and (3) quantitatively, whether the measured structure-factor amplitudes $|F_{obs}|$ match the calculated amplitudes $|F_{calc}|$ from the current model, judged by R_{work} and R_{free} values (Brünger, 1992; Rupp, 2009). In these final single conformer models, a few outliers in backbone and sidechain torsion angles persisted, but they are likely real protein features as the model matches the $2F_o - F_c$ map shape. For example, D39, a member of the catalytic triad of Proteinase K, appeared to have unfavorable backbone torsions, and this outlier is observed not only in our datasets, but also in previously published PDB models. These regions where intrinsic conformational preferences are potentially perturbed by surrounding forces may be of interest for further

[a] B-factors are also named thermal factors, temperature factors or atomic displacement parameters (ADP) and used interchangeably.

investigation when modeling is complete—as they may arise from structural constraints or represent features that are evolutionarily-selected and provide functional benefits. All final single conformer models have $R_{work} < 0.2$, indicating a reasonably high model quality (Fig. 4B). R values appear to be larger for higher temperature datasets, which is expected due to increased thermal motions that cannot be accounted for by single conformer models.

3.4 Modeling an unknown ligand appearing at high temperatures

Intriguingly, at the Proteinase K active site, some unexplained electron densities gradually appeared for datasets obtained at higher temperatures. Because Proteinase K binds peptide substrates, and the shape of these densities resembles a peptide chain, we reasoned that a short peptide may be able to bind better at higher temperatures, and the apparent increase in the peptide density may reflect a shifted equilibrium favoring the bound state (Fig. 2A). While all the datasets were derived from crystals with the same content, the compositions of bound and unbound species in ordered parts of the crystals were different, and generated different diffraction data and density maps, reflecting altered compositional heterogeneity across temperature. It is also possible that proteolysis occurred only when the crystals were exposed to higher temperatures, but less likely due to the short time period the crystal was placed at elevated temperature for data collection. It is impossible to distinguish whether shifted equilibrium or increased proteolysis is responsible for the peptide density, which we modeled by refining occupancies, as we described below.

To determine the sequence and the conformation of the unknown peptide, we used the 363 K dataset which contains the most complete densities for this peptide as a guide. A poly-alanine chain (chain B) was built based on the overall $2F_o - F_c$ density shape (using the *Add Terminal Residue* tool in *Coot*), followed by an automatic refinement round (using *phenix.refine*). Next, sidechain identities were estimated based on the shape of the $2F_o - F_c$ densities that were not explained by the poly-alanine model; the $F_o - F_c$ map further informs sidechain choices (e.g. an unmodeled valine sidechain would give a signature shape of two adjacent negative density blobs). In *Coot*, non-alanine residues were mutated (using *Mutate & Auto Fit*), with the final sequence determined to be AAASVK. In the 343 and 353 K datasets, we modeled the same peptide sequence with roughly the same conformation as modeled in the 363 K dataset while fitting to local densities which are less complete than those in the 363 K

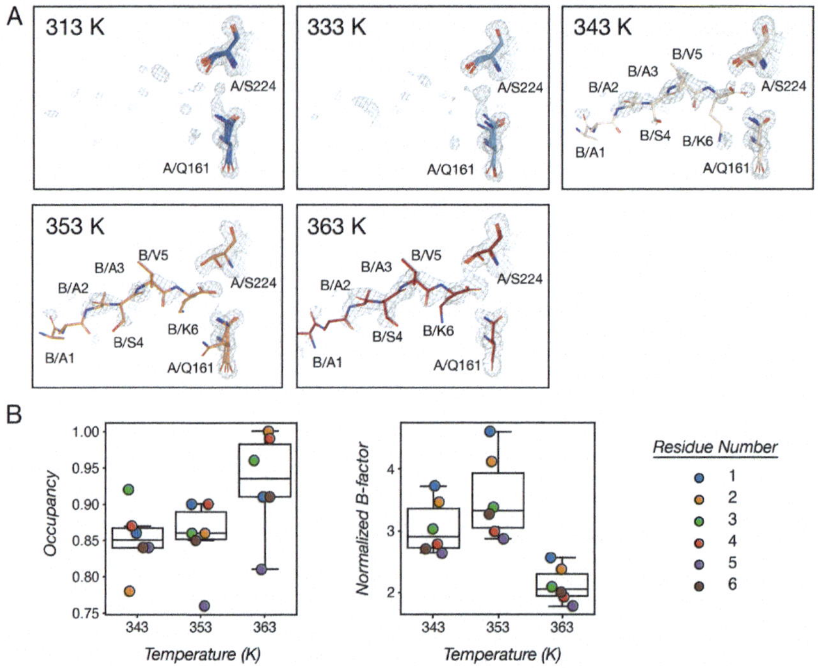

Fig. 2 An unknown peptide is bound at the Proteinase K active site at high temperatures. (A) $2F_o - F_c$ map and modeled residues for the binding site for datasets showed electron densities increasing for the bound peptide at higher temperatures. S224 from the Proteinase K (chain A) is the catalytic serine, while S224's backbone amide as well as the Q161 sidechain form the "oxyanion hole" hydrogen bond donors that interact with the carbonyl of the peptide ligand. The unknown peptides (chain B) were modeled for the 343, 353 and 363 K datasets. (B) Occupancies and normalized B-factors for Cα atoms of the unknown peptide residues.

dataset. Since the densities observed for this peptide are not complete, we set occupancies of chain B residues for all three datasets to a number below 1, which allowed the following automatic refinement step (using *phenix.refine*) to refine their partial occupancies (Fig. 2B).

Overall, the occupancies of the peptide residues continue to increase from 343 to 363 K, indicating higher bound species at higher temperatures. To compare the variability of these modeled positions across datasets, we calculated normalized B-factors by dividing the B-factors by the average B-factor of all atoms in each dataset. As expected, the peptide residues have higher-than-average B-factors due to incomplete densities (Fig. 2B). The normalized B-factors are lower for the 363 K dataset, consistent with higher ordering of the bound species (Fig. 2B). In practice, there is some degeneracy between occupancy and B-factor refinement, but the refined results here, obtained from high resolution data, are consistent with greater occupancy and higher order (decreased B-factors) as temperature increases.

4. Multiconformer model refinement

Conformational heterogeneity from diffraction data can be represented by different metrics and data formats, each with its own limitations. In single-conformer models, B-factors are typically used as a proxy for the degree of flexibility of a group, but they cannot be directly related to interpretable molecular geometries (e.g. atomic distances, bond angles and rotameric states) and involve contribution from other factors (e.g. crystallographic disorder) (Kuzmanic et al., 2014; Sun, Liu, Qu, Feng, & Reetz, 2019). Ensemble models generated using X-ray restrained molecular dynamics (MD) simulations provide 10–100 s of separate single conformer models, where the relative population of different conformers reflect their occupancies (Burnley & Gros, 2013; Burnley et al., 2012; Forneris et al., 2014; Pearce & Gros, 2021; Ploscariu, Burnley, Gros, & Pearce, 2021). However, because of the high parameter-to-observation ratio, discrete conformers modeled for areas with ambiguous electron densities can be a result of overfitting instead of real conformational heterogeneity (Burling & Brünger, 1994; Wankowicz & Fraser, 2020).

Recent attempts to represent heterogeneity also include the use of bond-based parameters (bond lengths, angles and torsion angles) instead of Cartesian coordinates; in this scheme, B-factors can be replaced by parameters describing the variation in torsion angles, which capture the physical nature of molecular motions more parsimoniously (Ginn, 2021). While promising in reducing the number of model parameters (and therefore reducing overfitting) and in improving the physical interpretability of X-ray models, a refinement method based on this scheme (*Vagabond*) is still under development and has not achieved the accuracies of traditional Cartesian-based models by conventional R_{free} metrics (Ginn, 2021). In addition, both MD-based ensemble models and bond-based models are incompatible with current software for further manual or automatic refinement. They therefore do not allow the fine-tuning of regions and structural features of interest, especially those detailing compositional heterogeneity that require more sophisticated refinement methods, which we describe below (Section 4.4).

To improve interpretability, accuracy, and compatibility while minimizing model complexity, we chose to refine each Proteinase K dataset into a multiconformer model, using the program *qFit* initially to sample and select alternative conformations (Keedy, Fraser et al., 2015; Riley et al., 2021; van den Bedem, Dhanik, Latombe, & Deacon, 2009; van Zundert et al., 2018). In these multiconformer models, each protein residue has one to five alternative conformers, as needed to explain local densities. Each conformer for a residue is assigned an "*altloc*" label (A, B, etc.), and each atom for that conformer has its coordinates, occupancies, and B-factors recorded in a separate line in the model file. Unlike MD-based ensemble refinement, the approach we took only introduces additional parameters as needed to explain the experimental data; these models would be less likely to overfit. Practically, multiconformer models describe ensemble information in a single model following the conventional PDB (or mmCIF) format; thus, they are compatible with all common structural biology tools for further structural refinement and manual adjustment (e.g. in *Coot*) (Fig. 1).

Both multiconformer models and MD-based ensemble models present visualization challenges. For example, in *PyMol* or *Chimera*, a multiconformer model is viewed in a single "state", and the alternative conformations are all visible. For visualization, coloring by *altLoc* label is helpful in interpreting coupled conformations while viewing all modeled conformations. In contrast, ensemble models contain multiple "states" with whole copies of the entire system. Scrolling through the states is helpful for visualization since viewing all models contained in the ensemble is often visually overwhelming. Further improvements in macromolecular visualization software for analyzing these complex model types will help further enable their use.

4.1 Automatic refinement using *qFit*

qFit is a Python-based software developed to automatically model and refine alternative conformers for protein residues and ligand molecules (Keedy, Fraser et al., 2015; Riley et al., 2021; van den Bedem et al., 2009; van Zundert et al., 2018). Here, we used *qFit* 3.0 to obtain initial multiconformer models for the Proteinase K datasets. To sample residue conformations, *qFit* first performs backbone sampling based on the anisotropic B-factors of the C_β atom (or O atom for Gly) which define the directionality of its potential motions, moving the atom around the ellipsoid defined by the anisotropic B-factors while adjusting adjacent atoms (within a 5-residue segment) such that the backbone linkages are closed (van den Bedem, Lotan, & Latombe, 2005). To sample sidechain conformations, *qFit* starts from the backbone conformations identified in the previous step and samples either around the C_α–C_β–C_γ bond for planar, aromatic sidechains or around the χ angles for other sidechains. At each χ angle and again once the entire sidechain is built, *qFit* evaluates the quality of sampled conformations and removes unnecessary and low occupancy (<0.09) conformers, keeping 1–5 optimal conformers for each residue whose positions, occupancies and B-factors best fit local densities.

After the optimal fitting of each individual residue, *qFit* reconnects the entire structural model taking account of conformer interconnections. Neighboring backbones residues with alternative conformations are split into segments, with each segment delimited by a residue with single-conformer backbone atoms. For each segment, *qFit* brings all residues to the same number of alternative conformations to avoid any "floating" conformers caused by missing backbones, and consistently assigns backbone occupancies and *altLoc* labels (*qFit-segment*). Next, *qFit* determines the coupling of alternative conformers within each segment using a simulated annealing algorithm, relabeling all alternative conformers so that the coupled conformers do not clash (*qFit-relabel*). Details of the *qFit* algorithm have been described previously (Keedy, Fraser et al., 2015; Riley et al., 2021; van den Bedem et al., 2009; van Zundert et al., 2018) and the open-source software is available at https://github.com/ExcitedStates/qfit-3.0.

To obtain a multiconformer model from *qFit*, we need a single conformer model (.pdb) of reasonably high quality and a composite omit map (.mtz) (Terwilliger et al., 2008). A composite omit map provides the advantage of reducing model bias. To build such a map, the asymmetric unit is segmented into contiguous regions, and for the iterative refinement of each map region, model atoms located within that region are given an occupancy of 0 so they do not bias structure factor calculations; the final "composite" map then combines all refined segments (Terwilliger et al., 2008). A composite omit map was obtained from the single conformer model and map refined in Section 3 using *phenix.composite_omit_map* with the *omit-type=refine* flag.

To sample conformers, we used the *qfit_protein* function with *-rmsd 0.1* setting, which removes redundant conformers when they have an all-atom root-mean-square deviation (RMSD) below 0.1 Å. This RMSD setting was determined by testing *qfit_protein* with the default setting (RMSD threshold = 0.01 Å) and increased thresholds of 0.1 Å and 0.2 Å.

Qualitatively, the 0.01 Å threshold produced the greatest number of redundant conformers; the 0.1 and 0.2 Å threshold produced similar models with less conformers but overall appropriate fits to the density maps. Therefore, we proceeded with the model generated from the 0.1 Å threshold. Setting an appropriate RMSD threshold that balances conformational fit and parsimony at this step reduces model parameters and helps minimize manual efforts to prune conformers in later steps.

qfit_protein produced a multiconformer model (*multiconformer_model2.pdb*) that was then refined using the *qfit_final_refine_xray.sh* script. To ensure a parsimonious model, this refinement protocol involves iterative refinement (using *phenix.refine* functionalities) of atomic positions, occupancies, and B-factors and removal of low occupancy (<0.09) conformers until no such conformers emerge. This step produces a refined multiconformer model and map (with suffixes *_qFit.pdb* and *_qFit.mtz*).

4.2 Manual pruning and refinement

Manual inspection and refinement of the model and map from Section 4.1 are required for two reasons: (1) *qFit* may produce spurious conformers fitted to densities from noise or the bulk solvent and (2) additional backbone conformations may need to be added, since the backbone sampling of *qFit* 3.0 depends on the anisotropy of C_β, which encodes backrub (Davis, Arendall, Richardson, & Richardson, 2006), crankshaft (Fadel, Jin, Montelione, & Levy, 1995; Fenwick, Orellana, Esteban-Martín, Orozco, & Salvatella, 2014), and shear (Hallen, Keedy, & Donald, 2013; Smith & Kortemme, 2008) motions, but does not report on large backbone rearrangements such as the 180° peptide flips (Keedy, Fraser et al., 2015).

In *Coot*, we inspected each residue to prune any spurious or unnecessary conformers, including those that do not fit to local densities, those that would cause strain or clashes with neighboring residues, and those that are too similar. While the criteria for similarity may be qualitative and ad hoc, we note that both sidechain and backbone atoms need to be compared to decide if a conformer needs to be pruned. For example, two conformers may have the same sidechain conformation but obviously different backbone positions. In this case, both sidechain conformers need to be kept in the model, because the current PDB format will not allow two sets of backbone atom positions linked to only one sidechain conformer (even though a single backbone conformation can spawn two sidechain conformations). The sidechains in solvent-exposed areas are more likely to show spurious conformers. In some cases, there were no $2F_o - F_c$ contours even at $<0.5\sigma$ around the spurious conformers and also no positive $F_o - F_c$ peaks, suggesting that these conformers may have been incorrectly fitted to densities resulting from noise or bulk solvent contributions. Only the conformers supported by the $2F_o - F_c$ map were kept in the model. In the meantime, we checked for any backbone conformations that needed to be rebuilt or sidechains that could be refined to fit the $2F_o - F_c$ map better, manually adjusting their positions as needed.

4.3 Automatic relabeling of structural segments

Manual pruning and refinement are essential to correct and improve the model, but may also introduce model inconsistencies that need to be resolved. First, because some conformers were deleted, the combined occupancies of the remaining conformers of a residue did not sum to one. Second, deletion of conformers resulted in breaks in peptide linkages. To redistribute occupancies and reconnect the peptide, we re-ran *qfit_protein* with the flag *–only-segment*. With this option, *qFit* does not re-sample and score residue conformers, but re-distributes the occupancies of the remaining conformations and performs the segmentation and labeling step as described in 4.1 (*qFit-segment* and *qFit-relabel*). This step is followed by another automatic refinement cycle using *qFit_segment_refine*.

This procedure generates connected backbones with consistent occupancies for coupled neighboring conformers, but at the cost of increased number of parameters, since it requires bringing in duplicate conformers. For example, if residue N has four alternative backbone conformations (A, B, C, D) and residue $N+1$ has two alternative conformations (A, B), this procedure will create C and D conformers for residue $N+1$ by duplicating its A and B conformers. This duplication may continue until we reach the end of a segment, so that all backbones have the same number of alternative conformations (A, B, C, D) and are therefore properly connected. The alternative to the duplication of conformers is to have "floating" backbone atoms, e.g. with residue N conformers C and D having no connection from the backbone carbonyl to the next alpha carbon. Ideally, we would like to have a nested model format where the C and D conformations can be "children" of the A and B conformations, but currently, neither the PDB nor CIF format allow for that representation (Hancock et al., 2022; Pearce, Krojer, & von Delft, 2017; Vallat et al., 2023).

4.4 Modeling coupled conformational preferences

To fine-tune the model for regions of interest where conformational preferences of multiple groups may be coupled to one another, we used a constrained group occupancy refinement approach, which we illustrate below using the example of the Proteinase K Ca^{2+} binding site. This binding site was identified in previous structural studies (Betzel, Pal, & Saenger, 1988), where a Ca^{2+} ion is coordinated by the sidechain of D200, the backbone O atoms of V177 and P175, and surrounding water molecules (Fig. 3A). Nevertheless, our diffraction data revealed a more complex picture for Ca^{2+} interactions in this binding site. The $2F_o - F_c$ map does not clearly indicate one unique position for the Ca^{2+}. Instead, for datasets obtained at 313–353 K, there are four spherical densities within this binding site. Two of these spheres are very close together, with their merged densities forming a dumbbell shape (Fig. 3C). Our density map and initial multiconformer model suggested that the Ca^{2+} can occupy these alternative positions in the binding site, for three reasons. First, the commonly modeled position where Ca^{2+} forms a bivalent interaction with the D200 sidechain lies within the overlapping dumbbell-shaped density. This overlapping density suggests alternative conformations of the same molecule instead of an additional coordinating water, as the interaction distance would be too close (<2 Å) and highly unfavorable (Fig. 3C). Second, the alternative conformers modeled for nearby residues such as D200 and V177 include those that orient towards positions other than the commonly modeled one, suggesting that these residues can stabilize Ca^{2+} when it occupies these other positions (Fig. 3C). Lastly, in the 363 K dataset, densities for the commonly modeled position disappeared and the dumbbell-shaped density shrunk to an elliptical shape, suggesting that alternative conformations are favored at high temperatures (Fig. 3C, D). To unambiguously determine possible positions of Ca^{2+}, future experiments can collect diffraction data at longer wavelengths to detect Ca^{2+} anomalous signals; here, we considered all possible alternative configurations as suggested by the electron density maps.

To model alternative positions of Ca^{2+} and to see how the interacting residues move accordingly, we manually set up alternative conformers of Ca^{2+} and its surrounding protein residues and water molecules as "groups" in *Coot* by creating multiple copies of the same atoms and assigning the same *altLoc* label to the atoms in the same configuration (Fig. 3B). Then, we used *phenix.refine* to perform automatic refinement with group

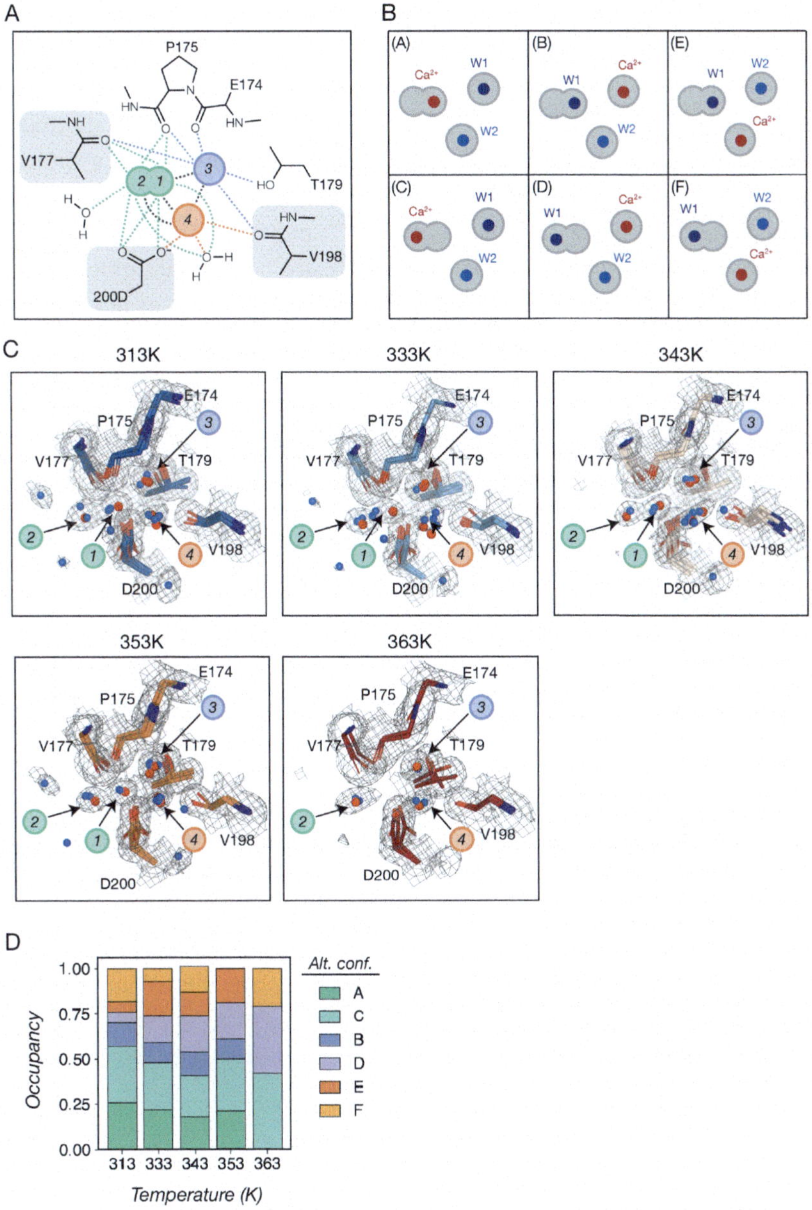

Fig. 3 Modeling and refinement of the Proteinase K Ca^{2+} binding site. (A) Four possible positions for Ca^{2+} as suggested by the $2F_o - F_c$ maps; each position appears to be stabilized by 5–7 metal-coordinating interactions with surrounding water molecules or protein residues. Position 1 corresponds to the Ca^{2+} position that is typically modeled. Protein residues that showed alternative conformers orienting towards different Ca^{2+} positions are indicated by gray boxes; these residues were included in

(Continued)

occupancy constraints that will produce consistent occupancies for chemical entities within a group.

To enumerate all possible Ca^{2+} and binding site residue configurations, we considered each of the four spherical densities as potential alternative positions for Ca^{2+}, and in each case assigned the other density blobs as water molecules (Fig. 3B). Because only one atom can occupy the dumbbell region at a time, we modeled one Ca^{2+} and two water molecules for each alternative conformation. In total, there are 6 different configurations for the Ca^{2+} and the two coordinating waters as a group, as illustrated in Fig. 3B; we therefore created alternative conformations A through F for these molecules accordingly. Next, we identified protein residues that showed correlated conformational preferences with these different Ca^{2+} positions, which include D200, V177, and V198 (Fig. 3A, B). We also modeled 6 alternative conformers (A through F) for each of these residues, and their alternative conformer labels were reassigned so that each conformer was in the correct group. For example, the conformer of D200 that is the closest to the A conformer of Ca^{2+} was labeled "A", et cetera. For the 363 K dataset, the dumbbell-shaped density observed for other datasets diminished into an ellipse with no clear indication for two separate configurations (Fig. 3C); therefore, duplicate configurations (A, B and E) were removed.

To model the positions, B-factors, and occupancies of these chemical entities as a group, we included group occupancy refinement strategies in our next cycle of *phenix.refine*, assigning each alternative configuration as a constrained group (e.g. group A was the A conformers of Ca^{2+}, waters, D200, V177, and V198) (see *Complex occupancy refinement strategy* in the *phenix* documentation: https://phenix-online.org/documentation/reference/refinement.html). Using this approach, all atoms in a group are refined to the same occupancy and each chemical entity will have a total occupancy of 1 summed over all its alternative conformers. The positions and B-factors were also allowed to further refine.

Nevertheless, occupancy refinement would be performed for the entire model, and some alternative conformers of protein residues outside the

Fig. 3—Cont'd the group occupancy refinement. (B) All possible configurations (conformers A through F) for the Ca^{2+} and water molecules. (C) $2F_o - F_c$ map and models for the Ca^{2+} binding site, including all alternative conformations for each model. Water molecules are shown as blue spheres, and Ca^{2+} as red spheres. (D) Changes in the refined occupancies of each alternative conformation across temperature.

Ca^{2+} binding site may drop below 0.09 again. Therefore, we created chimeric models that merged the pre-grouped model (from Section 4.3) with the updated positions, B-factors, and occupancies for grouped atoms in the grouped and refined model obtained here. Additional refinement runs for this chimeric model were then performed with fixed occupancies, allowing only the atomic positions and B-factors to fluctuate.

Increasing temperatures favor alternative Ca^{2+} binding configurations, as suggested by the refined group occupancies: at lower temperatures, Ca^{2+} mainly occupies the dumbbell region (configurations A and C); as temperature increases, the occupancy for Ca^{2+} at the more distal position (configurations B and D) increases (Fig. 3D). In addition, the elliptical instead of dumbbell shape for the center density at 363 K suggests less distinction and therefore higher mobility for occupying the A and C sites (Fig. 3C).

Overall, the final multiconformer models showed decreased R factors across all datasets (Fig. 4), indicating improved fit of the models to the underlying data after multiconformer refinement; in particular, the decrease in the cross-validation term R_{free} suggests that the improved accuracy does not arise from overfitting (Fig. 4A).

5. Identifying temperature-dependent conformational changes

Multiconformer models provide rich information for protein conformational ensembles, but it can be difficult to extract conformational changes that are significant, and relevant to functional aspects of interest. This difficulty arises from the fact that each residue may have a different number of alternative conformers modeled for different datasets, and each

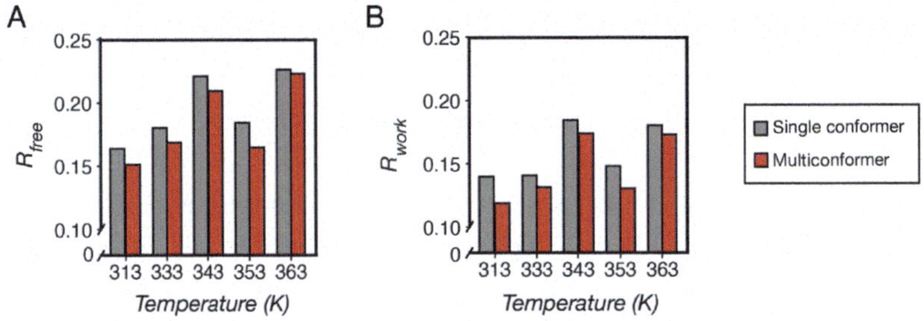

Fig. 4 R_{free} (A) and R_{work} (B) for the final single conformer versus multiconformer models indicate improved model accuracy after multiconformer refinement.

alternative conformer has its own modeled position, occupancy, and B-factors, preventing a matched statistical comparison across datasets. Here, we used the program *Ringer* (Lang et al., 2010) to guide our search for interesting conformational changes, and identified widespread changes of the proteinase K ensemble in response to temperature. The approach that we describe here can also be extended to study other structural perturbations, such as ligand binding and mutations.

5.1 Ringer analysis

The *Ringer* program systematically samples electron densities around sidechain rotamers, allowing for the detection of low-occupancy sidechain conformational states and the comparison of sidechain states across datasets at different temperatures. *Ringer* analysis complements multiconformer models, as it provides torsional electron density profiles for all sidechains at 5 or 10° intervals that can be systematically compared across datasets. Nevertheless, *Ringer* can only sample electron densities around sidechains based on backbone positions from a single conformer model; thus, the resulting profile reflects a mixture of sidechain and backbone motions. For example, a broad *Ringer* peak may result from a highly flexible sidechain attached to constrained backbone atoms, or the opposite, or a moderate level of flexibility from both. Therefore, to distinguish between these possibilities, one must return to the multiconformer model and electron density maps.

Ringer can be accessed via *phenix* using the *mmtbx.ringer* command with a model and a single conformer model supplied (*mmtbx.ringer model.pdb map.mtz*). We used the single conformer model from Section 3.4 and the final map after multiconformer refinement from Section 4.4, since the final map provides more accurate electron densities. This command produces a table of electron densities for each residue-rotamer in the model from 0° to 359° at specified intervals (default 5°).

The raw *Ringer* profiles are helpful for the interpretation of weak densities and further refinement of the multiconformer model. Crudely, any rotamer angles at $\geq 0.3\sigma$ are likely to be conformational features rather than noise from hydrogens (Lang et al., 2010). One may return to the multiconformer model to refine particular areas as informed by *Ringer*.

For the systematic comparison of rotamers dynamics across datasets, we need to normalize σ values (Eq. (1)), as the scale of electron density values can vary across datasets and obscure changes of σ.

$$\text{Normalized } \sigma = (\sigma - \sigma_{\min})/(\sigma_{\max} - \sigma_{\min}) \quad (1)$$

The normalized *Ringer* profiles revealed diverse patterns of sidechain conformational changes across temperatures. In the simplest scenario, we would expect high temperatures to favor higher-entropy states, and the distributions of sidechain rotamer angles are expected to broaden. One example of this pattern is the $\chi 1$ of Glu43, as indicated by the broader shoulders of the 363 K *Ringer* profile and its more dispersed $2F_o - F_c$ densities around the sidechain (Fig. 5A). In the second case, we observed the emergence of an alternative sidechain rotamer at higher temperatures, such as for the $\chi 1$ of the catalytic residue Ser224 (Fig. 5A). Unexpectedly, we also observed the disappearance of rotamer states at high temperatures, such as for the $\chi 1$ of Ser63, emphasizing the idiosyncrasy of temperature effects on individual rotamers, residues and regions, rather than a universally higher flexibility (Fig. 5A). Lastly, there are also highly positioned residues such as Asn163 whose $\chi 1$ profiles do not change across temperature (Fig. 5A).

To further quantify the similarities and differences between *Ringer* profiles, we calculated Pearson correlation coefficients (P_{CC}, or Pearson's *r*) for the *Ringer* profiles of each rotamer obtained from different datasets, using the *scipy.stats.pearsonr* function of the *SciPy* package (Virtanen et al., 2020). P_{CC} values lie between −1 and 1, and a higher positive P_{CC} value indicates a stronger positive correlation between the two *Ringer* profiles, and therefore more similar rotameric distributions. Across sidechain rotamers of the entire Proteinase K structure, P_{CC} values decrease when comparing the 313 K model to higher temperature models, and are especially low for the 363 K dataset (Fig. 5B). For the comparison of 313 K versus 363 K dataset, we identified 83 rotamers with $P_{CC} \leq 0.9$ among a total of 410 rotamers, suggesting widespread conformational changes in response to higher temperatures (Fig. 5C). As all datasets here were collected above the glass transition, these changes are mostly subtle, and we would expect more significant changes for comparisons of datasets below and above glass transition (Fraser et al., 2011; Halle, 2004; Keedy et al., 2014; Rasmussen et al., 1992; Tilton et al., 1992).

6. Summary and conclusions

Conformational ensembles, rather than static structures, are needed to deepen our understanding of protein functions and ultimately reach to

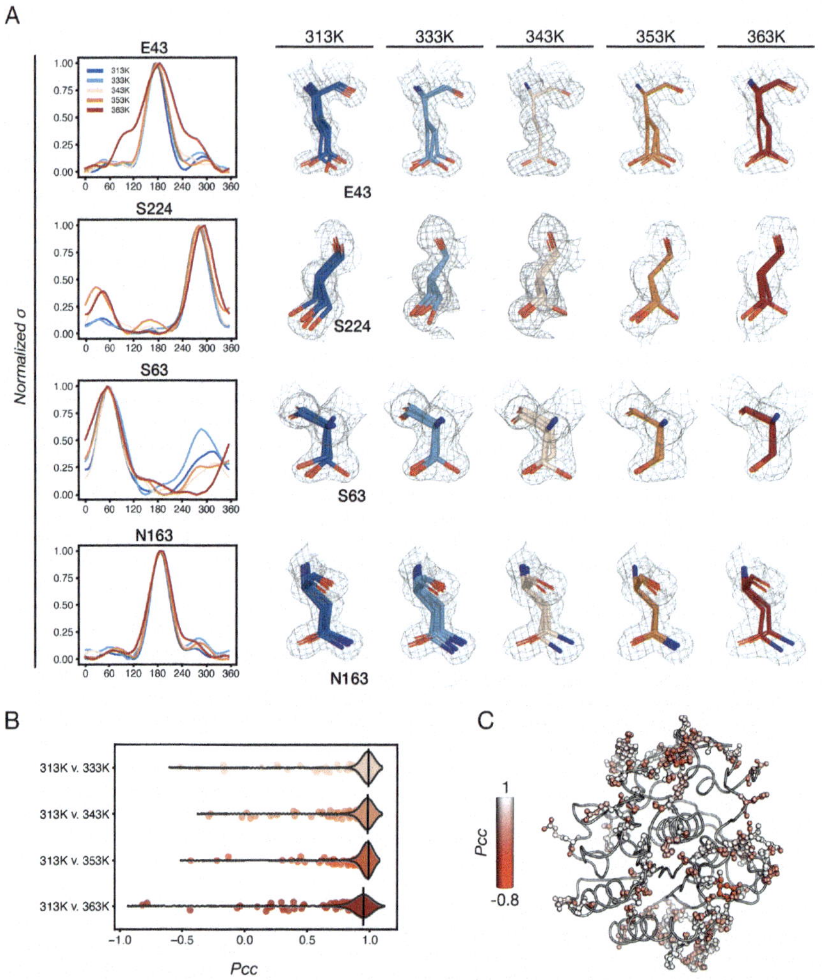

Fig. 5 Temperature-dependent rotamer changes in Proteinase K. (A) Examples for how rotamers change across temperatures and their *Ringer* profiles. $2F_o - F_c$ maps are contoured at 1σ. (B) Distributions of P_{CC} for comparisons of all rotamers across datasets. (C) Comparison of the 313 K v. 363 K *Ringer* profiles showed low P_{CC} values for sidechain rotamers throughout the Proteinase K structure. Only rotamers with $P_{CC} < 0.9$ shown.

the ability to derive quantitative, predictive models for protein functions (Austin et al., 1975; Benkovic & Hammes-Schiffer, 2003; Benkovic et al., 2008; Frauenfelder et al. 1988, 1991; Hammes et al., 2011; Mokhtari, Appel, Fordyce, & Herschlag, 2021). Nevertheless, X-ray derived ensemble data is limited due to experimental challenges (which we addressed in Doukov et al., 2020) and the requirement for specialized refinement approaches, which are not easily accessible. Here, we used a series of Proteinase K datasets collected at increasing temperatures to provide a practical and detailed tutorial for the refinement of multiconformer models and correlated conformational preferences within these models, and we discussed the rationale behind our refinement choices and their advantages and limitations.

We note that many of our refinement choices are limited by the PDB format and interpretations by refinement softwares. In particular, multiconformer models need to account for alternative conformations for each individual residue as well as the connections between conformers across the protein backbone, and this multidimensional information cannot be cleanly represented by the "flat" PDB format without duplicated model parameters. The mmCIF format could potentially represent multiconformer connectivities and interrelationships because of its more flexible formatting. Such future efforts will need to evolve with projects that have high compositional [e.g. fragment screening (Krojer, Fraser, & von Delft, 2020; Weiss et al., 2022)] and conformational [e.g. time-resolved serial femtosecond crystallography (Oda et al., 2021; Schmidt, 2021)] heterogeneity. Meeting these challenges will also help build molecular models compatible with increasingly complex 3D classification and heterogeneous map reconstruction methods in cryo-EM (Zhong, Bepler, Berger, & Davis, 2021).

In addition, we encountered issues during refinement and PDB deposition because many widely-used tools (e.g. *MolProbity* and *Reduce*) are not optimized for multiconformer models. We suggest that future efforts in improving the PDB/mmCIF format and structural biology tools to accommodate ensemble features will simplify the process of obtaining ensemble models and allow the database of conformational ensembles to grow.

Proteinase K appears to undergo widespread conformational changes across temperature. These observed changes are potentially linked to its stability, binding, and catalysis, such as the increased occupancies of the bound peptide ligand, changes in Ca^{2+} binding configurations, and altered distributions of rotameric angles for catalytic residues. While *qFit* automates the sampling of alternative conformations and provides a preliminary model, we emphasize that additional fine-tuning is needed to improve the accuracy of the model and to extract interesting local changes. For example, we showed that the Ca^{2+} binding site can be modeled by 6 different alternative configurations, and determined how the occupancies of each change across temperatures. This strategy may be extended to model other coupled conformations of interest, e.g. to determine if the motions of the active site groups are constrained or facilitated by surrounding residues, or if the binding of an allosteric ligand shifts the equilibrium of conformational states of a network of residues that move together. We expect that these modeled changes will lead to hypotheses that can be tested by additional experiments—for example, by introducing structural perturbations that change the magnitude or direction of these motions or disrupt their couplings.

Acknowledgments

This work was funded by a National Institutes of Health (NIH) grant (GM145238) to J. S. F., a National Science Foundation (NSF) grant (MCB-1714723) to D.H, and a long-term Human Frontiers Science Program postdoctoral fellowship to F.Y. Use of the Stanford Synchrotron Radiation Lightsource (SSRL), SLAC National Accelerator Laboratory is supported by the US Department of Energy, Office of Science and Office of Basic Energy Sciences under Contract No. DE-AC02-76SF00515. The SSRL Structural Molecular Biology Program is supported by the DOE Office of Biological and Environmental Research and by the National Institutes of Health (NIH), National Institute of General Medical Sciences (NIGMS, P41GM103393). The contents of this publication are solely the responsibility of the authors and do not necessarily represent the official views of NIH or NIGMS. We thank Lisa Dunn (SSRL) for help with scheduling experimental beam time.

References

Austin, R. H., Beeson, K. W., Eisenstein, L., Frauenfelder, H., & Gunsalus, I. C. (1975). Dynamics of ligand binding to myoglobin. *Biochemistry, 14*, 5355–5373.

Benkovic, S. J., Hammes, G. G., & Hammes-Schiffer, S. (2008). Free-energy landscape of enzyme catalysis. *Biochemistry, 47*, 3317–3321.

Benkovic, S. J., & Hammes-Schiffer, S. (2003). A perspective on enzyme catalysis. *Science (New York, N. Y.), 301*, 1196–1202.

Betzel, C., Pal, G. P., & Saenger, W. (1988). Three-dimensional structure of proteinase K at 0.15-nm resolution. *European Journal of Biochemistry/FEBS, 178*, 155–171.

Brünger, A. T. (1992). Free R value: A novel statistical quantity for assessing the accuracy of crystal structures. *Nature, 355*, 472–475.

Burling, F. T., & Brünger, A. T. (1994). Thermal motion and conformational disorder in protein crystal structures: Comparison of multi-conformer and time-averaging models. *Israel Journal of Chemistry, 34*, 165–175.

Burnley, B. T., Afonine, P. V., Adams, P. D., & Gros, P. (2012). Modelling dynamics in protein crystal structures by ensemble refinement. *Elife, 1*, e00311.

Burnley, B. T., & Gros, P. (2013). phenix.ensemble_refinement: a test study of apo and holo BACE1. *Computational crystallography newsletter, 4*, 51–58.

Bury, C. S., Brooks-Bartlett, J. C., Walsh, S. P., & Garman, E. F. (2018). Estimate your dose: RADDOSE-3D. *Protein Science: A Publication of the Protein Society, 27*, 217–228.

Casanas, A., Warshamanage, R., Finke, A. D., Panepucci, E., Olieric, V., Nöll, A., ... Wang, M. (2016). EIGER detector: Application in macromolecular crystallography. *Acta Crystallographica. Section D: Structural Biology, 72*, 1036–1048.

Davis, I. W., Arendall, W. B., 3rd, Richardson, D. C., & Richardson, J. S. (2006). The backrub motion: How protein backbone shrugs when a sidechain dances. *Structure (London, England: 1993), 14*, 265–274.
DePristo, M. A., de Bakker, P. I. W., & Blundell, T. L. (2004). Heterogeneity and inaccuracy in protein structures solved by X-ray crystallography. *Structure (London, England: 1993), 12*, 831–838.
Doukov, T., Herschlag, D., & Yabukarski, F. (2020). Instrumentation and experimental procedures for robust collection of X-ray diffraction data from protein crystals across physiological temperatures. *Journal of Applied Crystallography, 53*, 1493–1501.
Evans, P. (2006). Scaling and assessment of data quality. *Acta Crystallographica. Section D, Biological Crystallography, 62*, 72–82.
Evans, P. R., & Murshudov, G. N. (2013). How good are my data and what is the resolution? *Acta crystallographica. Section D, Biological Crystallography, 69*, 1204–1214.
Fadel, A. R., Jin, D. Q., Montelione, G. T., & Levy, R. M. (1995). Crankshaft motions of the polypeptide backbone in molecular dynamics simulations of human type-alpha transforming growth factor. *Journal of Biomolecular NMR, 6*, 221–226.
Fenwick, R. B., Orellana, L., Esteban-Martín, S., Orozco, M., & Salvatella, X. (2014). Correlated motions are a fundamental property of β-sheets. *Nature Communications, 5*, 4070.
Forneris, F., Burnley, B. T., & Gros, P. (2014). Ensemble refinement shows conformational flexibility in crystal structures of human complement factor D. *Acta Crystallographica. Section D, Biological Crystallography, 70*, 733–743.
Fraser, J. S., Clarkson, M. W., Degnan, S. C., Erion, R., Kern, D., & Alber, T. (2009). Hidden alternative structures of proline isomerase essential for catalysis. *Nature, 462*, 669–673.
Fraser, J. S., van den Bedem, H., Samelson, A. J., Lang, P. T., Holton, J. M., Echols, N., & Alber, T. (2011). Accessing protein conformational ensembles using room-temperature X-ray crystallography. *Proceedings of the National Academy of Sciences of the United States of America, 108*, 16247–16252.
Frauenfelder, H., Parak, F., & Young, R. D. (1988). Conformational substates in proteins. *Annual Review of Biophysics and Biophysical Chemistry, 17*, 451–479.
Frauenfelder, H., Petsko, G. A., & Tsernoglou, D. (1979). Temperature-dependent X-ray diffraction as a probe of protein structural dynamics. *Nature, 280*, 558–563.
Frauenfelder, H., Sligar, S. G., & Wolynes, P. G. (1991). The energy landscapes and motions of proteins. *Science (New York, N. Y.), 254*, 1598–1603.
Furnham, N., Blundell, T. L., DePristo, M. A., & Terwilliger, T. C. (2006). Is one solution good enough? *Nature Structural & Molecular Biology, 13*, 184–185.
Garman, E. (2003). Cool crystals: Macromolecular cryocrystallography and radiation damage. *Current Opinion in Structural Biology, 13*, 545–551.
Garman, E. F., & Owen, R. L. (2006). Cryocooling and radiation damage in macromolecular crystallography. *Acta Crystallographica. Section D, Biological Crystallography, 62*, 32–47.
Garman, E. F., & Weik, M. (2017). X-ray radiation damage to biological macromolecules: further insights. *Journal of Synchrotron Radiation, 24*, 1–6.
Ginn, H. M. (2021). Vagabond: Bond-based parametrization reduces overfitting for refinement of proteins. *Acta Crystallogr D Struct Biol, 77*, 424–437.
Gutermuth, T., Sieg, J., Stohn, T., & Rarey, M. (2023). Modeling with alternate locations in X-ray protein structures. *Journal of Chemical Information and Modeling, 63*, 2573–2585.
Halle, B. (2004). Biomolecular cryocrystallography: Structural changes during flash-cooling. *Proceedings of the National Academy of Sciences of the United States of America, 101*, 4793–4798.
Hallen, M. A., Keedy, D. A., & Donald, B. R. (2013). Dead-end elimination with perturbations (DEEPer): A provable protein design algorithm with continuous sidechain and backbone flexibility. *Proteins: Structure, Function, and Bioinformatics*. https://doi.org/10.1002/prot.24150

Hammes, G. G., Benkovic, S. J., & Hammes-Schiffer, S. (2011). Flexibility, diversity, and cooperativity: Pillars of enzyme catalysis. *Biochemistry, 50*, 10422–10430.

Hancock, M., Peulen, T.-O., Webb, B., Poon, B., Fraser, J. S., Adams, P., & Sali, A. (2022). Integration of software tools for integrative modeling of biomolecular systems. *Journal of Structural Biology, 214*, 107841.

Hope, H. (1990). Crystallography of biological macromolecules at ultra-low temperature. *Annual Review of Biophysics and Biophysical Chemistry, 19*, 107–126.

Indiani, C., & O'Donnell, M. (2006). The replication clamp-loading machine at work in the three domains of life. *Nature Reviews. Molecular Cell Biology, 7*, 751–761.

Ishima, R., & Torchia, D. A. (2000). Protein dynamics from NMR. *Nature Structural Biology, 7*, 740–743.

Kabsch, W. (2010). XDS. *Acta Crystallographica. Section D, Biological Crystallography*. https://doi.org/10.1107/s0907444909047337

Kato, Y., Miyakawa, T., & Tanokura, M. (2018). Overview of the mechanism of cytoskeletal motors based on structure. *Biophysical Reviews, 10*, 571–581.

Keedy, D. A., Fraser, J. S., & van den Bedem, H. (2015). Exposing hidden alternative backbone conformations in X-ray crystallography using qFit. *PLoS Computational Biology, 11*, e1004507.

Keedy, D. A., Kenner, L. R., Warkentin, M., Woldeyes, R. A., Hopkins, J. B., Thompson, M. C., ... Fraser, J. S. (2015). Mapping the conformational landscape of a dynamic enzyme by multitemperature and XFEL crystallography. *Elife, 4*, e07574. https://doi.org/10.7554/eLife.07574

Keedy, D. A., van den Bedem, H., Sivak, D. A., Petsko, G. A., Ringe, D., Wilson, M. A., & Fraser, J. S. (2014). Crystal cryocooling distorts conformational heterogeneity in a model Michaelis complex of DHFR. *Structure (London, England: 1993), 22*, 899–910.

Kempf, J. G., & Loria, J. P. (2003). Protein dynamics from solution NMR: Theory and applications. *Cell Biochemistry and Biophysics, 37*, 187–211.

Kleckner, I. R., & Foster, M. P. (2011). An introduction to NMR-based approaches for measuring protein dynamics. *Biochimica et Biophysica Acta, 1814*, 942–968.

Kovermann, M., Rogne, P., & Wolf-Watz, M. (2016). Protein dynamics and function from solution state NMR spectroscopy. *Quarterly Reviews of Biophysics, 49*, e6.

Krojer, T., Fraser, J. S., & von Delft, F. (2020). Discovery of allosteric binding sites by crystallographic fragment screening. *Current Opinion in Structural Biology, 65*, 209–216.

Kuriyan, J., Petsko, G. A., Levy, R. M., & Karplus, M. (1986). Effect of anisotropy and anharmonicity on protein crystallographic refinement. An evaluation by molecular dynamics. *Journal of Molecular Biology, 190*, 227–254.

Kuzmanic, A., Pannu, N. S., & Zagrovic, B. (2014). X-ray refinement significantly underestimates the level of microscopic heterogeneity in biomolecular crystals. *Nature Communications, 5*, 3220.

Lang, P. T., Ng, H.-L., Fraser, J. S., Corn, J. E., Echols, N., Sales, M., ... Alber, T. (2010). Automated electron-density sampling reveals widespread conformational polymorphism in proteins. *Protein Science: A Publication of the Protein Society, 19*, 1420–1431.

Mazal, H., & Haran, G. (2019). Single-molecule FRET methods to study the dynamics of proteins at work. *Current Opinion in Biomedical Engineering, 12*, 8–17.

McPhillips, T. M., McPhillips, S. E., Chiu, H.-J., Cohen, A. E., Deacon, A. M., Ellis, P. J., ... Kuhn, P. (2002). Blu-Ice and the Distributed Control System: Software for data acquisition and instrument control at macromolecular crystallography beamlines. *Journal of Synchrotron Radiation, 9*, 401–406.

Mittermaier, A., & Kay, L. E. (2006). New tools provide new insights in NMR studies of protein dynamics. *Science (New York, N. Y.), 312*, 224–228.

Mokhtari, D. A., Appel, M. J., Fordyce, P. M., & Herschlag, D. (2021). High throughput and quantitative enzymology in the genomic era. *Current Opinion in Structural Biology, 71*, 259–273.

Nave, C., & Garman, E. F. (2005). Towards an understanding of radiation damage in cryocooled macromolecular crystals. *Journal of Synchrotron Radiation, 12*, 257–260.
Oda, K., Nomura, T., Nakane, T., Yamashita, K., Inoue, K., Ito, S., ... Nureki, O. (2021). Time-resolved serial femtosecond crystallography reveals early structural changes in channelrhodopsin. *Elife, 10*, e62389. https://doi.org/10.7554/eLife.62389
Okamoto, K., & Sako, Y. (2017). Recent advances in FRET for the study of protein interactions and dynamics. *Current Opinion in Structural Biology, 46*, 16–23.
Pearce, N. M., & Gros, P. (2021). A method for intuitively extracting macromolecular dynamics from structural disorder. *Nature Communications, 12*, 5493.
Pearce, N. M., Krojer, T., & von Delft, F. (2017). Proper modelling of ligand binding requires an ensemble of bound and unbound states. *Acta Crystallographica. Section D: Structural Biology, 73*, 256–266.
Pflugrath, J. W. (2015). Practical macromolecular cryocrystallography. *Acta Crystallographica. Section F, Structural Biology and Crystallization Communications, 71*, 622–642.
Ploscariu, N., Burnley, T., Gros, P., & Pearce, N. M. (2021). Improving sampling of crystallographic disorder in ensemble refinement. *Acta Crystallographica. Section D: Structural Biology, 77*, 1357–1364.
Rasmussen, B. F., Stock, A. M., Ringe, D., & Petsko, G. A. (1992). Crystalline ribonuclease A loses function below the dynamical transition at 220 K. *Nature, 357*, 423–424.
Rejto, P. A., & Freer, S. T. (1996). Protein conformational substates from X-ray crystallography. *Progress in Biophysics and Molecular Biology, 66*, 167–196.
Riley, B. T., Wankowicz, S. A., de Oliveira, S. H. P., van Zundert, G. C. P., Hogan, D. W., Fraser, J. S., ... van den Bedem, H. (2021). qFit 3: Protein and ligand multiconformer modeling for X-ray crystallographic and single-particle cryo-EM density maps. *Protein Science: A Publication of the Protein Society, 30*, 270–285.
Robertus, J. D., Kraut, J., Alden, R. A., & Birktoft, J. J. (1972). Subtilisin; A stereochemical mechanism involving transition-state stabilization. *Biochemistry, 11*, 4293–4303.
Roedig, P., Duman, R., Sanchez-Weatherby, J., Vartiainen, I., Burkhardt, A., Warmer, M., ... Meents, A. (2016). Room-temperature macromolecular crystallography using a micro-patterned silicon chip with minimal background scattering. *Journal of Applied Crystallography, 49*, 968–975.
Rupp, B. (2009). *Biomolecular crystallography: Principles, practice, and application to structural biology*. Garland Science.
Schmidt, M. (2021). Macromolecular movies, storybooks written by nature. *Biophysical Reviews, 13*, 1191–1197.
Schuler, B., & Eaton, W. A. (2008). Protein folding studied by single-molecule FRET. *Current Opinion in Structural Biology, 18*, 16–26.
Smith, C. A., & Kortemme, T. (2008). Backrub-like backbone simulation recapitulates natural protein conformational variability and improves mutant side-chain prediction. *Journal of Molecular Biology, 380*, 742–756.
Smith, J. L., Hendrickson, W. A., Honzatko, R. B., & Sheriff, S. (1986). Structural heterogeneity in protein crystals. *Biochemistry, 25*, 5018–5027.
Southworth-Davies, R. J., Medina, M. A., Carmichael, I., & Garman, E. F. (2007). Observation of decreased radiation damage at higher dose rates in room temperature protein crystallography. *Structure (London, England: 1993), 15*, 1531–1541.
Sun, Z., Liu, Q., Qu, G., Feng, Y., & Reetz, M. T. (2019). Utility of B-factors in protein science: Interpreting rigidity, flexibility, and internal motion and engineering thermostability. *Chemical Reviews, 119*, 1626–1665.
Terwilliger, T. C., Grosse-Kunstleve, R. W., Afonine, P. V., Moriarty, N. W., Adams, P. D., Read, R. J., ... Hung, L.-W. (2008). Iterative-build OMIT maps: Map improvement by iterative model building and refinement without model bias. *Acta Crystallographica. Section D, Biological Crystallography, 64*, 515–524. https://doi.org/10.1107/s0907444908004319

Tilton, R. F., Jr, Dewan, J. C., & Petsko, G. A. (1992). Effects of temperature on protein structure and dynamics: X-ray crystallographic studies of the protein ribonuclease-A at nine different temperatures from 98 to 320 K. *Biochemistry, 31*, 2469–2481.

Tsukada, H., & Blow, D. M. (1985). Structure of alpha-chymotrypsin refined at 1.68 A resolution. *Journal of Molecular Biology, 184*, 703–711.

Vallat, B., Tauriello, G., Bienert, S., Haas, J., Webb, B. M., Žídek, A., ... Westbrook, J. D. (2023). ModelCIF: An extension of PDBx/mmCIF data representation for computed structure models. *Journal of Molecular Biology, 435*, 168021.

van den Bedem, H., Dhanik, A., Latombe, J. C., & Deacon, A. M. (2009). Modeling discrete heterogeneity in X-ray diffraction data by fitting multi-conformers. *Acta Crystallographica. Section D, Biological Crystallography, 65*, 1107–1117.

van den Bedem, H., Lotan, I., & Latombe, J. C. (2005). Real-space protein-model completion: An inverse-kinematics approach. *Section D: Biological ....*

van Zundert, G. C. P., Hudson, B. M., de Oliveira, S. H. P., Keedy, D. A., Fonseca, R., Heliou, A., ... van den Bedem, H. (2018). *qFit-ligand* reveals widespread conformational heterogeneity of drug-like molecules in X-ray electron density maps. *Journal of Medicinal Chemistry, 61*, 11183–11198. https://doi.org/10.1021/acs.jmedchem.8b01292

Virtanen, P., Gommers, R., Oliphant, T. E., Haberland, M., Reddy, T., Cournapeau, D., ... van Mulbregt, P. SciPy 1.0 Contributors. (2020). SciPy 1.0: Fundamental algorithms for scientific computing in Python. *Nature Methods, 17*, 261–272.

Wankowicz, S. A., & Fraser, J. S. (2020). Ensemble refinement produces consistent R-free values but smaller ensemble sizes than previously reported. *Computational Crystallography Newsletter, 11*, 11–22.

Warkentin, M., Badeau, R., Hopkins, J., & Thorne, R. E. (2011). Dark progression reveals slow timescales for radiation damage between $T = 180$ and 240 K. *Acta Crystallographica. Section D, Biological Crystallography*. https://doi.org/10.1107/s0907444911027600

Warkentin, M., & Thorne, R. E. (2010). Glass transition in thaumatin crystals revealed through temperature-dependent radiation-sensitivity measurements. *Acta Crystallographica. Section D, Biological Crystallography, 66*, 1092–1100.

Weik, M., & Colletier, J. P. (2010). Temperature-dependent macromolecular X-ray crystallography. *Acta Crystallographica. Section D, Biological Crystallography, 66*, 437–446.

Weik, M., Schreurs, A. M. M., Leiros, H. K. S., Zaccai, G., Ravelli, R. B. G., & Gros, P. (2005). Supercooled liquid-like solvent in trypsin crystals: Implications for crystal annealing and temperature-controlled X-ray radiation damage studies. *Journal of Synchrotron Radiation, 12*, 310–317.

Weiss, M. S., Wollenhaupt, J., Correy, G. J., Fraser, J. S., Heine, A., Klebe, G., ... Pearce, N. M. (2022). Of problems and opportunities-How to treat and how to not treat crystallographic fragment screening data. *Protein Science: A Publication of the Protein Society, 31*, e4391.

Yabukarski, F., Doukov, T., Pinney, M. M., Biel, J. T., Fraser, J. S., & Herschlag, D. (2022). Ensemble-function relationships to dissect mechanisms of enzyme catalysis. *Science Advances, 8*, eabn7738.

Zeldin, O. B., Gerstel, M., & Garman, E. F. (2013). RADDOSE-3D: Time- and space-resolved modelling of dose in macromolecular crystallography. *Journal of Applied Crystallography, 46*, 1225–1230.

Zhong, E. D., Bepler, T., Berger, B., & Davis, J. H. (2021). CryoDRGN: Reconstruction of heterogeneous cryo-EM structures using neural networks. *Nature Methods, 18*, 176–185.

CHAPTER NINE

Combining temperature perturbations with X-ray crystallography to study dynamic macromolecules: A thorough discussion of experimental methods

Michael C. Thompson*
Department of Chemistry and Biochemistry, University of California, Merced, Merced, CA, United States
*Corresponding author. e-mail address: mthompson30@ucmerced.edu

Contents

Abstract

Temperature is an important state variable that governs the behavior of microscopic systems, yet crystallographers rarely exploit temperature changes to study the structure and dynamics of biological macromolecules. In fact, approximately 90% of crystal structures in the Protein Data Bank were determined under cryogenic conditions, because sample cryocooling makes crystals robust to X-ray radiation damage and facilitates data collection. On the other hand, cryocooling can introduce artifacts into macromolecular structures, and can suppress conformational dynamics that are

Methods in Enzymology, Volume 688
ISSN 0076-6879, https://doi.org/10.1016/bs.mie.2023.07.008

critical for function. Fortunately, recent advances in X-ray detector technology, X-ray sources, and computational data processing algorithms make non-cryogenic X-ray crystallography easier and more broadly applicable than ever before. Without the reliance on cryocooling, high-resolution crystallography can be combined with various temperature perturbations to gain deep insight into the conformational landscapes of macromolecules. This Chapter reviews the historical reasons for the prevalence of cryocooling in macromolecular crystallography, and discusses its potential drawbacks. Next, the Chapter summarizes technological developments and methodologies that facilitate non-cryogenic crystallography experiments. Finally, the chapter discusses the theoretical underpinnings and practical aspects of multi-temperature and temperature-jump crystallography experiments, which are powerful tools for understanding the relationship between the structure, dynamics, and function of proteins and other biological macromolecules.

1. Introduction

X-ray crystallography is a cornerstone technique in structural biology. It is responsible for a large majority of experimentally determined structures in the PDB (175,602 of 205,451 total), and has enabled the elucidation of macromolecular structures and their functional implications. While cryogenic electron microscopy (cryoEM) has stepped into the limelight recently, X-ray crystallography continues to be the workhorse for many structural biology endeavors that rely on models with high coordinate precision, derived from high-resolution experimental data. This includes drug discovery, protein design, and studies of enzyme catalysis or ligand binding. Throughout the roughly six-decade history of macromolecular crystallographic structure determination, nearly 90% of all crystal structures have been determined at cryogenic temperatures of approximately 100 K (−173 °C). The temperature at which these experiments were performed is approximately the temperature on the surface of the planet Neptune, which is quite far from the physiological conditions that are relevant to these macromolecules while they function on Earth.

Of course, there are important reasons that sample cryocooling became standard practice in macromolecular crystallography. Furthermore, despite the decidedly non-physiological temperatures used in the determination of most macromolecular crystal structures, the results of these experiments generally reflect accurate molecular details. This has been validated by decades of structural observations that were followed by judicious biochemical validation, often employing site-directed mutagenesis to test hypotheses about the relationship between structure and function.

Nevertheless, crystal cryocooling can introduce perturbations to the atomic structures of macromolecules that can obscure the interpretation of biochemical mechanisms (Fraser et al., 2011; Keedy et al., 2014; Thorne, 2023). Also, while proteins are known to be dynamic (Henzler-Wildman & Kern, 2007), a traditional limitation of crystallography has been the inherently static nature of macromolecular crystal structures, notwithstanding the information contained in atomic displacement parameters (B-factors) and sparsely modeled alternative conformations (Ringe & Petsko, 1986; Thompson, Yeates, & Rodriguez, 2020). Collecting X-ray diffraction data from macromolecular crystals at non-cryogenic and near-physiological temperatures can give a more relevant view of their atomic structures, and can provide greater insight into the conformational dynamics that underlie many important protein functions, such as enzyme catalysis and allosteric communication (Fraser et al., 2011).

Performing crystallography at non-cryogenic temperatures is now easier than ever, because modern X-ray detectors and data processing software enable diffraction measurements using low X-ray doses (Ueno et al., 2019; Winter et al., 2018, 2019). Additionally, computational tools are available that can model alternative conformations or entire conformational ensembles from experimental diffraction data (Burnley, Afonine, Adams, & Gros, 2012; Lang et al., 2010; Riley et al., 2021; Van Den Bedem, Dhanik, Latombe, & Deacon, 2009), allowing crystallographers to maximize the amount of structural and biophysical information that can be extracted from experiments at near-physiological temperatures.

With the ability to collect non-cryogenic X-ray diffraction data easily, it becomes possible to use temperature as a perturbation to map the conformational landscapes of biological macromolecules and probe their energetics. This chapter will explore the use of temperature perturbation as a tool for macromolecular X-ray crystallography, from both a fundamental and practical perspective. First, the historical reason for the prevalence of cryocooling in macromolecular crystallography will be introduced, and a brief review of the literature will demonstrate some of the hazards of sample cryocooling in macromolecular crystallography. Next, modern developments in instrumentation and software that have made non-cryogenic crystallography experiments robust and accessible will be discussed, followed by a practical description of how best to use these experimental tools. Finally, theoretical and practical aspects of using temperature as a perturbation in both static (multi-temperature) and time-resolved (temperature-jump) crystallography experiments will be discussed, including a review of several examples that show how these experiments can be applied. This Chapter will serve as a guide for structural biologists who wish to apply these techniques in their own research.

2. Sample cryocooling in macromolecular crystallography

2.1 Cryocooling has become standard practice in macromolecular crystallography

In the late 1970s, the practice of cryocooling became standard operating procedure for the vast majority of macromolecular crystallography experiments (Haas, 2020; Pflugrath, 2015). Cryocooling involves the rapid vitrification of a crystal specimen by flash freezing in cryogenic liquid, most commonly liquid nitrogen, and maintaining the crystal at cryogenic temperature (~100 K) throughout the course of data collection. The widespread utilization of cryocooling in modern macromolecular crystallography results from a key advantage offered by the technique. Cryocooling increases the tolerance of macromolecular crystals to radiation damage, a common limiting factor for successful data collection, by approximately two orders of magnitude (Garman & Weik, 2017; Garman, 2003; Holton, 2009; Nave & Garman, 2005). Generally, the X-ray dose that is considered "safe" for a 100 μm^3 crystal at a photon energy of approximately 12 keV is approximately 0.2 MGy for crystals kept at ambient temperature during data collection, and that dose increases to approximately 20 MGy when crystals are held at 100 K. Crystals that are smaller, or contain radiation sensitive moieties, such as metal ions, are even more sensitive.

At the time cryocooling was introduced, the protective effects against radiation damage were critical for expanding the applicability of crystallography to increasingly challenging systems. For example, it enabled structural measurements for samples that yielded only small crystals, which required a large X-ray dose relative to their volume for successful data collection. Additionally, it led to more facile phasing experiments, where small anomalous differences might be sensitive to the rapid onset of radiation damage. This was especially true for multi-wavelength experiments that involved large X-ray doses and/or relatively low photon energies that exacerbate radiation damage. Importantly, it also opened the door to working with more radiation-sensitive samples, including crystals of metalloproteins and other redox-sensitive molecules that are acutely sensitive to damage resulting from photoreduction.

As synchrotron data collection became more commonplace, an additional advantage of cryocooling became critically important. As long as cryocooled crystals remain frozen, they are extremely robust to dehydration and mechanical damage, almost indefinitely. This allows for transportation of samples across the world with relative ease, and offers a straightforward solution for long-term storage. The significant advantages of cryocooling have led to its incorporation into almost every macromolecular crystallography workflow since the early 1980s

2.2 Cryocooling can cause lattice distortions that lead to practical challenges

While cryocooling provides substantial advantages described above, it also has several notable disadvantages. Practically speaking, cryocooling can be challenging. Protein crystals are, on average, about 50% solvent by volume, which leads to complications. When solvent channels and cavities in the protein crystal are large, it is possible for crystalline ice to form within these regions of the crystal (Juers & Matthews, 2004a; Moreau, Atakisi, & Thorne, 2019; Moreau, Atakisi, & Thorne, 2021; Weik et al., 2001). Even when solvent channels are relatively small, freezing of the mother liquor which bathes the crystal and keeps it hydrated during the harvesting and mounting process can form crystalline ice. When mother liquor solutions produce crystalline ice upon freezing, it is necessary to add cryoprotectants, additional solutes that promote the formation of vitreous (amorphous) ice (Pflugrath, 2015). Cryoprotection often involves high concentrations of small organic molecules, such as glycerol or ethylene glycol, salts, such as lithium chloride, or polymers, such as polyethylene glycol. The osmotic shock that results from transferring crystals from their mother liquor solution to a cryoprotectant solution can damage crystals, often leading to cracking (López-Jaramillo, Moraleda, González-Ramírez, Carazo, & García-Ruiz, 2002).

Even when cryoprotection can be effectively optimized, other physical phenomena that take place during flash cooling can damage the crystal lattice. In 2001, Juers and Matthews described a contraction of the crystal lattice upon flash cooling *Escherichia coli* β-galactosidase crystals, which was accompanied by a 50% increase in the buried surface area at crystal contacts (Juers & Matthews, 2001). Following on this observation, they performed a survey of 15 unique proteins for which both room temperature and cryogenic structures had been determined at the time, and found that on average, lattice contractions during cryocooling lead to a roughly 4% reduction in unit cell volume. In contrast, the protein volumes are reduced by only about 1%. In 2011, Fraser et al. performed a similar analysis with 30 pairs of cryogenic and room temperature structures and found nearly identical results (Fraser et al., 2011).

The non-uniform contraction of the protein and solvent can lead to flow of solvent out of the crystal volume during the freezing process (Juers & Matthews, 2004b; Moreau et al., 2019). These effects can lead to strain within the crystal, especially when cooling rates are non-uniform (Juers & Matthews, 2004a; Kriminski, Caylor, Nonato, Finkelstein, & Thorne, 2002; Kriminski, Kazmierczak, & Thorne, 2003). All of these phenomena associated with cryocooling are potential causes of disorder that can degrade the quality of X-ray diffraction from frozen crystals, often leading to non-isomorphism (Kriminski et al., 2002) and increased mosaicity (Dobrianov, Caylor, Lemay, Finkelstein, & Thorne, 1999; Juers & Matthews, 2004b; Shaikevitch & Kam, 1981; Vahedi-Faridi, Lovelace, Bellamy, Snell, & Borgstahl, 2003). Various approaches have been developed to mitigate these complications, including hyperquenching (Warkentin & Thorne, 2007) and annealing (Kriminski et al., 2002) strategies, however, development of an effective cryocooling protocol remains a major bottleneck in the standard X-ray crystallography pipeline.

2.3 Cryocooling perturbs the structures of macromolecules

In addition to potential technical challenges, a more troubling consequence of cryocooling is that it can alter the structure and dynamics of the crystallized molecules, sometimes affecting the mechanistic interpretation of results obtained from crystallographic measurements. Following the widespread adoption of cryocooling in macromolecular crystallography, a handful of studies noted the potential introduction of artifacts into protein structures because of the procedure. For example, in 1997, Deacon et al. reported significant differences in some side chain rotamers and disordered segments when comparing room temperature and cryogenic structures of concanavalin A (Fig. 1A), and noted the number of ordered solvent molecules that could be identified in the cryogenic structure was more than double what could be identified in the room temperature structure (Deacon et al., 1997).

The potential for these cryocooling artifacts to complicate our interpretation of structure-function relationships was later highlighted by studies of two enzymes. Sandalova, et al. observed that the active site of dethiobiotin synthetase undergoes a structural transition during cryocooling

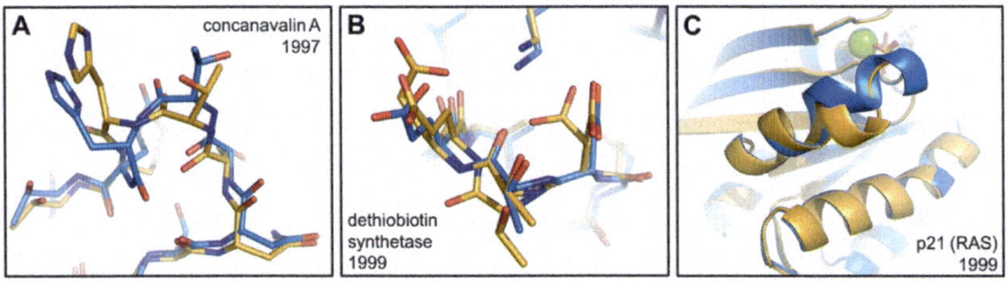

Fig. 1 Early examples of discrepancies between cryogenic and non-cryogenic protein structures. The figure summarizes findings from three papers published during the late 1990s. In each of the panels, a cryogenic structure is shown in blue, and a corresponding non-cryogenic structure is shown in orange. (A) Cryocooling results in small conformational changes in the structure of concanavalin A, including significant rearrangements of several loop regions (Deacon et al., 1997). (B) Cryocooling suppresses conformational heterogeneity and leads to other small conformational changes in dethiobiotin synthetase (Sandalova et al., 1999). (C) Cryocooling distorts the switch 2 helix in p21 (RAS) (Scheidig et al., 1999), a region that is critical for it's signaling function.

(Sandalova, Schneider, Käck, & Lindqvist, 1999) (Fig. 1B). Similarly, Scheidig, et al. noted differences in the structure of a key catalytic loop and associated active site water molecules in the switch II region of the p21(ras) GTPase (Scheidig, Burmester, & Goody, 1999) (Fig. 1C).

To address the prevalence of cryocooling artifacts, Fraser and coworkers systematically studied the distortion of side chain conformation upon cryocooling (Fraser et al., 2011). By analyzing 4008 residues across 30 proteins for which both cryogenic and room temperature structures were available, they found that 18.9% of all residues and 7.8% of buried residues change their $\chi 1$ rotamer angle significantly upon cryocooling. A growing body of work also suggests that protein–ligand interactions are potentially disrupted by crystal cryocooling (Fischer, Shoichet, & Fraser, 2015; Skaist Mehlman et al., 2023), which has important implications for drug discovery.

In addition to the structural distortions noted above, cryocooling also negates a major strength of X-ray crystallography with respect to the study of dynamic macromolecules like proteins. In a crystallographic experiment, the observed electron density is a spatiotemporal average of all conformations that are collectively explored by molecules in the irradiated crystal volume during the course of data collection. Cryocooling lowers the temperature of the sample below the solvent glass transition temperature (near 200 K) (Halle, 2004), thereby arresting protein dynamics, which are intimately coupled to the dynamics of the solvent (Fenimore et al., 2013; Frauenfelder et al., 2009).

Ideally, crystal cryocooling would be rapid enough to capture the ensemble of conformations present at ambient temperature, but in practice, cooling rates for macromolecular crystals are in the range of 0.1–1 s (Warkentin & Thorne, 2007), which is much slower than most molecular motions.

Therefore, the conformational ensemble is constantly relaxing toward a new thermal equilibrium during the flash-cooling process, with a tendency for the conformational ensemble to collapse toward compact, low-enthalpy states (Halle, 2004) that eliminate void space in the molecule (Fraser et al., 2011; Frauenfelder et al., 1987). It has already been noted that flash cooling reduces the volumes of crystallized proteins by approximately 1–1.5%. Because protein functions often involve exchange among alternative conformations, this loss of conformational heterogeneity can obscure our interpretation of molecular mechanisms from crystal structures, as demonstrated in a 2009 study of proline isomerase (cyclophilin A) by Fraser et al. (2009). In this work, the authors demonstrate that alternative conformations of the enzyme, which are critical for catalysis, were invisible in electron density maps calculated from cryogenic X-ray data but could be clearly modeled using maps derived from data collected at ambient temperature.

This study provided a powerful demonstration of the extent to which functional information can be lost during cryocooling. Additional experiments that included careful controls for X-ray dose highlighted that the increase in alternative conformations observed in non-cryogenic experiments is not due to enhanced radiation damage, and confirmed that it represents the true underlying conformational ensemble (Russi et al., 2017; Yabukarski, Doukov, Mokhtari, Du, & Herschlag, 2021).

3. Non-cryogenic macromolecular crystallography

3.1 Modern hardware and software obviate the need for cryocooling

After the introduction of cryocooling into macromolecular crystallography, experiments performed at non-cryogenic temperatures came to be viewed by most as "boutique" experiments that could be applied only to uniquely well-behaved crystal systems that yielded very large (hundreds of μm) and well-diffracting crystals, such as lysozyme, myoglobin, and RNase. This notion is rooted in the history of X-ray detector capability, and is now less true thanks to developments in both hardware and software. Given the crystal pathologies and artifacts that can arise from cryocooling (Section 2.2), it is valuable to consider how bottlenecks to performing non-cryogenic experiments have been removed, and to understand how to utilize new technology to perform these experiments effectively.

Cryocooling is an important tool; however, one can argue that technique is overused in modern macromolecular crystallography, especially given the potential for distortion of the underlying crystal structure. There are still valid uses for sample cryocooling in a variety of experimental contexts where radiation damage is a concern (see Section 2.1). Additionally, many crystallographers have heard anecdotes or had their own experience in which only a single crystal was ever obtained for a protein of interest, yet it was enough to determine the structure. When faced with this type of scenario, it would likely be prudent to freeze your one-and-only crystal. Outside of those specific situations, most samples that can be crystallized reproducibly and are not acutely sensitive to radiation damage are suitable for non-cryogenic data collection, owing to technological developments in both hardware and software.

3.1.1 Breakthroughs in X-ray detector technology

When cryocooling was originally introduced in the late 1970s, measuring X-ray diffraction was difficult, because detector technology was still somewhat primitive. Film and image plate (Amemiya, 2008) detectors used prior to the 1990s are poorly suited for data collection at low X-ray doses needed to mitigate radiation damage, leading to the introduction of cryocooling as an enabling technology. Detectors developed throughout the 1990s, based on charge-coupled devices (CCDs) offered a variety of advantages for synchrotron data collection due to their relative speed, however they still had typically low detective quantum efficiency (DQE), typically in the range of 0.3 (Phillips et al., 1993). DQE is a measure of how much the signal-to-noise ratio is degraded by the detector electronics, with a perfect ("noiseless") detector having a DQE of 1 (Gruner, Eikenberry, & Tate, 2006). Detectors with lower DQE require more photons to be measured to achieve the desired measurement precision. Because of these detector limitations, data collection with very low X-ray dose was impossible, so sample cryocooling was revolutionary. In the ensuing decades cryocooling became a standard practice that enabled many new experiments and an explosion of structural data.

The mid-2000s ushered in a technological breakthrough in commercial X-ray detector technology that makes non-cryogenic data collection routine. Specifically, during this period, the pixel-array detector (PAD) became part of the standard instrumentation available at many macromolecular crystallography beamlines across the world. These devices have several advantages over their predecessors, namely CCDs, in terms of their speed and detection properties. Critical to their utility in non-cryogenic experiments is the fact that PADs are direct photon counting detectors, which can have single-photon sensitivity and DQE values approaching 1 (Förster, Brandstetter, & Schulze-Briese, 2019). This allows for measurements with high signal-to-noise, even at very low X-ray doses. With the ability to measure weak but reliable diffraction using doses as low as tens of kGy, radiation sensitivity is a lesser problem for many macromolecular crystal systems, even when crystals are small or otherwise sensitive. The weak signal resulting from low dose per image can be overcome with high data multiplicity and merging of the integrated reflection intensities.

3.1.2 Integration of weak reflection intensities with modern software

The ability to reliably measure X-ray diffraction patterns with single-photon sensitivity required data processing algorithms that could operate with very small signals. Specifically, crystallographic data reduction software needs to reliably find reflections that can be used for indexing, and then to integrate them accurately (with background subtraction) prior to merging. When working with very low X-ray doses, data reduction programs that fit pixel intensities against an empirical reflection profile, such as DIALS (Winter et al., 2018) and XDS (Kabsch, 2010), are known to give better estimates of weak reflection intensities (Diamond, 1969). One might recommend that one of these data processing programs be used for indexing and integration of weak data resulting from low X-ray doses. In 2012, a groundbreaking paper by Karplus and Diederichs (Karplus & Diederichs, 2012) demonstrated that useful signal can be extracted from much weaker (lower $I/\sigma(I)$) diffraction data than was previously appreciated (and introduced the useful statistic CC1/2). This revelation further validated the use of low X-ray doses in macromolecular crystallography, especially for non-cryogenic measurements.

3.2 Practical considerations for non-cryogenic data collection

When performing non-cryogenic crystallography experiments, there are two major challenges that the experimenter must be aware of and work to mitigate—avoiding crystal dehydration and not exceeding the relatively low X-ray dose threshold for radiation damage. These two key considerations, relevant at the sample preparation and data collection stages of the experiment, respectively, are critical for success and also require procedures that are markedly different from what one would do during a cryocrystallography experiment.

3.2.1 Sample preparation

The first branch point where the non-cryogenic macromolecular crystallography workflow deviates from the common procedure for working with frozen crystals is at the stage of crystal harvesting and mounting. This stage of sample preparation is critical for protecting the crystal from dehydration during data collection, which is more challenging when crystals and their surrounding mother liquor are not frozen, which we noted above as a major advantage of cryocooling. An additional benefit of cryocooling, which has not been mentioned yet, was the introduction of the cryoloop (Fig. 2A), a small polymer lasso that can be used to capture a crystal from its growth solution with little excess solvent, and then to plunge it directly into a cryogen such as liquid nitrogen (Pflugrath, 2015).

Before introduction of the cryoloop, the traditional method of preparing macromolecular crystals for X-ray diffraction experiments involved the assembly of complicated "capillary mounts" (First Analysis of Macromolecular Crystals | Springer Nature Experiments, n.d.) (Fig. 2B). In this procedure, an experimenter carefully manipulates a crystal into a thin-walled glass capillary tube using a pipette or syringe, and then uses a small wicking device (a tiny sliver of filter paper) to remove just the right amount of excess mother liquor from the crystal to keep the specimen hydrated without producing excessive background scattering. A small volume of excess mother liquor is left inside the tube, and the ends are sealed with wax or glue before affixing to a goniometer base for data collection. Mounting protein crystals, which are delicate and sensitive to humidity, is far easier to do with a cryoloop than with a capillary mount. Current best practices for preparing samples for non-cryogenic X-ray diffraction experiments combine the simplicity of cryoloop mounting with various physical or chemical barriers that protect the crystal from dehydration, rather than with plunge freezing.

The most common approach to preparing crystals for non-cryogenic data collection is essentially a hybrid of the cryoloop and capillary methods (Fig. 2C). This approach is straightforward, and the required materials are available commercially. Specifically, the method requires a standard cryoloop with a magnetic goniometer base and a 2 mm internal

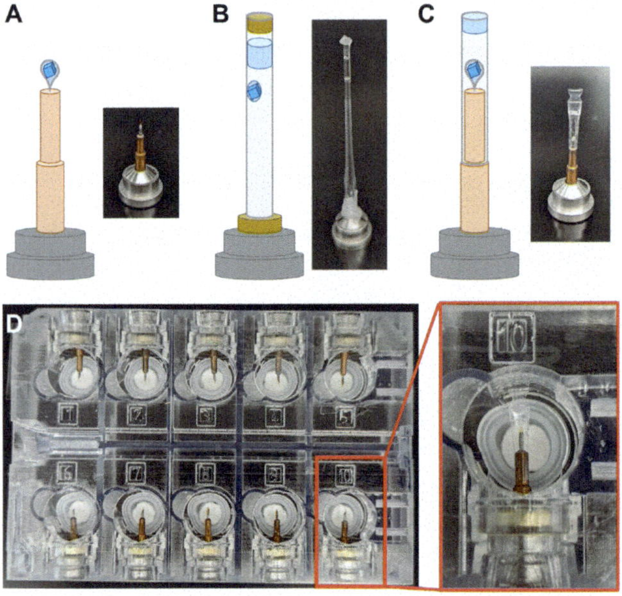

Fig. 2 Crystal handling for non-cryogenic experiments. Different crystal mounting methods are shown as cartoon alongside a photo of a prepared mount. (A) The cryoloop, used commonly for cryogenic crystallography experiments consists of a polymer loop at the end of a metal pin that is affixed to a magnetic goniometer base. The loop can be used to isolate a crystal and then plunged into cryogen to freeze the crystal. Ideally, the size of the crystal (bright blue) is well-matched to the size of the loop as shown to reduce background scattering. (B) The original method for preparing macromolecular crystals for non-cryogenic data collected is known as a "capillary mount." A thin glass capillary contains a hydrated crystal stuck to the inside surface of the capillary, as well as a small volume of crystal mother liquor to keep the crystal hydrated (light blue). The capillary is sealed at both ends with wax (yellow), and affixed to a goniometer base. The preparation of capillary mounts is laborious compared to freezing crystals in cryoloops. (C) A hybrid system allows crystals to be prepared for non-cryogenic data collection by harvesting them in a cryoloop, and then sealing the cryoloop with a thin polycarbonate capillary that is closed at one end. The plastic capillary also contains a small volume of crystal mother liquor to keep the sample hydrated, like a capillary mount. (D) A commercial plate system is available for shipping crystals prepared for non-cryogenic data collection. Each well contains a disposable fiber plug that can be soaked with crystal mother liquor to maintain a humid environment. Crystals are inserted into the wells of the plate through a hole in the side that grips the magnetic base of the cryoloop. The top of the well is sealed with a water impermeable tape.

diameter, thin-walled polyester capillary, which is sealed on one end. Such cryoloops have become ubiquitous in macromolecular crystallography, and the tubing is available commercially. The procedure begins by cutting the capillary at the open end to roughly match its length to that of the cryoloop, and then pipetting a small amount (~4 μL) of crystal mother liquor into the sealed end of the capillary. Next, a crystal is harvested from its growth environment using a standard cryoloop with a magnetic goniometer base. The process of harvesting the crystal from its growth environment mirrors the way in which crystals are typically harvested for cryocooling. However, rather than plunging the harvested crystal into cryogen, the loop is placed upright, and the polyester capillary is drawn down over the sample and sealed to the goniometer base using a small amount of vacuum grease. The assembly retains the advantages of a traditional capillary mount—the water-impermeable polyester capillary and vacuum grease create a dehydration barrier, while the small plug of mother liquor maintains an appropriately humid environment to keep the crystal hydrated—but does not require challenging crystal manipulations that can damage the specimen. The thin polyester tube produces a relatively low amount of background scattering, which is beneficial for data quality.

Another useful method for preparing crystals for non-cryogenic data collection involves the combination of a cryoloop with chemical barriers to dehydration, usually some type of highly viscous and water impermeable oil. Commonly used oils are proprietary mixtures (chemical compositions unavailable) known as "NVH Oil" and "Parabar" (or "Paratone"), which are available commercially. The idea is simple in principle—the crystal is mounted in a cryoloop and coated in oil to prevent dehydration. This method can be highly effective, but the procedure can be challenging and requires practice for most experimenters. Once again, a crystal is harvested from its growth environment using a standard cryoloop with a magnetic goniometer base, and it is transferred into a small droplet of oil. The crystal must be fully immersed in the oil, and carefully manipulated to strip away excess aqueous mother liquor from around the crystal. Once this has been done successfully, the cryoloop can be used to remove the oil-coated crystal for data collection.

Although the process seems straightforward, there are several significant challenges. First, the oil is very viscous, and maintaining a coating around the crystal at non-cryogenic temperatures while manipulating the crystal in the oil can result in mechanical damage if not done carefully. Next, the refractive indices of the useful oils are generally very close to the refractive indices of protein crystals, which means that most crystals become invisible when they are well-coated in oil. Using a microscope with a cross-polarizer can help overcome this challenge. Finally, this method is quite unforgiving, and it is essential to make sure that the mounted crystal is completely covered in oil, otherwise it will not be sufficiently protected from dehydration. For this reason, some crystal morphologies are not ideal for this method, especially those with sharp vertices and edges that tend to poke out of the oil and create an "Achilles heel" for crystal dehydration. Although it can be difficult to execute, the high-viscosity oil method works extremely well if executed properly, because the oil creates a barrier that is completely impermeable to water vapor. This maintains the crystal under the same hydration conditions it was grown in, unlike the capillary method, in which vapor diffusion can occur within the capillary. Adaptations of this method include sealing the cryoloop-mounted crystal with more exotic materials, including polymer glues (Baba, Hoshino, Ito, & Kumasaka, 2013) and graphene (Warren et al., 2015), but these procedures require specialized materials and/or equipment.

Regardless of the sample mounting method (capillaries or oils), there are three important and general considerations for preparing samples for non-cryogenic data collection. First, when preparing crystals for non-cryogenic data collection, it is obviously not necessary to add cryoprotectants to prevent the formation of ice. However, if the goal of an experiment is to compare structures at cryogenic versus non-cryogenic temperatures (see Section 4), it can be useful to add any cryoprotectants that were utilized for the cryogenic data collection. Cryoprotection often involves large changes to the solution conditions (López-Jaramillo et al., 2002), and treating crystals in an identical manner prior to data collection at different temperatures can circumvent uncertainty about whether any observed structural differences are due to temperature or differing solution conditions.

Second, it is important to select an appropriately-sized cryoloop for the crystal specimen. This is not really necessary in cryocrystallography, because after the crystal is mounted and frozen, it is virtually immobile on the loop; it can be larger or smaller than the loop. In a non-cryogenic experiment, mother liquors and oils that surround the crystal remain liquid, and therefore if the crystal is small relative to the volume of liquid in the loop, it can move within the loop as it rotates during data collection, resulting in data that are difficult to interpret. On the other hand, crystals become more prone to drying out in capillaries or poking out of oil coatings if the loop is too small relative to the size of the crystal.

Third, non-cryogenic crystallography experiments often involve transportation of crystal trays to the X-ray facility, or crystal growth on-site, followed by mounting of the samples immediately before data collection. This can be logistically difficult for researchers that are not close to an X-ray facility. It would reduce throughput if samples must be prepared during the experiment, rather than in advance. Recently, shippable humidity chambers were developed and commercialized that enable the transportation of mounted crystals at ambient temperatures (Fig. 2D). These chambers can be pre-loaded with samples (prepared using either capillary or oil methods) and shipped or transported in a thermally-insulated container to the experimental facility for data collection. The development of these tools will likely be transformative for the widespread adoption of non-cryogenic crystallography because it lowers a major barrier. This will be especially true if these tools for non-cryogenic experiments are integrated into remote data collection programs at synchrotron facilities.

3.2.2 Data collection

In a reference to the Wild West, the pioneering protein crystallographer Michael Rossmann often joked about the "American Method" of crystallography, in which one could "shoot first, and ask questions later" (Wu & Arnold, 2019). Initially, this phrase referred to the benefits of autoindexing algorithms [developed by Rossmann himself (Steller, Bolotovsky, & Rossmann, 1997)] that could identify the unit cell regardless of crystal orientation relative to the beam. The haphazard approach to data collection was only reinforced by cryocooling procedures that made samples robust to large X-ray doses that yielded visibly strong diffraction patterns.

Removing some of the barriers to successful data collection enabled high-throughput experiments, and made crystallography accessible to the non-expert, both of which were critical for the development of structural biology as a field. Indeed, in 2023 most data sets are collected in this manner, with a relatively high X-ray dose (20–30 MGy) being distributed over a large (e.g. 360°) wedge of reciprocal space, ensuring satisfactory completeness of the data regardless of the space group. Despite the advantages of the "American Method," collecting data at non-cryogenic temperatures requires a more thoughtful approach so as not to exceed the relatively low X-ray dose threshold for radiation damage under non-cryogenic conditions.

The approximate dose threshold for radiation damage under non-cryogenic conditions, which is approximately 0.2 MGy for a 100 μm^3 crystal at a photon energy of approximately 12 keV (Garman & Weik, 2017), has already been discussed (Section 2.1). For crystals that are larger or smaller than 100 μm^3, it will be necessary to adjust the dose. A variety of tools are available for determining the appropriate X-ray dose for non-cryogenic data collection, the most comprehensive of which is RADDOSE-3D (Bury, Brooks-Bartlett, Walsh, & Garman, 2018). These tools take various crystal, X-ray beam, and experiment parameters as input, and calculate the required X-ray dose and ideal data collection strategy for non-cryogenic experiments. This total X-ray dose can be converted to a dose rate in terms of Gy/s or photons/s for a given size X-ray beam, so that the source parameters can be adjusted appropriately. Please note that the X-ray dose used for non-cryogenic data collection is approximately two orders of magnitude less than what is used for cryogenic experiments (Garman & Weik, 2017). This can make the diffraction images appear very sparse, and sometimes nearly blank, but the detector will be able to measure reflections that are too weak be visualized on the computer screen (Fig. 3).

To get a sense for how well crystals actually diffract, it is often useful to sacrifice a sample and collect a single image with 10–20 times the planned dose, so that high-resolution diffraction is clearly visible. This will provide

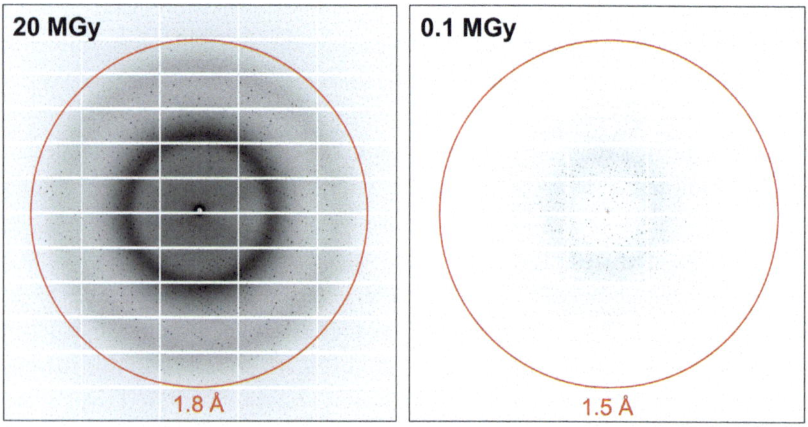

Fig. 3 X-ray diffraction under high and low dose regimes. X-ray diffraction images are shown that are collected from identical protein crystals, using a different X-ray dose. The images are displayed on the same intensity scale. The image on the left was collected with a high (20 MGy) dose, and the overall signal appears as most crystallographers would expect from a crystal that diffracts well. The image on the right was collected with 200-fold less dose than the image on the left (0.1 MGy), and only the very strongest reflections are visible at low resolution. Nevertheless, the data set corresponding to the image on the right can be integrated and merged to a resolution of 1.7 Å using only 180 images collected using this low dose.

feedback on crystal quality and will help position the detector at the correct distance from the sample. Note that the higher dose per image does not make the crystals diffract better, it simply makes the reflections more visible to the human eye on a computer screen. With a highly sensitive PAD, the weak reflections can be accurately integrated (Ueno et al., 2019; Winter et al., 2019). As described in Section 3.1.2, the weak individual images may lead to a lower-than-expected overall ratio of $I/\sigma(I)$ for the integrated and merged data, however the correlations between unique reflection measurements (as measured by $CC_{1/2}$, for example) should remain strong. Certain data reduction programs (Kabsch, 2010; Winter et al., 2018) are better-suited to handle these weak reflections (Diamond, 1969).

After determining the desired X-ray dose for a particular crystal, one must consider how the dose will be distributed evenly throughout the crystal volume. The simplest strategy for doing this is to closely match the beam size and the crystal size, so the entire crystal volume remains bathed in the beam throughout the entire rotation series (Fig. 4A). Most X-ray beams are focused and subsequently shaped with apertures so that their cross-section is roughly circular. Many beams have adjustable sizes via the insertion of different apertures or via adjustable focusing (Evans, Axford, & Owen, 2011; Fuchs et al., 2014). Beams that can be adjusted via focusing are advantageous because their size is the most variable, but due to the behavior of synchrotron radiation these beams are often non-circular, with dimensions that are broader in the horizontal direction than in the vertical, leading to non-circular shapes. When possible, instrument adjustments should be used to closely match the crystal size and beam size, so the dose is inherently distributed across the entire crystal.

Distributing the X-ray dose evenly across the crystal volume is the most straightforward when the crystal is roughly the same size in all three dimensions; this is rarely the case. Crystals that are substantially longer in one or two dimensions require translation of the sample to distribute the dose throughout the crystal volume, and can also require other compromises. In order to facilitate crystal translation during data collection, it is desirable to mount the crystals so that their longest physical dimension is roughly (but not necessarily exactly) parallel to the goniometer rotation axis (Fig. 4B,C). Because most diffractometers allow crystal translation along the goniometer rotation axis, samples that are mounted in this way can be effectively moved through the X-ray beam during data collection, to distribute the dose throughout the largest possible crystal volume.

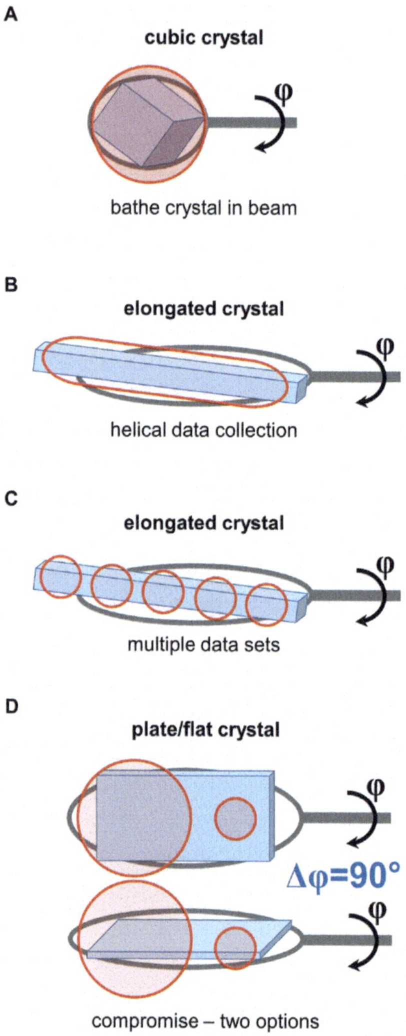

Fig. 4 Strategies for low-dose X-ray data collection. A key aspect of mitigating radiation damage during X-ray data collection, especially for non-cryogenic samples, is to distribute the X-ray dose throughout the crystal. The cartoons depict differently shaped crystals held in cryoloops that are rotatable about a goniometer φ axis. In each case, the footprint of the X-ray beam during data collection is outlined in red, and shaded according to the approximate dose deposited over the underlying area. (A) If the crystal is roughly the same size in all dimensions (cubic) then one can simply match the size of the X-ray beam to that of the crystal. Elongated, rod-shaped crystals can be treated in one of two ways. (B) Helical data collection can be implemented, in which the crystal is translated during the measurements, so that the entire crystal volume is measured probed during the collection of a single data set. (C) Another strategy for collecting data from elongated crystals is to collect a full data set from each of several locations on the crystal, using a lower overall dose for each individual data set. (D) When crystals are small in a single dimension only (flat/plate crystals), then one must make a compromise on whether to match the beam to the larger or smaller dimension of the crystal about the rotation axis.

There are two distinct approaches to data collection from an elongated crystal via sample translation, and both will be considered in light of the following example. Consider a crystal that is 1000 μm × 100 μm × 100 μm (a rod shape), mounted roughly parallel to the goniometer rotation axis, a desired data collection wedge that includes 180° of reciprocal space with 1° oscillation per image and a total dose of 180 kGy. Assume an X-ray beam with a diameter of 100 μm, and hard edges (n.b. most X-ray beams do not have perfect edges, but this simplification makes the following considerations simpler).

The first, and most common, strategy for doing this is referred to as "helical" data collection (Polsinelli et al., 2017; Rajashankar & Dauter, 2014). In this approach, the crystal is translated simultaneously in all three directions as it is rotated during data collection, so that fresh crystal volume is continuously brought into the beam, and expended crystal volume is removed from the beam (Fig. 4B). Many synchrotron beamlines have implemented helical data collection into their control software. Assuming the above example, a helical data collection strategy will yield 180 images, each covering 1° of reciprocal space. If the total dose is 180 kGy, then the dose per image is 1 kGy. If we translate along the length of this crystal, a particular region will be exposed for 18 consecutive images, and receive a dose of 18 kGy. When the data are collected in this way, each region of the crystal is sampled across 18° of reciprocal space.

The second, and less commonly implemented strategy, involves collecting non-overlapping and individually complete data sets across the length of the crystal, each utilizing only a fraction of the total desired dose (Fig. 4C). Assuming the above example, this data collection strategy will yield 10 unique, non-overlapping data sets, each containing 180 images and collected with 18 kGy total dose. Each of the 10 data sets can be integrated individually, and then merged appropriately. In this case, the dose per image is only 0.1 kGy. The composite data set will contain 1800 images, and 180° of reciprocal space have been sampled at each position on the crystal.

Given the two strategies outlined above, the one which is most appropriate to use for a particular experiment depends on several factors. First, if the diffractometer control software is not capable of helical data collection, then there is no choice but to implement the second strategy, where multiple unique data sets are collected. Next, the type of detector becomes a consideration. The differences between older CCD detectors and newer PAD devices have already been mentioned (Section 3.1.1), and the important difference between the two is the DQE, which is much higher for PADs than for CCDs, which makes them much more sensitive.

A consequence of this is that PADs are much more effective at measuring weak diffraction signals. If a CCD is available, it might be desirable to have a high per-image X-ray dose of 1 kGy that results from the helical strategy, rather than the low per-image X-ray dose of 0.1 kGy that results from the multiple data set approach. On the other hand, if a PAD is available, then it should be possible to detect the diffraction signal in the weaker, low dose images, and effectively merge the integrated reflections.

Finally, crystal quality is an important consideration. In the case of a perfect crystal, there should be no difference between the two strategies; however, crystals are never perfect. One can demonstrate that even within a single macromolecular crystal, there can be domains that are non-isomorphous with one another (Gallagher-Jones et al., 2020; Thompson, Cascio, & Yeates, 2018). The helical data collection strategy results in different regions of reciprocal of reciprocal space being sampled from different regions of the crystal. Therefore, if a particular region of the crystal is not isomorphous with the rest, the composite dataset can be rendered useless due to incompleteness. When using the multiple data set approach, complete data sets are collected for each region of the crystal, and if one or more regions are non-isomorphous, they can simply be excluded from the final, composite data set. Unit cell clustering algorithms are available to aid in this process (Foadi et al., 2013; Gildea et al., 2022; Soares et al., 2022; Thompson et al., 2018; Zeldin et al., 2015).

In the case of plate-like crystal morphologies, in which the crystals are large in two dimensions, but small in the third, a compromise must be made (Fig. 4D). The crystal can still be translated along its longest dimension, but a choice must be made with respect to the beam size. On one hand, data can be collected with a beam that matches the second largest dimension of the crystal. In this case, the same volume of crystal remains in the beam during rotation, but there will be more background air scatter due to the larger beam. On the other hand, the beam can be made to match more closely to the smallest crystal dimension. This reduces the background scattering, but the diffraction will be stronger for some crystal orientations than for others, because different crystal volumes will be exposed to the beam as a function of rotation angle.

3.2.3 Multi-crystal approaches and serial crystallography

Section 3.2.2 above described a data collection strategy in which 10 essentially unique data sets were collected from a single crystal and merged, based on their isomorphism. This same strategy can be applied to data collection from 10 (or more) different crystals. By collecting multiple data sets from many different crystals, it becomes possible to collect high-multiplicity data with essentially infinitely high X-ray doses, because the dose can be spread over any number of individual samples. The individual data sets are indexed and integrated as independent single-crystal data sets, and then merged to improve signal-to-noise.

Well-established data reduction programs, such as XDS (Kabsch, 2010) or POINTLESS/AIMLESS/TRUNCATE (Evans, 2006), have options to merge multiple data sets. Once again, before merging, algorithms that cluster data sets based on unit cell isomorphism are important for determining which data sets can be combined (Foadi et al., 2013; Thompson et al., 2018). In a multi-crystal experiment, full data sets (i.e. 180° of reciprocal space) or partial data sets can be collected from each individual crystal. If partial data sets are collected, it is important that the multiple crystals measured are in different orientations, so that the merged data set is complete.

An advantage to collecting partial data sets is that the individual images can be measured using a higher X-ray dose, because only a fraction of reciprocal space needs to be sampled before a given crystal succumbs to radiation damage. To facilitate multi-crystal experiments, a variety of different mounting supports have been developed that allow the preparation of multiple crystals on a single device, to minimize the need for sample changes. These include a variety of polymer grids (Baxter et al., 2016), meshes (Nam, Kim, & Cho, 2021; Park, Choi, Eo, Cho, & Nam, 2020), and chips (Gilbile et al., 2021; Lyubimov et al., 2015; Murray et al., 2015; Perry et al., 2014; Saha et al., 2023; Sui et al., 2021) that can be affixed to a standard magnetic goniometer base. Additionally, plate goniometers are available that allow data collection from crystals directly in the crystallization plate, allowing the *in situ* measurement of multiple crystals without removing them from their growth environment (Hargreaves, 2012; Okumura et al., 2022). Because it is impossible to cryocool an entire crystallization plate efficiently, these in-situ measurements are always performed at ambient temperature. Solid sample supports and plate goniometers are generally limited in the accessible rotation angles during data collection, so these strategies rely on the random orientation of different crystals on the support or in the tray.

The most extreme case of multi-crystal X-ray crystallography experiments are those which are referred to as "serial crystallography" (Gruner & Lattman, 2015; Mehrabi et al., 2021; Schlichting, 2015). In these experiments, rather than using relatively large, single crystals that are tens to hundreds of μm in each dimension, researchers prepare slurries of micro-crystals, that are on the order of 1–20 μm in all dimensions. In contrast to rotation crystallography, where a single crystal rotated in the X-ray beam to obtain a complete (or partial) data set, in the serial crystallography experiment, each crystal produces just a single diffraction image representing a random slice through reciprocal space. One reconstructs a complete diffraction pattern by measuring, processing, and scaling together thousands of randomly oriented crystals (Boutet et al., 2012; Chapman et al., 2011), producing a random sampling of reciprocal space. This approach allows a large X-ray dose to be used for each image, effectively destroying the crystal in a single shot.

When performed at a synchrotron (Martin-Garcia et al., 2017; Owen et al., 2017; Schulz, Yorke, Pearson, & Mehrabi, 2022; Weinert et al., 2017), one measures individual diffraction images with tens of kGy, typically delivered in milliseconds or less. This constitutes a relatively low dose for the crystal, but a high dose for an individual image (De La Mora et al., 2020). When the experiment is performed at an X-ray free electron laser (XFEL) (Boutet et al., 2012; Chapman et al., 2011), the data are collected using ultrafast X-ray pulses that are roughly 30 fs in duration, but deliver approximately 10^{12} photons per pulse. The peak power of these pulses is so high that they cause the crystals to undergo coulombic explosion, however they are also so fast that the diffraction patterns are collected before the onset of any damage. This approach, which has been referred to as "diffraction before destruction," provides structural data that is ostensibly free of the effects of radiation damage, even at non-cryogenic temperatures, provided that the X-ray pulses used in the experiment remain short (<50 fs). The nature of these experiments, in which each crystal specimen contributes a single diffraction image to the data set prior to its destruction, requires that the sample be continually refreshed in the X-ray beam. This requirement is the defining feature of the technique that inspired the name "serial" crystallography.

From a practical standpoint, serial crystallography experiments differ from standard rotation experiments in two main ways, alluded to above. The first has to do with sample preparation and delivery of crystals into the X-ray beam. As noted, serial crystallography requires a slurry of micro-crystals suspended in mother liquor. In favorable cases, tens to hundreds of microliters of crystal slurry are sufficient to determine a crystal structure in a serial experiment (Weierstall et al., 2014), however some experiments can consume milliliters of this slurry. Therefore, it is necessary to have a reliable method for growing large volumes of microcrystals, preferably in a batch format. A straightforward and reliable protocol for adapting traditional crystallization conditions, used for growing large single crystals, to a batch microcrystallization format was developed by Wolff, et al. (Wolff et al., 2020). Once a slurry of satisfactory microcrystals has been obtained, one must decide how they should be delivered to the X-ray beam to be measured. Various methods are employed for this purpose, each tailored to overcome different challenges. A variety of microfluidic jets are available that create stream of liquid or other viscous medium carrying the crystals, which continuously flows through the X-ray beam (DePonte et al., 2008; Shimazu et al., 2019; Sierra et al., 2016; Vakili et al., 2020; Weierstall et al., 2014) (Fig. 5A).

Another approach is the use of fixed-target systems (Hunter et al., 2014; Lahey-Rudolph et al., 2021; Roedig et al., 2017; Zarrine-Afsar et al., 2012), where microcrystals are mounted at high density on a solid support, such as a thin film or a microfabricated chip (Gilbile et al., 2021; Roedig et al., 2015; Roedig et al., 2016; Sui et al., 2021) (Fig. 5B). With fixed-target approaches, microcrystals are positioned in an ordered array, allowing for rapid, automated data collection from each individual specimen. Acoustic droplet ejection is yet another approach, which utilizes sound waves to precisely manipulate and eject individual droplets containing crystals into the

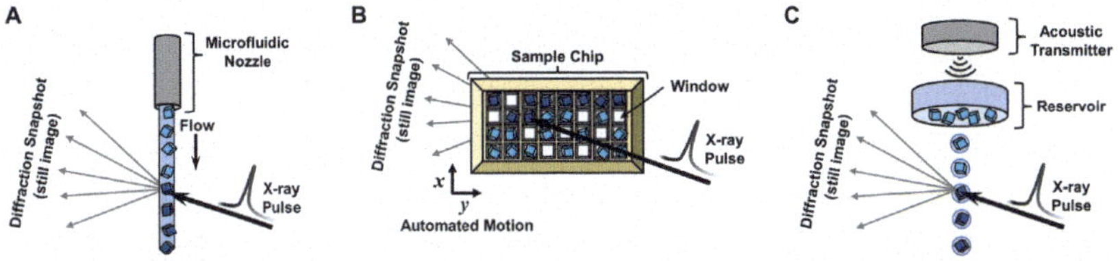

Fig. 5 Sample delivery strategies for serial crystallography. In a serial crystallography experiment, only a single diffraction image is collected from each crystal, destroying the crystal in the process and necessitating the replenishment of crystals in the beam. (A) A variety of microfluidic devices have been developed that create a freestanding liquid jet that flow crystals through the interaction region where they intersect with the (pulsed) X-ray beam. (B) Fixed-target strategies involve the deposition of hydrated crystals on some type of solid support chip, which is then rapidly translated in the horizontal and vertical directions to move crystals through the X-ray beam. (C) Acoustic methods involve the use of sound waves to eject small droplets of mother liquor containing crystals into the interaction region where they intersect with the X-ray beam. All of these methods rely on the random orientation of crystals relative to the X-ray beam. In the figure, crystals that have not yet been probed with the X-ray beam are colored bright blue, while crystals that have already been probed and are damaged are colored dark blue.

X-ray beam (Roessler et al., 2016) (Fig. 5C). A unifying feature of all these different sample delivery methods is that the many crystals measured during the experiment should be delivered to the X-ray beam in random orientations to ensure complete sampling of reciprocal space.

The second way in which serial crystallography experiments differ from standard rotation experiments relates to the data processing. In a rotation crystallography experiment, the crystal orientation only needs to be determined once, because all the images in the dataset correspond to crystal orientations that are deterministically related to one another by a known rotation about the goniometer axis. This powerful constraint is absent in serial crystallography, because every image in the data set is produced by a randomly oriented crystal. Therefore, every image must be indexed and integrated separately before merging (Boutet et al., 2012; Chapman et al., 2011). An additional problem is that because the crystals are not oscillated in the beam during the collection of individual images, full reflections are not swept through the Ewald sphere, and therefore all measurements are effectively recorded as partial intensities. This problem can be handled in several ways.

One option is to measure a large number of reflections, and simply take the mean, so that errors due to partiality average out (White, 2014). A second approach is to utilize a geometric postrefinement procedure when scaling and merging the data, which provides an estimate of the partiality of each measured reflection, and can improve the estimation of the underlying true reflection intensities (Uervirojnangkoorn et al., 2015; White, 2014). Such procedures are especially valuable when data sets only contain a modest number of images. These postrefinement procedures have their roots in the field of virus crystallography (Rossmann, Leslie, Abdel-Meguid, & Tsukihara, 1979; Winkler, Schutt, & Harrison, 1979); here, massive unit cells prohibited sizeable crystal oscillations during data collection due to the introduction of reflection overlap, resulting in a similar partiality problem.

If the X-ray source permits, the partiality problem can also be addressed by broadening the energy spectrum of the beam, creating what is known as a "pink beam" (Meents et al., 2017). When the energy spectrum of the X-ray beam is broadened, the Ewald sphere effectively becomes thicker, allowing more interaction between the sphere and reciprocal lattice points, and a better estimation of the true reflection intensity. The idiosyncrasies of processing serial crystallographic data require purpose-built computer programs that differ from the ones that are used to process data from rotation experiments. The two primary computational tools for processing serial crystallographic data are the cctbx.xfel (Brewster et al., 2018) and CrystFEL (White, 2019) suites of programs.

3.3 Structural modeling with non-cryogenic data

In principle, modeling macromolecular structures from non-cryogenic data is no different than for cryogenic data, and identical refinement tools (Afonine et al., 2012; Murshudov et al., 2011) can be used in both cases. One notable feature of electron density maps derived from non-cryogenic experiments is that they contain fewer water molecules than those derived from cryogenic experiments (Deacon et al., 1997; Keedy et al., 2014). This observation speaks to the effect of temperature on the behavior solvent on the protein surface, which is critical for function (Fenimore, Frauenfelder, McMahon, & Parak, 2002). On the other hand, as discussed in Section 2.3, a major advantage of non-cryogenic experiments is that the resulting electron density often contains additional evidence for alternative conformations of the crystallized molecules (Fraser et al., 2009; Fraser et al., 2011).

To capitalize on this additional information, it is helpful to employ computational tools that are capable of systematically modeling alternative conformations throughout the molecule. Other Chapters within this Volume provide a detailed discussion of best practices for modeling conformational heterogeneity from crystallographic data. Therefore, this section will not discuss this topic in technical depth, but will instead simply introduce two relevant model types (Fig. 6) and tools for creating them. The first type of model is commonly referred to as a "multi-conformer" model, which has alternative conformations for some, but not all, atoms in the structure. These models can be created in an automated and systematic manner using the program qFit (Riley et al., 2021), which takes a standard model of a single protein conformation as input along with a crystallographic electron density map, and algorithmically identifies and models alternative conformations. The second type of model is an ensemble (or multi-member) model, where the entire macromolecule is represented in alternative conformations. This type of model can be generated using the PHENIX (Zwart et al., 2008) suite of software [phenix.ensemble_refinement (Burnley et al., 2012)].

A major challenge with ensemble models, and multi-conformer models to a lesser extent, is that they are prone to overfitting, because of the large number of parameters needed to represent the ensemble (Babcock, Keedy, Fraser, & Sivak, 2018). When implementing these modeling strategies, to

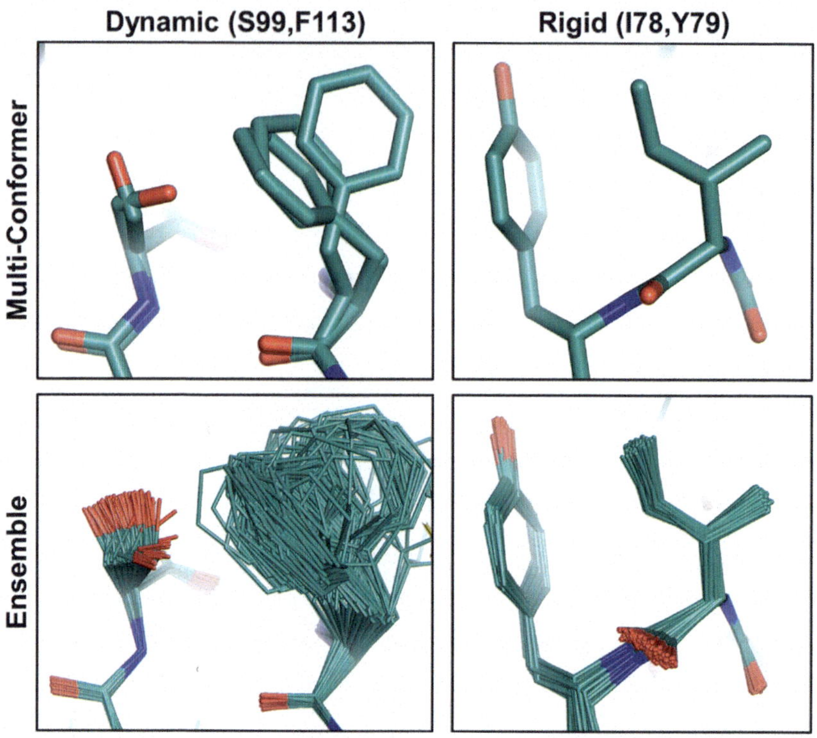

Fig. 6 Modeling conformational heterogeneity in crystal structures. The images depict the differences between two model types that are commonly used to represent conformational heterogeneity in proteins. The enzyme cyclophilin A (CypA) serves as an example, and two regions are highlighted; the sidechains of residues Ser99 and Phe113 are relatively dynamic, while the sidechains of residues Ile78 and Tyr79 are relatively rigid. The multi-conformer model (Keedy et al., 2015) includes multiple conformations for residues that display significant conformational heterogeneity, but not for residues that can be modeled well using a single conformation. On the other hand, every residue in the ensemble model (Burnley et al., 2012) contains the same number of conformations. In the case of the more dynamic residues, the side chain conformations are spread over different rotamers, whereas all the side chain conformations appear to correspond to the same rotamers for the more rigid residues.

guard against overfitting, one must watch the R-work/R-free gap as models are made increasingly complex. A recently-developed program called Vagabond also addresses the overfitting problem by reparameterizing the macromolecular structure in terms of bonds rather than atomic coordinates, and creating ensembles from deterministically related alternative conformations (Ginn, 2021).

To conclude this brief section, one might note that non-cryogenic diffraction experiments also enable the collection of diffuse scattering data, which can be used to model correlated motions of the lattice and within the crystallized molecules (Meisburger, Thomas, Watkins, & Ando, 2017; Meisburger, Case, & Ando, 2020; Meisburger, Case, & Ando, 2023; Wall, Wolff, & Fraser, 2018). This is another topic that is reviewed in great depth in other Chapters of this Volume, and therefore will only be mentioned briefly here as a potential opportunity arising from non-cryogenic macromolecular crystallography.

4. Temperature as a tool for studying conformational heterogeneity and dynamics

4.1 Multi-temperature crystallography

Temperature is a key state-variable that helps one understand the behavior of a dynamical molecular system. Specifically, when attempting to understand the behavior of a complex system, chemists and physicists have long implemented temperature perturbations to gain insight into energetics and kinetics. While this idea is commonplace in other scientific fields, it has been scarcely exploited by structural biologists attempting to understand the structures and dynamic properties of complex macromolecules, likely because the main techniques that are used to study the atomic structures of macromolecules, namely X-ray crystallography and cryoEM, have become reliant on cryocooling of samples.

The need for samples to be held in vacuum during cryoEM data collection will likely remain a prohibitive factor for performing electron microscopy at non-cryogenic temperatures; however, as described in Section 3.1, modern hardware and software developments in X-ray crystallography make non-cryogenic data collection easier than ever. This opens the door to performing high-resolution structural measurements of biological macromolecules as a function of temperature, allowing the structural biologist to explore their conformational ensembles with the tools of physical chemistry (Schmidt, Graber, Henning, & Srajer, 2010; Weik & Colletier, 2010). Experiments in which molecular structures are analyzed as a function of data collection temperature are referred to as "multi-temperature crystallography" experiments.

4.1.1 Practical considerations for multi-temperature crystallography

To perform a multi temperature crystallography experiment, a series of X-ray diffraction data sets are collected at varying temperatures. Each temperature that the investigator wishes to sample is essentially treated as an independent rotation crystallography experiment, using a separate crystal (or multiple crystals which then get merged). Because each temperature in the series is measured from a different crystal, it is important to develop a highly reproducible crystallization protocol. Additionally, care should be taken to treat all crystals the same during sample preparation in order to ensure that temperature is, in fact, the only difference between data sets in a multi-temperature series.

Many of the important practical considerations for performing multi-temperature data collection have been discussed already in this Chapter (Section 3.2). For data sets collected below 260 K, sample prep can be managed as if one is collecting "cryogenic" data, because the crystals will remain frozen during data collection. For data sets collected above 260 K, samples should be prepared for non-cryogenic data collection (see Section 3.2.1). Multi-temperature data collection requires a stable and adjustable cryojet device that can be set to maintain temperatures across the broad range of 100–300 K and higher if possible. These devices are standard at most macromolecular crystallography beamlines. Additionally, some cryojets are also equipped with a humidity stream, which offers yet another option for protecting crystals from dehydration at non-cryogenic temperatures.

An important consideration when performing multi-temperature experiments is that crystals should be equilibrated to their data collection temperatures in the cryojet stream. After mounting a crystal at ambient temperature, the crystal should be transferred to the goniometer in the cryojet stream at the desired data collection temperature. For example, if the goal is to collect data at 150 K, the experimenter should not plunge freeze the crystal in liquid nitrogen and then transfer to the diffractometer at 150 K, because then the crystal would be subjected to a cooling and warming process that other crystals (e.g. those measured at temperatures above the freezing point) did not experience. Instead, the crystal should be mounted and transferred to the diffractometer and allowed to freeze in the cryojet stream at 150 K.

This practice has a major advantage and a major drawback. The advantage is that all crystals will be subjected to the same temperature equilibration protocol, which sees them transferred from ambient temperature to data collection temperature. The disadvantage is that the cooling rate of crystals in the cryojet stream is not as fast as in liquid cryogen, and so the crystals can be prone to developing some of the cryocooling artifacts described in Section 2.2. This can be especially problematic at intermediate temperatures, between the solvent glass transition (~200 K) and the freezing temperature of water (273 K). The problem can be mitigated by using smaller crystals that freeze faster, although this is a delicate compromise, because smaller crystals typically yield lower resolution data, and the structural modeling protocols commonly used to analyze multi-temperature experiments (see below) benefit greatly from high-resolution information. An additional strategy for improving the freezing of macromolecular crystals is slow cooling (Warkentin & Thorne, 2009).

Once crystals are successfully mounted on the goniometer and equilibrated to the data collection temperature (within seconds), the data can be collected. It is important to collect all data sets in a multi-temperature series using the same X-ray dose. This means that crystals measured at low temperatures, which may be robust to radiation damage, should receive the same dose as crystals measured at non-cryogenic temperatures. Once recorded, the data can be processed as they would for any typical rotation crystallography experiment (see Section 3.2.2).

4.1.2 Studying the conformational ensemble with multi-temperature crystallography

The idea that temperature can be used as a physical perturbation in macromolecular crystallography to probe conformational ensembles is motivated by fundamental statistical mechanics—the Boltzmann distribution dictates that increasing the temperature of a molecular ensemble will increase the population of molecules in high energy states. The first multi-temperature crystallography experiments were performed decades ago on myoglobin (Frauenfelder, Petsko, & Tsernoglou, 1979; Frauenfelder et al. 1987) and ribonuclease-A (Tilton, Dewan, & Petsko, 1992). These studies demonstrated that proteins undergo a small thermal expansion, and that B-factors also increase as a function of temperature, suggesting an increase in dynamics of the crystallized molecules.

The observed B-factor increase in these early experiments is consistent with the idea that higher energy conformational states are being additionally populated at higher temperatures. Much more recently, in 2015, Keedy, et al. reported an extremely detailed multi-temperature crystallographic study of the proline isomerase enzyme, cyclophilin A (CypA), which demonstrated the power of the technique for mapping the conformational ensembles of proteins (Keedy et al., 2015). In this work, the authors collected eight distinct X-ray diffraction data sets across a broad range of temperatures, spanning 100–310 K. These data sets were used to algorithmically generate multi-conformer models of the enzyme with qFit (Riley et al., 2021) (see Section 3.3), which were then compared across temperatures using a composite, per-residue analysis of both B-factors and alternative conformations.

Above the solvent glass transition, most parts of CypA tended to become more heterogeneous as a function of temperature, as expected. Interestingly, the response to increasing temperature was not uniform across the enzyme, with some residues showing large increases in conformational heterogeneity, and others showing only modest increases. Additionally, there were several regions of the enzyme that appeared less dynamic with increasing temperature. This analysis yielded detailed insight into how conformational equilibria change as a function of temperature in proteins. The authors of this work also suggested that multi-conformer models derived from multi-temperature data can be used to assess which dynamic features of a protein are thermodynamically coupled to one another, by correlating their temperature-dependent behavior. Hypothetically, regions of the protein that have similar changes in heterogeneity as a function of temperature may be dynamically coupled to one another.

A powerful aspect of multi-temperature macromolecular crystallography is that it opens the door to experimental measurement of the thermodynamics that underlie observed conformational changes. When discreet alternative conformations of a macromolecule are modeled in a crystal structure and their occupancies are refined against the diffraction data, the refined occupancies give direct insight into equilibrium populations of the modeled conformations. These populations can provide estimates of equilibrium constants for conformational interconversions. Thus, for conformationally heterogeneous regions of a molecular structure, a multi-temperature crystallography experiment provides equilibrium constant information as a function of temperature.

These data can form the basis of a Van't Hoff analysis, which allows the calculation of enthalpy changes (ΔH) and entropy changes (ΔS) associated with a conformational transition, and their temperature-dependence. Although the theoretical basis for this type of analysis is valid, a detailed Van't Hoff analysis has not yet been performed using multi-temperature macromolecular crystallography data, and this represents a frontier in the field.

At present, a major limitation is likely the accuracy of occupancy values that are refined against the diffraction data. An additional frontier area of research in multi-temperature crystallography is to co-refine structures simultaneously across all temperatures, with occupancies restrained to vary smoothly across a temperature series. This type of refinement has not been performed in macromolecular crystallography; however, a related technique known as "parametric refinement" has been applied to analogous problems in materials science (Stinton & Evans, 2007). Implementing this type of refinement approach would likely enable detailed thermodynamic analysis from multi-temperature, multi-conformer models of macromolecules.

4.1.3 Examples of multi-temperature crystallography to study protein function

Multi-temperature crystallography can provide insight into the thermodynamics of a macromolecule's conformational ensemble, and this information can lead to a greater understanding of how conformational changes drive protein function. This capability is highlighted by an increasingly large body of work. In the previous section, a multi-temperature crystallographic study of CypA by Keedy, et al. (Keedy et al., 2015) was introduced, and used to describe the results that could be derived from such an experiment. Note that the observations in this study were consequential for understanding the functional mechanism of this enzyme.

In CypA, there is a network of amino acid sidechains (Arg55, Met61, Ser99, Phe113) that are known to undergo rotamer interconversions that are critical for catalytic turnover (Fraser et al., 2009). Analysis of rotamer occupancies as a function of temperature for these residues suggested that two of the four (Ser99, Phe113), which are buried deeper in the protein core, form a more tightly-coupled sub-network. This observation refined a previous model based on NMR spectroscopy, where a single exchange rate was fit to all residues in this network (Eisenmesser et al., 2005). Since 2015, several additional examples in which multi-temperature crystallography has been used to study the role of conformational heterogeneity in defining protein function have been reported. We summarize three examples here.

In 2018, in a different study, Keedy, et al. used multi-temperature crystallography to investigate allostery in protein tyrosine phosphatase 1B (Ptp1B) (Keedy et al., 2018). Ptp1B is an important drug target in diabetes (Elchebly et al., 1999) and cancer (Krishnan et al., 2014). However, most orthosteric inhibitors of Ptp1B suffer from off-target effects as a result of inhibiting other homologs of the enzyme. In an effort to advance the design of allosteric inhibitors, the authors used multi-temperature crystallography to characterize the motion of the WPD loop, which opens and closes over the active site. Additionally, they searched for other regions of the protein that moved in concert with the WPD loop, identifying two distinct allosteric sites on the protein that were coupled to the active site. The ability of these allosteric sites, identified through multi-temperature crystallography, to perturb active site function was validated using a chemical tethering approach. This work demonstrated the potential value of multi-temperature crystallography for drug discovery.

In 2020, Otten, et al. implemented an interesting use of non-cryogenic X-ray crystallography in a study of how directed evolution reshaped the energy landscape in an enzyme to improve catalytic function (Otten et al., 2020). Specifically, the authors used crystallography and NMR to describe how successive mutations altered the conformational heterogeneity of the enzyme, to populate a narrow, highly active ensemble. This study is not a full multi-temperature series, but remarkably, the authors were able to collect crystallographic data at 70 °C (343 K). Equilibrating the crystal at this high temperature populated a hidden active conformation of the enzyme that was so sparsely occupied it could not be modeled previously. The large temperature perturbation employed in this study shows an extreme case of what is possible using temperature as a perturbation in macromolecular crystallography. Indeed, it may be possible to regularly determine crystal structures of folded proteins at temperatures that exceed their melting temperatures in solution, because the crystal lattice will provide additional stabilization to the folded structure.

In 2022, in response to the COVID-19 pandemic, Ebrahim, et al. performed a multi-temperature crystallographic study of the SARS-CoV-2 main protease (Mpro) (Ebrahim et al., 2022), with the goal of characterizing the conformational landscape and elucidating conformational states that could be targeted with novel orthosteric and allosteric inhibitors. They collected data at five temperatures spanning a range of 100–300 K, and performed a detailed analysis of conformational heterogeneity, following in the footsteps of Keedy, et al. (Keedy et al., 2015). Analysis of multi-conformer and ensemble models across the temperature series revealed coupled conformational changes across the active site, substrate-binding pocket, interdomain interface, and parts of the broad dimer interface, which offer intriguing possibilities for the design of small molecules that disrupt function by interacting with these structural elements. Additionally, in this work the authors explored the role of relative humidity as a perturbation in addition to temperature, and found that altering humidity led to an observable change in the protein hydration shell. This idea, and its relationship to temperature perturbation was also explored by Atakisi, et al. in a study of lysozyme crystals (Atakisi, Moreau, & Thorne, 2018).

4.2 Time-resolved crystallography using temperature-jump

Electron density from multi-temperature X-ray crystallography often reveals multiple conformations of macromolecules, as described extensively above, and their occupancies represent the equilibrium populations of the observed conformational states. Experiments performed at multiple temperatures cannot reveal conformations that are not significantly populated at equilibrium, notably, the intermediate states that are transiently populated along a particular trajectory of motion (Fig. 7). Additionally, because molecular motion is very fast relative to the experimental measurement time, all kinetic information is lost, and we cannot retrieve any information about the kinematic details (dynamics) of the observed conformational changes. NMR and other spectroscopic measurements are ideal for studying the kinetics of molecular motions, but it can be difficult to correlate spectroscopic observables with the atomic details of specific conformational transitions that are observed crystallographically. Simulations

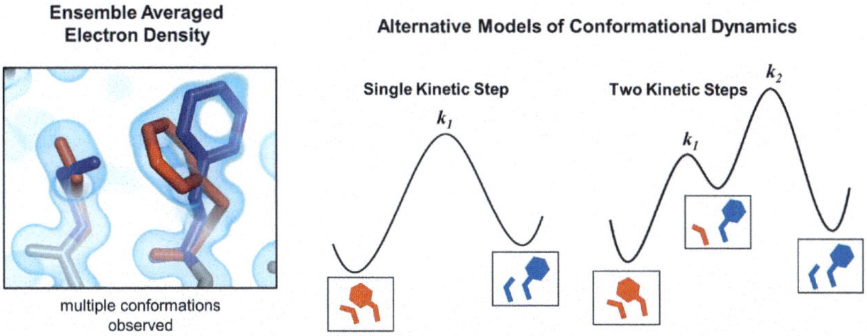

Fig. 7 Limitations of spatiotemporally-averaged electron density. This simple example shows how the electron density derived from traditional crystallographic experiments, which represents an ensemble average over space and time, is insufficient to describe the dynamics that underlie any observed conformational heterogeneity. The electron density on the left shows clear alternative conformations for the serine and phenylalanine residues shown [S99 and F113 of CypA (Fraser et al., 2009)]; however, multiple different models of conformational exchange (shown) could result in the approximately the same observed density. Time-resolved measurements have the potential ability to disambiguate these models.

can provide additional insight, but computational studies are not feasible for large systems with dynamics that occur on the microsecond to millisecond time scale.

Overcoming the challenges highlighted above, time-resolved crystallography is a powerful tool for studying functional protein motions (Moffat, 2001; Orville, 2020; Šrajer & Schmidt, 2017; Thompson et al., 2020; Wilson, 2022). This method utilizes brilliant, ultrafast X-ray pulses to provide simultaneous information about the structure and dynamics of macromolecules at high spatial and temporal resolution, thus acting like a "molecular high-speed camera." The approach allows direct measurement of structural dynamics in real time using a pump-probe strategy, where the molecules in the sample are rapidly perturbed to initiate synchronous conformational changes, and then measured at defined time delays following the perturbation to map the resulting structural changes.

A major technical bottleneck for time-resolved crystallography experiments is the need to rapidly synchronize conformational changes in the billions of molecules that constitute the sample. Historically, time-resolved crystallography has been reserved as a niche technique for studying photoactive proteins, which are easy to perturb through optical excitation of a protein-bound chromophore using a pulsed laser. The multi-temperature crystallography experiments described above (Section 4.1) demonstrate across multiple protein systems that increasing the temperature of crystallized macromolecules generally enhances populations of higher-energy conformations that are invisible at lower temperatures, suggesting temperature as viable perturbation for time-resolved measurements. Because temperature is a universal perturbation—the conformational ensembles of all complex macromolecules are sensitive to changes of this state variable—the use of rapid temperature change as a perturbation of conformational dynamics has the potential to make time-resolved crystallography a general tool for structural biology.

4.2.1 Infrared laser-induced temperature-jump

Rapid temperature change as a perturbation of chemical dynamics, known as a temperature-jump (T-jump), has been applied for decades to study chemical kinetics in a variety of different contexts (Crooks, 1983). In biochemistry, the method has been used to study protein folding (Gruebele, Sabelko, Ballew, & Ervin, 1998; Gruebele, 1999) and enzyme catalyzed reactions (Meadows, Balakrishnan, Kier, Spiro, & Klinman, 2015; Phillips et al., 2008). The idea of a T-jump centers on a temperature

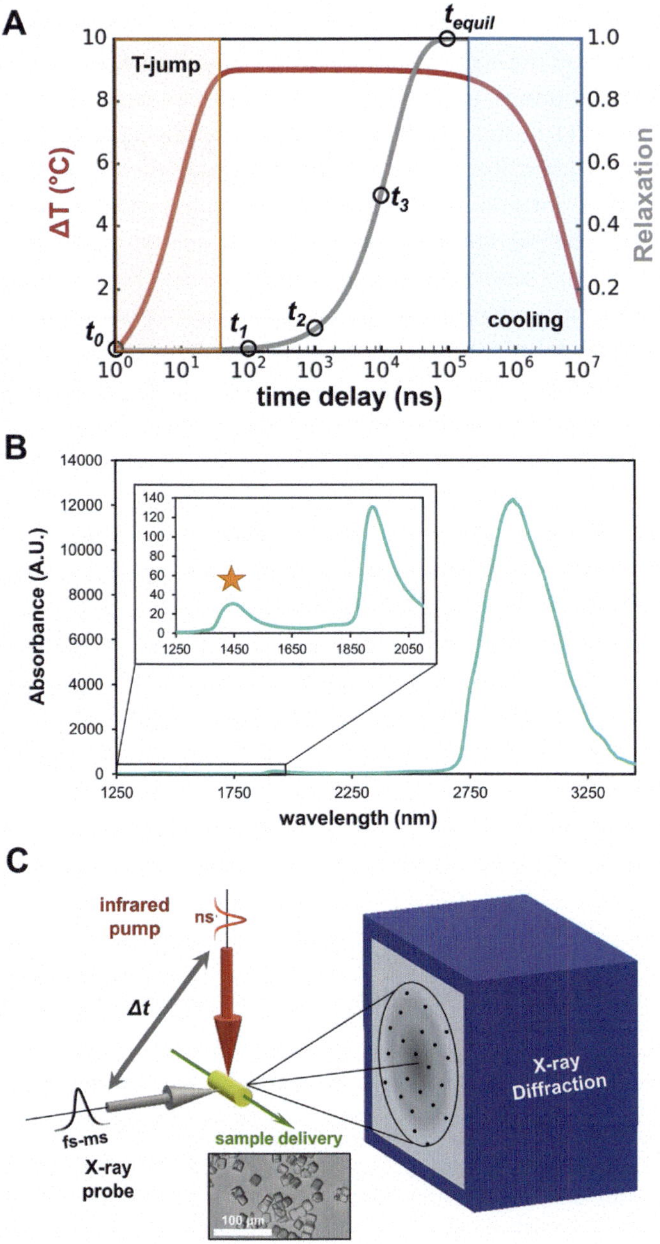

Fig. 8 Summary of T-jump crystallography. (A) In a T-jump experiment, an infrared laser pulse is used to rapidly heat an aqueous solution (red line), leading to nearly instantaneous heating, which removes any macromolecules in the solution from thermal equilibrium. On timescales that are slower than the heating pulse, the

(Continued)

change being so fast relative to the chemical process of interest that the perturbation takes the system out of thermal equilibrium, forcing it to relax to a new equilibrium (Fig. 8A). During this relaxation process, many molecules should be undergoing the same motions at roughly the same time, which allows for non-equilibrium measurements of the system. T-jumps can be introduced in a solution through excitation of molecular vibrations in the solvent molecules, which converts light energy into kinetic energy that results in a temperature increase. Because water is the common solvent for biological macromolecules, vibrational excitation of water is a universal method for increasing their temperature in solution.

Liquid water has relatively strong infrared absorbance properties, which relate to the excitation of molecular vibrations. Specifically, the absorption spectrum has peaks at approximately 1.4 and 1.9 μm, and relatively broad absorbance in the range of 2.5–4 μm (Palmer & Williams, 1974) (Fig. 8B). For most biological T-jump experiments, the absorbance peak at 1.4 μm, corresponding to the OH stretching overtone, is utilized to introduce T-jump. Absorbance at this wavelength is substantially weaker than at some other wavelengths. By utilizing another wavelength where absorbance is stronger, a larger T-jump could be achieved with the same amount of light; however, the increased absorbance would lead to a more significant temperature gradient within the sample, because the incident light intensity decreases as the beam passes through and energy is absorbed. The weaker absorbance at 1.4 μm provides an ideal compromise between the magnitude of the T-jump and attenuation through the sample.

In addition to being the correct wavelength to excite vibrations in the solvent, the incident light used for T-jump must also be delivered significantly quickly for the system to be taken out of thermal equilibrium. For studying biological macromolecules, laser sources that can deliver

Fig. 8—Cont'd macromolecules undergo conformational changes as they relax to a new thermal equilibrium (gray line), and rapid structural measurements are performed during the relaxation process (black circles). Note the logarithmic scaling of the time axis in the plot. (B) The absorbance spectrum of water (Palmer & Williams, 1974) shows several peaks in the infrared, including a large peak in the region of 2.5–3.5 μm (2500–3500 nm), and two smaller peaks centered at 1.9 μm (1900 nm) and 1.4 μm (1400 nm). The absorbance peak that is most commonly used for biological T-jump experiments is the peak at 1.4 μm (1400 nm) because it provides a good compromise between absorbance and penetration, and it is accessible with many tunable laser systems. (C) T-jump crystallography experiments are performed in a pump-probe serial format, with precise control of the delay between the infrared heating pulses and the X-ray pulses used to measure molecular structure.

several mJ of energy in 10 ns or less are ideal, because T-jumps of 10 K or more can typically be achieved and the duration of the heating pulse is faster than many functional motions, which take place in the microsecond to millisecond time scale. Additionally, because thermal equilibration of molecules typically occurs on the order of picoseconds (Rubtsov & Burin, 2019), one assumes that sample heating is homogenous over the duration of the nanosecond laser pulse. This assumption holds true for crystalline macromolecules as well, because their crystals tend to have large channels that are permeated with bulk solvent that also undergoes infrared irradiation and experiences a T-jump.

4.2.2 Practical considerations for T-jump crystallography experiments

Infrared laser-induced T-jump can be utilized as a rapid perturbation for time-resolved crystallography measurements of macromolecular structural dynamics (Wolff et al., 2022). T-jump crystallography experiments occur in a unique pump-probe format (Fig. 8C) that relies on the use of serial crystallography (see Section 3.2.3). In the serial crystallography experiment, a single X-ray diffraction snapshot is measured from each of many crystals using an X-ray pulse. To incorporate the T-jump, a pulsed infrared laser tuned to a wavelength of approximately 1.4 μm is focused on the "interaction region" where X-rays impinge on the sample. The laser pulses rapidly heat microcrystals as they pass through the interaction region (in either a microfluidic or fixed-target format, see Section 3.2.3), and the timing of infrared and X-ray pulses is coordinated so that the heating occurs at a defined time delay prior to structural measurement with an X-ray pulse. The heating is effectively instantaneous (typically nanoseconds, depending on the laser source) relative to slower (microsecond to millisecond) molecular motions, and the macromolecules are removed from thermal equilibrium. By sampling multiple pump-probe time delays, it is possible to map both the spatial and temporal details (structure and kinetics) of molecular motion.

An important aspect of this experiment is that it requires the ability to generate suitable X-ray pulses, which is only possible at XFEL sources or specialized synchrotron beamlines (Cammarata et al., 2009; Graber et al., 2011). The temporal resolution of the experiment is limited by the duration of either the pump (infrared) or probe (X-ray) pulses, whichever is longer in duration. Alternatively, time-resolved experiments can be performed without a pulsed X-ray source, if a fast frame rate detector is available (Singh et al., 2013). In this case, the temporal resolution is set by the detector frame rate.

There are several critical aspects of the experimental setup described above that must be considered carefully when designing a T-jump crystallography experiment. First, because the T-jump relies on the presence of water, crystals should be prepared for serial crystallography in an aqueous solution or carrier medium. If high sample viscosity is desired, hydroxyethyl cellulose is an ideal material (Sugahara et al., 2017), because it is aqueous and does not undergo unusual phase transitions like many of its hydrophobic counterparts including LCP (Czeslik, Winter, Rapp, & Bartels, 1995). Next, the properties of the infrared pump laser are crucial for success. The role of the wavelength in creating a uniform temperature jump and of the pulse duration in defining the temporal resolution of the experiment have already been mentioned. Additionally, the overall fluence (mJ/mm^2) of the infrared beam determines the magnitude of the T-jump that is introduced in the sample. High fluence at the sample can be achieved by producing many photons per pulse (high pulse energy), and/or by focusing the beam to a smaller area. Special cameras that are sensitive to the mid-infrared region of the electromagnetic spectrum are also critical for aligning the pump and probe beams with one another and with the sample.

A major question when performing T-jump experiments relates to the instantaneous temperature of the sample during the X-ray measurement. Fortunately, samples of macromolecules contain a sensitive internal thermometer, because the X-ray scattering of liquid water is highly sensitive to temperature. By analyzing the radial average of the solvent scattering background that is present in the diffraction images, it is possible to detect the presence or absence of a temperature jump, and in favorable cases, to estimate its magnitude (Cammarata et al., 2006; Cho et al., 2018; Thompson et al., 2019; Wolff et al., 2022).

The ability to use the solvent scattering signal as a proxy for temperature also allows the experimenter to monitor the system for cooling. In a T-jump experiment, the longest pump-probe time delays that can be sampled are limited by sample cooling, so it is valuable to have a reliable estimate of when the effect becomes limiting. Empirically, using current experimental setups, the longest possible pump-probe time delays that can be achieved are hundreds of microseconds (Thompson et al., 2019). As of this writing, T-jump crystallography is a new technique, and one might note that new hardware and software to facilitate these experiments are in rapid development.

Extracting meaningful structural changes from the results of a time-resolved crystallography experiment can be challenging, and this is especially true for T-jump experiments. Typically, the modest T-jumps that can be practically introduced only shift the populations of conformational states by a very small amount. Therefore, direct refinement of atomic coordinates against structure factor data in the presence and absence of a T-jump will likely produce identical models. In order to isolate the time-resolved signals that arise from the T-jump, it is necessary to analyze difference structure factors and difference electron density maps calculated between data measured with and without introduction of a T-jump. Difference electron density is sensitive to subtle differences between the underlying structure factors, and the small signals that represent protein motions are visible in time-resolved difference maps (Ursby & Bourgeois, 1997). Time-resolved difference density can be quantified for direct analysis (Wickstrand et al., 2020), or it can be used to create structural models of high-energy conformational states that can be refined against the raw or extrapolated structure factor data (De Zitter, Coquelle, Oeser, Barends, & Colletier, 2022), analysis methods that have become commonplace in time-resolved crystallography. It is worth noting that T-jump generally leads to widespread structural changes throughout the unit cell, which can be problematic for structure factor extrapolation methods that rely on scalar approximations (Genick, 2007), so this method should be approached with caution for T-jump experiments.

4.2.3 Studying macromolecular dynamics with T-jump

T-jump has only recently been combined with X-ray scattering to be used as a tool for structural biology. Consequently, there are only limited examples of its use, and they will be reviewed briefly here. The majority of time-resolved T-jump X-ray experiments in structural biology (3/4) have been performed using small- and wide-angle X-ray scattering (SAXS/WAXS) as a probe, rather than crystallography.

The first proof-of-principle that T-jump could be combined with X-ray scattering to study macromolecular dynamics came from a series of papers published by Rimmerman and coworkers (Rimmerman et al., 2017, 2018). In this work, the authors used T-jump to induce changes in insulin oligomerization state paired with SAXS/WAXS to follow the dynamics from nanoseconds to milliseconds. Analysis of the time-resolved scattering curves by singular value decomposition (SVD) revealed several transient intermediates involved in the association and dissociation of oligomers.

Later in 2018, Cho, et al. also performed T-jump SAXS/WAXS on the R-state of tetrameric hemoglobin, to study the effect of R-state heterogeneity on the transition from the R-state to the T-state (Cho et al., 2018). In 2019, Thompson et al. again used T-jump paired with SAXS/WAXS to study the internal dynamics of the enzyme cyclophilin A (CypA) (Thompson et al., 2019). In this work, two distinct functional motions could be mapped, one corresponding to catalysis and the other to substrate recognition. Importantly, in this study, mutants of the enzyme that are known to have altered catalytic and substrate binding functions were used as controls to confirm the functional relevance of the observed time-resolved signals.

More recently, Wolff and coworkers demonstrated the first application of T-jump to time-resolved macromolecular crystallography (Wolff et al., 2022). Here the authors used microcrystals of the enzyme lysozyme in a pump-probe serial T-jump crystallography experiment, as described in detail above (Section 4.2.2). The authors measured data corresponding to 20 ns, 20 μs, and 200 μs pump-probe time delays. Analysis of time-resolved difference electron density maps revealed signals related to heating of the enzyme, manifesting as increased harmonic motions (B-factors) and side chain rotations at 20 ns after the introduction of T-jump, followed by dissipation of those short amplitude motions into larger, correlated loop motions at longer pump-probe time delays. These larger amplitude motions, which appeared to be involved in the well-known hinge-bending motion of lysozyme (Mccammon, Gelin, Karplus, & Wolynes, 1976), could be included in multi-conformer models and refined against the time-resolved data. Similar to the SAXS/WAXS experiments by Thompson, et al. (Thompson et al., 2019), the T-jump crystallography study by Wolff, et al. included an important functional control. T-jump crystallography experiments were repeated in the presence of chitobiose, an enzyme inhibitor that arrests the hinge-bending motion of lysozyme. In the presence of chitobiose, the time-resolved signals do not dissipate in the same manner as they do for the apo enzyme, which corroborated the connection between the time-resolved signal changes and the hinge-bending motion of the enzyme.

5. Conclusion

Temperature is a key variable that governs the dynamic behavior of complex macromolecules, but it has been underutilized as a tool for structural biology. Given the traditional reliance of high-resolution structural techniques on sample cryocooling, this is somewhat unsurprising. Fortunately, modern advances in X-ray crystallography eliminate the need for cryocooling in many cases. Using sensitive detectors and modern data-processing software, it is possible to collect high-resolution X-ray diffraction with a very low X-ray dose, mitigating the ill effects of radiation damage. The ability to collect diffraction data near ambient/physiological temperature eliminates artifacts that are associated with cryocooling, and enables the use of temperature as a perturbation to study structure and dynamics using multi-temperature and time-resolved T-jump crystallography. Experiments that integrate temperature perturbations with high-resolution X-ray crystallography can provide detailed insight into the correlated motions that drive the functions of biological macromolecules.

References

Afonine, P. V., Grosse-Kunstleve, R. W., Echols, N., Headd, J. J., Moriarty, N. W., Mustyakimov, M., ... Adams, P. D. (2012). Towards automated crystallographic structure refinement with phenix.refine. *Acta Crystallographica. Section D, Biological Crystallography, 68*(Pt 4), 352–367. https://doi.org/10.1107/S0907444912001308.

Amemiya, Y. (2008). Imaging plate - X-ray area detector based on photostimulable phosphor. *Synchrotron Radiation News, 3*, 21–26. https://doi.org/10.1080/08940889008602552.

Atakisi, H., Moreau, D. W., & Thorne, R. E. (2018). Effects of protein-crystal hydration and temperature on side-chain conformational heterogeneity in monoclinic lysozyme crystals. *Acta Crystallographica. Section D, Structural Biology, 74*(Pt 4), 264–278. https://doi.org/10.1107/S2059798318000207.

Baba, S., Hoshino, T., Ito, L., & Kumasaka, T. (2013). Humidity control and hydrophilic glue coating applied to mounted protein crystals improves X-ray diffraction experiments. *Acta Crystallographica. Section D, Biological Crystallography, 69*(Pt 9), 1839–1849. https://doi.org/10.1107/S0907444913018027.

Babcock, N. S., Keedy, D. A., Fraser, J. S., & Sivak, D. A. (2018). Model selection for biological crystallography. *bioRxiv*. 448795. ⟨https://doi.org/10.1101/448795⟩.

Baxter, E. L., Aguila, L., Alonso-Mori, R., Barnes, C. O., Bonagura, C. A., Brehmer, W., ... Cohen, A. E. (2016). High-density grids for efficient data collection from multiple crystals. *Acta Crystallographica. Section D, Structural Biology, 72*(1), 1. https://doi.org/10.1107/S2059798315020847.

Boutet, S., Lomb, L., Williams, G. J., Barends, T. R. M., Aquila, A., Doak, R. B., ... Schlichting, I. (2012). High-resolution protein structure determination by serial femtosecond crystallography. *Science (New York, N. Y.), 337*(6092), 362–364. https://doi.org/10.1126/science.1217737.

Brewster, A. S., Waterman, D. G., Parkhurst, J. M., Gildea, R. J., Young, I. D., O'Riordan, L. J., ... Sauter, N. K. (2018). Improving signal strength in serial crystallography with DIALS geometry refinement. *Acta Crystallographica. Section D, Structural Biology, 74*(9), 9. https://doi.org/10.1107/S2059798318009191.

Burnley, B. T., Afonine, P. V., Adams, P. D., & Gros, P. (2012). Modelling dynamics in protein crystal structures by ensemble refinement. *ELife, 1*, e00311. https://doi.org/10.7554/eLife.00311.

Bury, C. S., Brooks-Bartlett, J. C., Walsh, S. P., & Garman, E. F. (2018). Estimate your dose: RADDOSE-3D. *Protein Science: A Publication of the Protein Society, 27*(1), 217–228. https://doi.org/10.1002/pro.3302.

Cammarata, M., Eybert, L., Ewald, F., Reichenbach, W., Wulff, M., Anfinrud, P., ... Polachowski, S. (2009). Chopper system for time resolved experiments with synchrotron radiation. *The Review of Scientific Instruments, 80*(1), 015101. https://doi.org/10.1063/1.3036983.

Cammarata, M., Lorenc, M., Kim, T. K., Lee, J. H., Kong, Q. Y., Pontecorvo, E., ... Ihee, H. (2006). Impulsive solvent heating probed by picosecond x-ray diffraction. *The Journal of Chemical Physics, 124*(12), 124504. https://doi.org/10.1063/1.2176617.

Chapman, H. N., Fromme, P., Barty, A., White, T. A., Kirian, R. A., Aquila, A., ... Spence, J. C. H. (2011). Femtosecond X-ray protein nanocrystallography. *Nature, 470*(7332), 73–77. https://doi.org/10.1038/nature09750.

Cho, H. S., Schotte, F., Stadnytskyi, V., DiChiara, A., Henning, R., & Anfinrud, P. (2018). Dynamics of quaternary structure transitions in R-state carbonmonoxyhemoglobin unveiled in time-resolved X-ray scattering patterns following a temperature jump. *The Journal of Physical Chemistry. B, 122*(49), 11488–11496. https://doi.org/10.1021/acs.jpcb.8b07414.

Crooks, J. E. (1983). The temperature-jump technique for the study of fast reactions in solution. *Journal of Physics E: Scientific Instruments, 16*(12), 1142–1147. https://doi.org/10.1088/0022-3735/16/12/003.

Czeslik, C., Winter, R., Rapp, G., & Bartels, K. (1995). Temperature- and pressure-dependent phase behavior of monoacylglycerides monoolein and monoelaidin. *Biophysical Journal, 68*(4), 1423–1429. https://doi.org/10.1016/S0006-3495(95)80315-2.

De La Mora, E., Coquelle, N., Bury, C. S., Rosenthal, M., Holton, J. M., Carmichael, I., ... Weik, M. (2020). Radiation damage and dose limits in serial synchrotron crystallography at cryo- and room temperatures. *Proceedings of the National Academy of Sciences of the United States of America, 117*(8), 4142–4151. https://doi.org/10.1073/pnas.1821522117.

De Zitter, E., Coquelle, N., Oeser, P., Barends, T. R. M., & Colletier, J.-P. (2022). Xtrapol8 enables automatic elucidation of low-occupancy intermediate-states in crystallographic studies. *Communications Biology, 5*(1), 1. https://doi.org/10.1038/s42003-022-03575-7.

Deacon, A., Gleichmann, T., Kalb (Gilboa), A. J., Price, H., Raftery, J., Bradbrook, G., ... Helliwell, J. R. (1997). The structure of concanavalin A and its bound solvent determined with small-molecule accuracy at 0.94 [Aring]resolution. *Journal of the Chemical Society, Faraday Transactions, 93*(24), 4305–4312. https://doi.org/10.1039/A704140C.

DePonte, D. P., Weierstall, U., Schmidt, K., Warner, J., Starodub, D., Spence, J. C. H., & Doak, R. B. (2008). Gas dynamic virtual nozzle for generation of microscopic droplet streams. *Journal of Physics D: Applied Physics, 41*(19), 195505. https://doi.org/10.1088/0022-3727/41/19/195505.

Diamond, R. (1969). Profile analysis in single crystal diffractometry. *Acta Crystallographica. Section A, Crystal Physics, Diffraction, Theoretical and General Crystallography, 25*(1), 43–55. https://doi.org/10.1107/s0567739469000064.

Dobrianov, I., Caylor, C., Lemay, S. G., Finkelstein, K. D., & Thorne, R. E. (1999). X-ray diffraction studies of protein crystal disorder. *Journal of Crystal Growth, 196*(2), 511–523. https://doi.org/10.1016/S0022-0248(98)00833-1.

Ebrahim, A., Riley, B. T., Kumaran, D., Andi, B., Fuchs, M. R., McSweeney, S., & Keedy, D. A. (2022). The temperature-dependent conformational ensemble of SARS-CoV-2 main protease (Mpro). *IUCrJ, 9*(5), 5. https://doi.org/10.1107/S2052252522007497.

Eisenmesser, E. Z., Millet, O., Labeikovsky, W., Korzhnev, D. M., Wolf-Watz, M., Bosco, D. A., ... Kern, D. (2005). Intrinsic dynamics of an enzyme underlies catalysis. *Nature, 438*(7064), 117–121. https://doi.org/10.1038/nature04105.

Elchebly, M., Payette, P., Michaliszyn, E., Cromlish, W., Collins, S., Loy, A. L., ... Kennedy, B. P. (1999). Increased insulin sensitivity and obesity resistance in mice lacking the protein tyrosine phosphatase-1B gene. *Science (New York, N. Y.), 283*(5407), 1544–1548. https://doi.org/10.1126/science.283.5407.1544.

Evans, G., Axford, D., & Owen, R. L. (2011). The design of macromolecular crystallography diffraction experiments. *Acta Crystallographica. Section D, Biological Crystallography, 67*(4), 261–270. https://doi.org/10.1107/S0907444911007608.

Evans, P. (2006). Scaling and assessment of data quality. *Acta Crystallographica. Section D, Biological Crystallography, 62*(Pt 1), 72–82. https://doi.org/10.1107/S0907444905036693.

Fenimore, P. W., Frauenfelder, H., Magazù, S., McMahon, B. H., Mezei, F., Migliardo, F., ... Stroe, I. (2013). Concepts and problems in protein dynamics. *Chemical Physics, 424*, 2–6. https://doi.org/10.1016/j.chemphys.2013.06.023.

Fenimore, P. W., Frauenfelder, H., McMahon, B. H., & Parak, F. G. (2002). Slaving: Solvent fluctuations dominate protein dynamics and functions. *Proceedings of the National Academy of Sciences of the United States of America, 99*(25), 16047–16051. https://doi.org/10.1073/pnas.212637899.

First Analysis of Macromolecular Crystals | Springer Nature Experiments. (n.d.). Retrieved June 1, 2023, from ⟨https://experiments.springernature.com/articles/10.1385/1–59745-266–1:43⟩.

Fischer, M., Shoichet, B. K., & Fraser, J. S. (2015). One crystal, two temperatures: Cryocooling penalties alter ligand binding to transient protein sites. *Chembiochem: A European Journal of Chemical Biology, 16*(11), 1560–1564. https://doi.org/10.1002/cbic.201500196.

Foadi, J., Aller, P., Alguel, Y., Cameron, A., Axford, D., Owen, R. L., ... Evans, G. (2013). Clustering procedures for the optimal selection of data sets from multiple crystals in macromolecular crystallography. *Acta Crystallographica Section D: Biological Crystallography, 69*(8), 1617–1632. https://doi.org/10.1107/S0907444913012274.

Förster, A., Brandstetter, S., & Schulze-Briese, C. (2019). Transforming X-ray detection with hybrid photon counting detectors. *Philosophical Transactions of the Royal Society A: Mathematical, Physical and Engineering Sciences, 377*(2147), 20180241. https://doi.org/10.1098/rsta.2018.0241.

Fraser, J. S., Clarkson, M. W., Degnan, S. C., Erion, R., Kern, D., & Alber, T. (2009). Hidden alternative structures of proline isomerase essential for catalysis. *Nature, 462*(7273), 669–673. https://doi.org/10.1038/nature08615.

Fraser, J. S., Van Den Bedem, H., Samelson, A. J., Lang, P. T., Holton, J. M., Echols, N., & Alber, T. (2011). Accessing protein conformational ensembles using room-temperature X-ray crystallography. *Proceedings of the National Academy of Sciences of the United States of America, 108*(39), 16247–16252. https://doi.org/10.1073/pnas.1111325108.

Frauenfelder, H., Chen, G., Berendzen, J., Fenimore, P. W., Jansson, H., McMahon, B. H., ... Young, R. D. (2009). A unified model of protein dynamics. *Proceedings of the National Academy of Sciences of the United States of America, 106*(13), 5129–5134. https://doi.org/10.1073/pnas.0900336106.

Frauenfelder, H., Hartmann, H., Karplus, M., Kuntz, I. D., Kuriyan, J., Parak, F., ... Connolly, M. L. (1987). Thermal expansion of a protein. *Biochemistry, 26*(1), 254–261. https://doi.org/10.1021/bi00375a035.

Frauenfelder, H., Petsko, G. A., & Tsernoglou, D. (1979). Temperature-dependent X-ray diffraction as a probe of protein structural dynamics. *Nature, 280*(5723), 558–563. https://doi.org/10.1038/280558a0.

Fuchs, M. R., Pradervand, C., Thominet, V., Schneider, R., Panepucci, E., Grunder, M., ... Wang, M. (2014). D3, the new diffractometer for the macromolecular crystallography beamlines of the Swiss Light Source. *Journal of Synchrotron Radiation, 21*(Pt 2), 340–351. https://doi.org/10.1107/S160057751400006X.

Gallagher-Jones, M., Bastillo, K. C., Ophus, C., Richards, L. S., Ciston, J., Lee, S., ... Rodriguez, J. A. (2020). Atomic structures determined from digitally defined nanocrystalline regions. *IUCrJ, 7*(3), 490–499. https://doi.org/10.1107/S2052252520004030.

Garman, E. (2003). "Cool" crystals: Macromolecular cryocrystallography and radiation damage. *Current Opinion in Structural Biology, 13*(5), 545–551. https://doi.org/10.1016/j.sbi.2003.09.013.

Garman, E. F., & Weik, M. (2017). Radiation damage in macromolecular crystallography. *Methods in Molecular Biology (Clifton, N. J.), 1607*, 467–489. https://doi.org/10.1007/978-1-4939-7000-1_20.

Genick, U. K. (2007). Structure-factor extrapolation using the scalar approximation: Theory, applications and limitations. *Acta Crystallographica. Section D, Biological Crystallography, 63*(Pt 10), 1029–1041. https://doi.org/10.1107/S0907444907038164.

Gilbile, D., Shelby, M. L., Lyubimov, A. Y., Wierman, J. L., Monteiro, D. C. F., Cohen, A. E., ... Kuhl, T. L. (2021). Plug-and-play polymer microfluidic chips for hydrated, room temperature, fixed-target serial crystallography. *Lab on a Chip, 21*(24), 4831–4845. https://doi.org/10.1039/D1LC00810B.

Gildea, R. J., Beilsten-Edmands, J., Axford, D., Horrell, S., Aller, P., Sandy, J., ... Winter, G. (2022). xia2.multiplex: A multi-crystal data-analysis pipeline. *Acta Crystallographica. Section D, Structural Biology, 78*(6), 752–769. https://doi.org/10.1107/S2059798322004399.

Ginn, H. M. (2021). Vagabond: Bond-based parametrization reduces overfitting for refinement of proteins. *Acta Crystallographica. Section D, Structural Biology,* 77(4), 4. https://doi.org/10.1107/S2059798321000826.

Graber, T., Anderson, S., Brewer, H., Chen, Y. S., Cho, H. S., Dashdorj, N., ... Moffat, K. (2011). BioCARS: A synchrotron resource for time-resolved X-ray science. *Journal of Synchrotron Radiation, 18*(Pt 4), 658–670. https://doi.org/009423>10.1107/S0909049511
009423.

Gruebele, M. (1999). The fast protein folding problem. *Annual Review of Physical Chemistry, 50*(1), 485–516. https://doi.org/10.1146/annurev.physchem.50.1.485.

Gruebele, M., Sabelko, J., Ballew, R., & Ervin, J. (1998). Laser temperature jump induced protein refolding. *Accounts of Chemical Research, 31*(11), 699–707. https://doi.org/10.1021/ar970083x.

Gruner, S. M., Eikenberry, E. F., & Tate, M. W. (2006). *Comparison of X-ray detectors* [Text]. Urn:Isbn:978–0-7923–6857-1; International Union of Crystallography. ⟨https://doi.org/10.1107/97809553602060000667⟩.

Gruner, S. M., & Lattman, E. E. (2015). Biostructural science inspired by next-generation X-ray sources. *Annual Review of Biophysics, 44*, 33–51. https://doi.org/10.1146/annurev-biophys-060414-033813.

Haas, D. J. (2020). The early history of cryo-cooling for macromolecular crystallography. *IUCrJ,* 7(Pt 2), 148–157. https://doi.org/10.1107/S2052252519016993.

Halle, B. (2004). Biomolecular cryocrystallography: Structural changes during flash-cooling. *Proceedings of the National Academy of Sciences, 101*(14), 4793–4798. https://doi.org/10.1073/pnas.0308315101.

Hargreaves, D. (2012). A manual low-cost protein-crystallization plate jig for in situ diffraction in the home laboratory. *Journal of Applied Crystallography, 45*(Pt 1), 138–140. https://doi.org/10.1107/S0021889811052654.

Henzler-Wildman, K., & Kern, D. (2007). Dynamic personalities of proteins. *Nature, 450*(7172), 964–972. https://doi.org/10.1038/nature06522.

Holton, J. M. (2009). A beginner's guide to radiation damage. *Journal of Synchrotron Radiation, 16*(Pt 2), 133–142. https://doi.org/10.1107/S0909049509004361.

Hunter, M. S., Segelke, B., Messerschmidt, M., Williams, G. J., Zatsepin, N. A., Barty, A., ... Frank, M. (2014). Fixed-target protein serial microcrystallography with an x-ray free electron laser. *Scientific Reports, 4*(1), 1. https://doi.org/10.1038/srep06026.

Juers, D. H., & Matthews, B. W. (2001). Reversible lattice repacking illustrates the temperature dependence of macromolecular interactions. *Journal of Molecular Biology, 311*(4), 851–862. https://doi.org/10.1006/jmbi.2001.4891.

Juers, D. H., & Matthews, B. W. (2004a). Cryo-cooling in macromolecular crystallography: Advantages, disadvantages and optimization. *Quarterly Reviews of Biophysics, 37*(2), 105–119. https://doi.org/10.1017/s0033583504004007.

Juers, D. H., & Matthews, B. W. (2004b). The role of solvent transport in cryo-annealing of macromolecular crystals. *Acta Crystallographica. Section D, Biological Crystallography, 60*(Pt 3), 412–421. https://doi.org/10.1107/S0907444903027938.

Kabsch, W. (2010). XDS. *Acta Crystallographica. Section D, Biological Crystallography, 66*(Pt 2), 125–132. https://doi.org/10.1107/S0907444909047337.

Karplus, P. A., & Diederichs, K. (2012). Linking crystallographic model and data quality. *Science (New York, N. Y.), 336*(6084), 1030–1033. https://doi.org/10.1126/science.1218231.

Keedy, D. A., Hill, Z. B., Biel, J. T., Kang, E., Rettenmaier, T. J., Brandão-Neto, J., ... Fraser, J. S. (2018). An expanded allosteric network in PTP1B by multitemperature crystallography, fragment screening, and covalent tethering. *ELife, 7*, e36307. https://doi.org/10.7554/eLife.36307.

Keedy, D. A., Kenner, L. R., Warkentin, M., Woldeyes, R. A., Hopkins, J. B., Thompson, M. C., ... Fraser, J. S. (2015). Mapping the conformational landscape of a dynamic enzyme by multitemperature and XFEL crystallography. *ELife, 4*, e07574. https://doi.org/10.7554/eLife.07574.

Keedy, D. A., Van Den Bedem, H., Sivak, D. A., Petsko, G. A., Ringe, D., Wilson, M. A., & Fraser, J. S. (2014). Crystal cryocooling distorts conformational heterogeneity in a model Michaelis complex of DHFR. *Structure (London, England: 1993), 22*(6), 899–910. https://doi.org/10.1016/j.str.2014.04.016.

Kriminski, S., Caylor, C. L., Nonato, M. C., Finkelstein, K. D., & Thorne, R. E. (2002). Flash-cooling and annealing of protein crystals. *Acta Crystallographica. Section D, Biological Crystallography, 58*(3), 3. https://doi.org/10.1107/S0907444902000112.

Kriminski, S., Kazmierczak, M., & Thorne, R. E. (2003). Heat transfer from protein crystals: Implications for flash-cooling and X-ray beam heating. *Acta Crystallographica. Section D, 59*(4), 697–708. https://doi.org/10.1107/S0907444903002713.

Krishnan, N., Koveal, D., Miller, D. H., Xue, B., Akshinthala, S. D., Kragelj, J., ... Tonks, N. K. (2014). Targeting the disordered C terminus of PTP1B with an allosteric inhibitor. *Nature Chemical Biology, 10*(7), 558–566. https://doi.org/10.1038/nchembio.1528.

Lahey-Rudolph, J. M., Schönherr, R., Barthelmess, M., Fischer, P., Seuring, C., Wagner, A., ... Redecke, L. (2021). Fixed-target serial femtosecond crystallography using in cellulo grown microcrystals. *IUCrJ, 8*(Pt 4), 665–677. https://doi.org/10.1107/S2052252521005297.

Lang, P. T., Ng, H.-L., Fraser, J. S., Corn, J. E., Echols, N., Sales, M., ... Alber, T. (2010). Automated electron-density sampling reveals widespread conformational polymorphism in proteins. *Protein Science: A Publication of the Protein Society, 19*(7), 1420–1431. https://doi.org/10.1002/pro.423.

López-Jaramillo, F. J., Moraleda, A. B., González-Ramírez, L. A., Carazo, A., & García-Ruiz, J. M. (2002). Soaking: The effect of osmotic shock on tetragonal lysozyme crystals. *Acta Crystallographica. Section D, Biological Crystallography, 58*(Pt 2), 209–214. https://doi.org/10.1107/s090744490101914x.

Lyubimov, A. Y., Murray, T. D., Koehl, A., Araci, I. E., Uervirojnangkoorn, M., Zeldin, O. B., ... Berger, J. M. (2015). Capture and X-ray diffraction studies of protein microcrystals in a microfluidic trap array. *Acta Crystallographica. Section D, Biological Crystallography, 71*(Pt 4), 928–940. https://doi.org/10.1107/S1399004715002308.

Martin-Garcia, J. M., Conrad, C. E., Nelson, G., Stander, N., Zatsepin, N. A., Zook, J., ... Liu, W. (2017). Serial millisecond crystallography of membrane and soluble protein microcrystals using synchrotron radiation. *IUCrJ, 4*(4), 439–454. https://doi.org/10.1107/S205225251700570X.

Mccammon, J. A., Gelin, B. R., Karplus, M., & Wolynes, P. G. (1976). The hinge-bending mode in lysozyme. *Nature, 262*(5566), 5566. https://doi.org/10.1038/262325a0.

Meadows, C. W., Balakrishnan, G., Kier, B. L., Spiro, T. G., & Klinman, J. P. (2015). Temperature-jump fluorescence provides evidence for fully reversible microsecond dynamics in a thermophilic alcohol dehydrogenase. *Journal of the American Chemical Society, 137*(32), 10060–10063. https://doi.org/10.1021/jacs.5b04413.

Meents, A., Wiedorn, M. O., Srajer, V., Henning, R., Sarrou, I., Bergtholdt, J., ... Chapman, H. N. (2017). Pink-beam serial crystallography. *Nature Communications, 8*(1), 1. https://doi.org/10.1038/s41467-017-01417-3.

Mehrabi, P., Bücker, R., Bourenkov, G., Ginn, H. M., von Stetten, D., Müller-Werkmeister, H. M., ... Miller, R. J. D. (2021). Serial femtosecond and serial synchrotron crystallography can yield data of equivalent quality: A systematic comparison. *Science Advances,* 7(12), eabf1380. https://doi.org/10.1126/sciadv.abf1380.

Meisburger, S. P., Case, D. A., & Ando, N. (2020). Diffuse X-ray scattering from correlated motions in a protein crystal. *Nature Communications, 11*(1), 1271. https://doi.org/10.1038/s41467-020-14933-6.

Meisburger, S. P., Case, D. A., & Ando, N. (2023). Robust total X-ray scattering workflow to study correlated motion of proteins in crystals. *Nature Communications, 14*(1), 1. https://doi.org/10.1038/s41467-023-36734-3.

Meisburger, S. P., Thomas, W. C., Watkins, M. B., & Ando, N. (2017). X-ray scattering studies of protein structural dynamics. *Chemical Reviews, 117*(12), 7615–7672. https://doi.org/10.1021/acs.chemrev.6b00790.

Moffat, K. (2001). Time-resolved biochemical crystallography: A mechanistic perspective. *Chemical Reviews, 101*(6), 1569–1582. https://doi.org/10.1021/cr990039q.

Moreau, D. W., Atakisi, H., & Thorne, R. E. (2019). Solvent flows, conformation changes and lattice reordering in a cold protein crystal. *Acta Crystallographica. Section D, Structural Biology, 75*(11), 11. https://doi.org/10.1107/S2059798319013822.

Moreau, D. W., Atakisi, H., & Thorne, R. E. (2021). Ice in biomolecular cryocrystallography. *Acta Crystallographica. Section D, Structural Biology,* 77(4), 4. https://doi.org/10.1107/S2059798321001170.

Murray, T. D., Lyubimov, A. Y., Ogata, C. M., Vo, H., Uervirojnangkoorn, M., Brunger, A. T., & Berger, J. M. (2015). A high-transparency, micro-patternable chip for X-ray diffraction analysis of microcrystals under native growth conditions. *Acta Crystallographica. Section D, Biological Crystallography, 71*(Pt 10), 1987–1997. https://doi.org/10.1107/S1399004715015011.

Murshudov, G. N., Skubák, P., Lebedev, A. A., Pannu, N. S., Steiner, R. A., Nicholls, R. A., ... Vagin, A. A. (2011). REFMAC5 for the refinement of macromolecular crystal structures. *Acta Crystallographica Section D: Biological Crystallography, 67*(4), 4. https://doi.org/10.1107/S0907444911001314.

Nam, K. H., Kim, J., & Cho, Y. (2021). Polyimide mesh-based sample holder with irregular crystal mounting holes for fixed-target serial crystallography. *Scientific Reports, 11*(1), 1. https://doi.org/10.1038/s41598-021-92687-x.

Nave, C., & Garman, E. F. (2005). Towards an understanding of radiation damage in cryocooled macromolecular crystals. *Journal of Synchrotron Radiation, 12*(Pt 3), 257–260. https://doi.org/10.1107/S0909049505007132.

Okumura, H., Sakai, N., Murakami, H., Mizuno, N., Nakamura, Y., Ueno, G., ... Kumasaka, T. (2022). In situ crystal data-collection and ligand-screening system at SPring-8. *Acta Crystallographica. Section F, Structural Biology Communications, 78*(Pt 6), 241–251. https://doi.org/10.1107/S2053230X22005283.

Orville, A. M. (2020). Recent results in time resolved serial femtosecond crystallography at XFELs. *Current Opinion in Structural Biology, 65*, 193–208. https://doi.org/10.1016/j.sbi.2020.08.011.

Otten, R., Pádua, R. A. P., Bunzel, H. A., Nguyen, V., Pitsawong, W., Patterson, M., ... Kern, D. (2020). How directed evolution reshapes the energy landscape in an enzyme to boost catalysis. *Science (New York, N. Y.), 370*(6523), 1442–1446. https://doi.org/10.1126/science.abd3623.

Owen, R. L., Axford, D., Sherrell, D. A., Kuo, A., Ernst, O. P., Schulz, E. C., ... Mueller-Werkmeister, H. M. (2017). Low-dose fixed-target serial synchrotron crystallography. *Acta Crystallographica. Section D, Structural Biology, 73*(Pt 4), 373–378. https://doi.org/10.1107/S2059798317002996.

Palmer, K. F., & Williams, D. (1974). Optical properties of water in the near infrared. *JOSA, 64*(8), 1107–1110. https://doi.org/10.1364/JOSA.64.001107.

Park, S.-Y., Choi, H., Eo, C., Cho, Y., & Nam, K. H. (2020). Fixed-target serial synchrotron crystallography using Nylon mesh and enclosed film-based sample holder. *Crystals, 10*(9), 9. https://doi.org/10.3390/cryst10090803.

Perry, S. L., Guha, S., Pawate, A. S., Henning, R., Kosheleva, I., Srajer, V., ... Ren, Z. (2014). In situ serial Laue diffraction on a microfluidic crystallization device. *Journal of Applied Crystallography, 47*(Pt 6), 1975–1982. https://doi.org/10.1107/S1600576714023322.

Pflugrath, J. W. (2015). Practical macromolecular cryocrystallography. *Acta Crystallographica. Section F, Structural Biology Communications, 71*(Pt 6), 622–642. https://doi.org/10.1107/S2053230X15008304.

Phillips, R. S., Miles, E. W., McPhie, P., Marchal, S., Georges, C., Dupont, Y., & Lange, R. (2008). Pressure and temperature jump relaxation kinetics of the conformational change in Salmonella typhimurium Tryptophan synthase l-serine complex: Large activation compressibility and heat capacity changes demonstrate the contribution of solvation. *Journal of the American Chemical Society, 130*(41), 13580–13588. https://doi.org/10.1021/ja8018466.

Phillips, W. C., Li, Y., Stanton, M., Xie, J., O'Mara, D., & Kalata, K. (1993). A CCD-based area detector for X-ray crystallography using synchrotron and laboratory sources. *Nuclear Instruments and Methods in Physics Research Section A: Accelerators, Spectrometers, Detectors and Associated Equipment, 334*(2), 621–630. https://doi.org/10.1016/0168-9002(93)90830-B.

Polsinelli, I., Savko, M., Rouanet-Mehouas, C., Ciccone, L., Nencetti, S., Orlandini, E., ... Shepard, W. (2017). Comparison of helical scan and standard rotation methods in single-crystal X-ray data collection strategies. *Journal of Synchrotron Radiation, 24*(1), 1. https://doi.org/10.1107/S1600577516018488.

Rajashankar, K., & Dauter, Z. (2014). Data collection for crystallographic structure determination. *Methods in Molecular Biology (Clifton, N. J.), 1140*, 211–237. https://doi.org/10.1007/978-1-4939-0354-2_17.

Riley, B. T., Wankowicz, S. A., De Oliveira, S. H. P., Van Zundert, G. C. P., Hogan, D. W., Fraser, J. S., ... Van Den Bedem, H. (2021). qFit 3: Protein and ligand multi-conformer modeling for X-ray crystallographic and single-particle cryo-EM density maps. *Protein Science: A Publication of the Protein Society, 30*(1), 270–285. https://doi.org/10.1002/pro.4001.

Rimmerman, D., Leshchev, D., Hsu, D. J., Hong, J., Abraham, B., Kosheleva, I., ... Chen, L. X. (2018). Insulin hexamer dissociation dynamics revealed by photoinduced T-jumps and time-resolved X-ray solution scattering. *Photochemical & Photobiological Sciences, 17*(7), 874–882. https://doi.org/10.1039/c8pp00034d.

Rimmerman, D., Leshchev, D., Hsu, D. J., Hong, J., Kosheleva, I., & Chen, L. X. (2017). Direct observation of insulin association dynamics with time-resolved X-ray scattering. *The Journal of Physical Chemistry Letters, 8*(18), 4413–4418. https://doi.org/10.1021/acs.jpclett.7b01720.

Ringe, D., & Petsko, G. A. (1986). *[19]Study of protein dynamics by X-ray diffraction. Methods in Enzymology, 131*, Academic Press, 389–433. ⟨https://doi.org/10.1016/0076-6879(86)31050-4⟩.

Roedig, P., Duman, R., Sanchez-Weatherby, J., Vartiainen, I., Burkhardt, A., Warmer, M., ... Meents, A. (2016). Room-temperature macromolecular crystallography using a micro-patterned silicon chip with minimal background scattering. *Journal of Applied Crystallography, 49*(Pt 3), 968–975. https://doi.org/10.1107/S1600576716006348.

Roedig, P., Ginn, H. M., Pakendorf, T., Sutton, G., Harlos, K., Walter, T. S., ... Meents, A. (2017). High-speed fixed-target serial virus crystallography. *Nature Methods, 14*(8), 805–810. https://doi.org/10.1038/nmeth.4335.

Roedig, P., Vartiainen, I., Duman, R., Panneerselvam, S., Stübe, N., Lorbeer, O., ... Meents, A. (2015). A micro-patterned silicon chip as sample holder for macromolecular crystallography experiments with minimal background scattering. *Scientific Reports, 5*, 10451. https://doi.org/10.1038/srep10451.

Roessler, C. G., Agarwal, R., Allaire, M., Alonso-Mori, R., Andi, B., Bachega, J. F. R., ... Zouni, A. (2016). Acoustic injectors for drop-on-demand serial femtosecond crystallography. *Structure (London, England: 1993), 24*(4), 631–640. https://doi.org/10.1016/j.str.2016.02.007.

Rossmann, M. G., Leslie, A. G. W., Abdel-Meguid, S. S., & Tsukihara, T. (1979). Processing and post-refinement of oscillation camera data. *Journal of Applied Crystallography, 12*(6), 6. https://doi.org/10.1107/S0021889879013273.

Rubtsov, I. V., & Burin, A. L. (2019). Ballistic and diffusive vibrational energy transport in molecules. *The Journal of Chemical Physics, 150*(2), 020901. https://doi.org/10.1063/1.5055670.

Russi, S., González, A., Kenner, L. R., Keedy, D. A., Fraser, J. S., & van den Bedem, H. (2017). Conformational variation of proteins at room temperature is not dominated by radiation damage. *Journal of Synchrotron Radiation, 24*(Pt 1), 73–82. https://doi.org/10.1107/S1600577516017343.

Saha, S., Özden, C., Samkutty, A., Russi, S., Cohen, A., Stratton, M. M., & Perry, S. L. (2023). Polymer-based microfluidic device for on-chip counter-diffusive crystallization and in situ X-ray crystallography at room temperature. *Lab on a Chip, 23*(8), 2075–2090. https://doi.org/10.1039/d2lc01194h.

Sandalova, T., Schneider, G., Käck, H., & Lindqvist, Y. (1999). Structure of dethiobiotin synthetase at 0.97 A resolution. *Acta Crystallographica. Section D, Biological Crystallography, 55*(Pt 3), 610–624. https://doi.org/10.1107/s090744499801381x.

Scheidig, A. J., Burmester, C., & Goody, R. S. (1999). The pre-hydrolysis state of p21(ras) in complex with GTP: New insights into the role of water molecules in the GTP hydrolysis reaction of ras-like proteins. *Structure (London, England: 1993)*, 7(11), 1311–1324. https://doi.org/10.1016/s0969-2126(00)80021-0.

Schlichting, I. (2015). Serial femtosecond crystallography: The first five years. *IUCrJ, 2*(Pt 2), 246–255. https://doi.org/10.1107/S205225251402702X.

Schmidt, M., Graber, T., Henning, R., & Srajer, V. (2010). Five-dimensional crystallography. *Acta Crystallographica. Section A, Foundations of Crystallography, 66*(2), 198–206. https://doi.org/10.1107/S0108767309054166.

Schulz, E. C., Yorke, B. A., Pearson, A. R., & Mehrabi, P. (2022). Best practices for time-resolved serial synchrotron crystallography. *Acta Crystallographica. Section D, Structural Biology, 78*(1), 14–29. https://doi.org/10.1107/S2059798321011621.

Shaikevitch, A., & Kam, Z. (1981). Investigation of long-range order in protein crystals by X-ray diffraction. *Acta Crystallographica. Section A, Crystal Physics, Diffraction, Theoretical and General Crystallography, 37*(6), 6. https://doi.org/10.1107/S0567739481001897.

Shimazu, Y., Tono, K., Tanaka, T., Yamanaka, Y., Nakane, T., Mori, C., ... Yabashi, M. (2019). High-viscosity sample-injection device for serial femtosecond crystallography at atmospheric pressure. *Journal of Applied Crystallography, 52*(6), 1280–1288. https://doi.org/10.1107/S1600576719012846.

Sierra, R. G., Gati, C., Laksmono, H., Dao, E. H., Gul, S., Fuller, F., ... DeMirci, H. (2016). Concentric-flow electrokinetic injector enables serial crystallography of ribosome and photosystem-II. *Nature Methods, 13*(1), 59–62. https://doi.org/10.1038/nmeth.3667.

Singh, B., Miller, S. R., Bhandari, H. B., Graceffa, R., Irving, T. C., & Nagarkar, V. V. (2013). High-speed detector for time-resolved diffraction studies. *Journal of Physics. Conference Series, 425*(9), 092005. https://doi.org/10.1088/1742-6596/425/9/092005.

Skaist Mehlman, T., Biel, J. T., Azeem, S. M., Nelson, E. R., Hossain, S., Dunnett, L., ... Keedy, D. A. (2023). Room-temperature crystallography reveals altered binding of small-molecule fragments to PTP1B. *ELife, 12*, e84632. https://doi.org/10.7554/eLife.84632.

Soares, A. S., Yamada, Y., Jakoncic, J., McSweeney, S., Sweet, R. M., Skinner, J., ... Bernstein, H. J. (2022). Serial crystallography with multi-stage merging of thousands of images. *Acta Crystallographica Section F: Structural Biology Communications, 78*(7), 7. https://doi.org/10.1107/S2053230X22006422.

Šrajer, V., & Schmidt, M. (2017). Watching proteins function with time-resolved X-ray crystallography. *Journal of Physics D: Applied Physics, 50*(37), 373001. https://doi.org/10.1088/1361-6463/aa7d32.

Steller, I., Bolotovsky, R., & Rossmann, M. G. (1997). An algorithm for automatic indexing of oscillation images using Fourier analysis. *Journal of Applied Crystallography, 30*(6), 6. https://doi.org/10.1107/S0021889897008777.

Stinton, G. W., & Evans, J. S. O. (2007). Parametric Rietveld refinement. *Journal of Applied Crystallography, 40*(Pt 1), 87–95. https://doi.org/10.1107/S0021889806043275.

Sugahara, M., Nakane, T., Masuda, T., Suzuki, M., Inoue, S., Song, C., ... Iwata, S. (2017). Hydroxyethyl cellulose matrix applied to serial crystallography. *Scientific Reports,* 7(1), 703. https://doi.org/10.1038/s41598-017-00761-0.

Sui, S., Mulichak, A., Kulathila, R., McGee, J., Filiatreault, D., Saha, S., ... Perry, S. L. (2021). A capillary-based microfluidic device enables primary high-throughput room-temperature crystallographic screening. *Journal of Applied Crystallography, 54*(4), 1034–1046. https://doi.org/10.1107/S1600576721004155.

Thompson, M. C., Barad, B. A., Wolff, A. M., Sun Cho, H., Schotte, F., Schwarz, D. M. C., ... Fraser, J. S. (2019). Temperature-jump solution X-ray scattering reveals distinct motions in a dynamic enzyme. *Nature Chemistry, 11*(11), 1058–1066. https://doi.org/10.1038/s41557-019-0329-3.

Thompson, M. C., Cascio, D., & Yeates, T. O. (2018). Microfocus diffraction from different regions of a protein crystal: Structural variations and unit-cell polymorphism. *Acta Crystallographica. Section D, Structural Biology,* 74(Pt 5), 411–421. https://doi.org/10.1107/S2059798318003479.

Thompson, M. C., Yeates, T. O., & Rodriguez, J. A. (2020). Advances in methods for atomic resolution macromolecular structure determination. *F1000Research, 9*, F1000 Faculty Rev-667. ⟨https://doi.org/10.12688/f1000research.25097.1⟩.

Thorne, R. E. (2023). Determining biomolecular structures near room temperature using X-ray crystallography: Concepts, methods and future optimization. *Acta Crystallographica. Section D, Structural Biology, 79*(1), 1. https://doi.org/10.1107/S2059798322011652.

Tilton, R. F., Jr., Dewan, J. C., & Petsko, G. A. (1992). Effects of temperature on protein structure and dynamics: X-ray crystallographic studies of the protein ribonuclease-A at nine different temperatures from 98 to 320K. *Biochemistry, 31*(9), 2469–2481. https://doi.org/10.1021/bi00124a006.

Ueno, G., Shimada, A., Yamashita, E., Hasegawa, K., Kumasaka, T., Shinzawa-Itoh, K., ... Yamamoto, M. (2019). Low-dose X-ray structure analysis of cytochrome c oxidase utilizing high-energy X-rays. *Journal of Synchrotron Radiation, 26*(Pt 4), 912–921. https://doi.org/10.1107/S1600577519006805.

Uervirojnangkoorn, M., Zeldin, O. B., Lyubimov, A. Y., Hattne, J., Brewster, A. S., Sauter, N. K., ... Weis, W. I. (2015). Enabling X-ray free electron laser crystallography for challenging biological systems from a limited number of crystals. *ELife, 4*, e05421. https://doi.org/10.7554/eLife.05421.

Ursby, T., & Bourgeois, D. (1997). Improved estimation of structure-factor difference amplitudes from poorly accurate data. *Acta Crystallographica. Section A, Foundations of Crystallography, 53*(5), 5. https://doi.org/10.1107/S0108767397004522.

Vahedi-Faridi, A., Lovelace, J., Bellamy, H. D., Snell, E. H., & Borgstahl, G. E. O. (2003). Physical and structural studies on the cryocooling of insulin crystals. *Acta Crystallographica. Section D, Biological Crystallography, 59*(Pt 12), 2169–2182. https://doi.org/10.1107/s0907444903019668.

Vakili, M., Vasireddi, R., Gwozdz, P. V., Monteiro, D. C. F., Heymann, M., Blick, R. H., & Trebbin, M. (2020). Microfluidic polyimide gas dynamic virtual nozzles for serial crystallography. *Review of Scientific Instruments, 91*(8), 085108. https://doi.org/10.1063/5.0012806.

Van Den Bedem, H., Dhanik, A., Latombe, J. C., & Deacon, A. M. (2009). Modeling discrete heterogeneity in X-ray diffraction data by fitting multi-conformers. *Acta Crystallographica. Section D, Biological Crystallography, 65*(Pt 10), 1107–1117. https://doi.org/10.1107/S0907444909030613.

Wall, M. E., Wolff, A. M., & Fraser, J. S. (2018). Bringing diffuse X-ray scattering into focus. *Current Opinion in Structural Biology, 50*, 109–116. https://doi.org/10.1016/j.sbi.2018.01.009.

Warkentin, M., & Thorne, R. E. (2007). A general method for hyperquenching protein crystals. *Journal of Structural and Functional Genomics, 8*(4), 141–144. https://doi.org/10.1007/s10969-007-9029-0.

Warkentin, M., & Thorne, R. E. (2009). Slow cooling of protein crystals. *Journal of Applied Crystallography, 42*(Pt 5), 944–952. https://doi.org/10.1107/S0021889809023553.

Warren, A. J., Crawshaw, A. D., Trincao, J., Aller, P., Alcock, S., Nistea, I., ... Evans, G. (2015). In vacuo X-ray data collection from graphene-wrapped protein crystals. *Acta Crystallographica. Section D, Biological Crystallography, 71*(Pt 10), 2079–2088. https://doi.org/10.1107/S1399004715014194.

Weierstall, U., James, D., Wang, C., White, T. A., Wang, D., Liu, W., ... Cherezov, V. (2014). Lipidic cubic phase injector facilitates membrane protein serial femtosecond crystallography. *Nature Communications, 5*(1), 3309. https://doi.org/10.1038/ncomms4309.

Weik, M., & Colletier, J.-P. (2010). Temperature-dependent macromolecular X-ray crystallography. *Acta Crystallographica. Section D, Biological Crystallography, 66*(Pt 4), 437–446. https://doi.org/10.1107/S0907444910002702.

Weik, M., Kryger, G., Schreurs, A. M., Bouma, B., Silman, I., Sussman, J. L., ... Kroon, J. (2001). Solvent behaviour in flash-cooled protein crystals at cryogenic temperatures. *Acta Crystallographica. Section D, Biological Crystallography, 57*(Pt 4), 566–573. https://doi.org/10.1107/s0907444901001196.

Weinert, T., Olieric, N., Cheng, R., Brünle, S., James, D., Ozerov, D., ... Standfuss, J. (2017). Serial millisecond crystallography for routine room-temperature structure determination at synchrotrons. *Nature Communications, 8*(1), 1. https://doi.org/10.1038/s41467-017-00630-4.

White, T. A. (2014). Post-refinement method for snapshot serial crystallography. *Philosophical Transactions of the Royal Society B: Biological Sciences, 369*(1647), 20130330. https://doi.org/10.1098/rstb.2013.0330.

White, T. A. (2019). Processing serial crystallography data with CrystFEL: A step-by-step guide. *Acta Crystallographica. Section D, Structural Biology, 75*(2), 219–233. https://doi.org/10.1107/S205979831801238X.

Wickstrand, C., Katona, G., Nakane, T., Nogly, P., Standfuss, J., Nango, E., & Neutze, R. (2020). A tool for visualizing protein motions in time-resolved crystallography. *Structural Dynamics (Melville, N. Y.)*, 7(2), 024701. https://doi.org/10.1063/1.5126921.

Wilson, M. A. (2022). Mapping enzyme landscapes by time-resolved crystallography with synchrotron and X-ray free electron laser light. *Annual Review of Biophysics, 51*(1), 79–98. https://doi.org/10.1146/annurev-biophys-100421-110959.

Winkler, F. K., Schutt, C. E., & Harrison, S. C. (1979). The oscillation method for crystals with very large unit cells. *Acta Crystallographica. Section A, Crystal Physics, Diffraction, Theoretical and General Crystallography, 35*(6), 6. https://doi.org/10.1107/S0567739479002035.

Winter, G., Gildea, R. J., Paterson, N. G., Beale, J., Gerstel, M., Axford, D., ... Hall, D. R. (2019). How best to use photons. *Acta Crystallographica. Section D, Structural Biology, 75*(Pt 3), 242–261. https://doi.org/10.1107/S2059798319003528.

Winter, G., Waterman, D. G., Parkhurst, J. M., Brewster, A. S., Gildea, R. J., Gerstel, M., ... Evans, G. (2018). DIALS: Implementation and evaluation of a new integration package. *Acta Crystallographica. Section D, Structural Biology, 74*(Pt 2), 85–97. https://doi.org/10.1107/S2059798317017235.

Wolff, A. M., Nango, E., Young, I. D., Brewster, A. S., Kubo, M., Nomura, T., ... Thompson, M.C. (2022). Mapping protein dynamics at high-resolution with temperature-jump X-ray crystallography. *bioRxiv*. 2022.06.10.495662. ⟨https://doi.org/10.1101/2022.06.10.495662⟩.

Wolff, A. M., Young, I. D., Sierra, R. G., Brewster, A. S., Martynowycz, M. W., Nango, E., ... Thompson, M. C. (2020). Comparing serial X-ray crystallography and microcrystal electron diffraction (MicroED) as methods for routine structure determination from small macromolecular crystals. *IUCrJ*, 7(Pt 2), 306–323. https://doi.org/10.1107/S205225252000072X.

Wu, H., & Arnold, E. (2019). Michael G. Rossmann (1930-2019). *Nature Structural & Molecular Biology, 26*(8), 660–662. https://doi.org/10.1038/s41594-019-0271-5.

Yabukarski, F., Doukov, T., Mokhtari, D. A., Du, S., & Herschlag, D. (2021). Damaged goods? Evaluating the impact of X-ray damage on conformational heterogeneity in room temperature and cryo-cooled protein crystals. *bioRxiv*. 2021.06.27.450091. ⟨https://doi.org/10.1101/2021.06.27.450091⟩.

Zarrine-Afsar, A., Barends, T. R. M., Müller, C., Fuchs, M. R., Lomb, L., Schlichting, I., & Miller, R. J. D. (2012). Crystallography on a chip. *Acta Crystallographica. Section D, Biological Crystallography, 68*(Pt 3), 321–323. https://doi.org/10.1107/S0907444911055296.

Zeldin, O. B., Brewster, A. S., Hattne, J., Uervirojnangkoorn, M., Lyubimov, A. Y., Zhou, Q., ... Brunger, A. T. (2015). Data Exploration Toolkit for serial diffraction experiments. *Acta Crystallographica. Section D, Biological Crystallography, 71*(Pt 2), 352–356. https://doi.org/10.1107/S1399004714025875.

Zwart, P. H., Afonine, P. V., Grosse-Kunstleve, R. W., Hung, L.-W., Ioerger, T. R., McCoy, A. J., ... Adams, P. D. (2008). Automated structure solution with the PHENIX suite. *Methods in Molecular Biology (Clifton, N. J.), 426*, 419–435. https://doi.org/10.1007/978-1-60327-058-8_28.

CHAPTER TEN

Room temperature crystallography and X-ray spectroscopy of metalloenzymes

Hiroki Makita, Miao Zhang, Junko Yano*, and Jan Kern*
Molecular Biophysics and Integrated Bioimaging Division, Lawrence Berkeley National Laboratory, Berkeley, CA, United States
*Corresponding authors. e-mail address: JYano@lbl.gov; JFKern@lbl.gov

Contents

Abstract

The ultrashort (10s of femtoseconds) X-ray pulses generated by X-ray free electron lasers enable the measurement of X-ray diffraction and spectroscopic data from radiation-sensitive metalloenzymes at room temperature while mostly avoiding the effects of radiation damage usually encountered when performing such experiments at synchrotron sources. Here we discuss an approach to measure both X-ray emission and X-ray crystallographic data at the same time from the same sample volume. The droplet-on-tape setup described allows for efficient sample use and the integration of different reaction triggering options in order to conduct time-resolved studies with limited sample amounts. The approach is illustrated by two examples, photosystem II

Methods in Enzymology, Volume 688
ISSN 0076-6879, https://doi.org/10.1016/bs.mie.2023.07.009

that catalyzes the light-driven oxidation of water to oxygen, and isopenicillin N synthase, an enzyme that catalyzes the double ring cyclization of a tripeptide precursor into the β-lactam isopenicillin and can be activated by oxygen exposure. We describe the necessary steps to obtain microcrystals of both proteins as well as the operation procedure for the drop-on-tape setup and details of the data acquisition and processing involved in this experiment. At the end, we present how the combination of time-resolved X-ray emission spectra and diffraction data can be used to improve the knowledge about the enzyme reaction mechanism.

1. Introduction

Metalloenzymes are central players in many biosynthetic and energy converting pathways in Nature. They are used to perform challenging chemical reactions with efficiency, speed, and complexity under ambient conditions and in a highly controlled manner in both time and space, by utilizing the flexibility of the protein, cofactors, and water network. To understand the mechanism of such enzymes at the atomic level, solving their macromolecular structures and combining these with spectroscopic studies of the catalytic site has been an important approach. X-ray crystallography (X-ray diffraction, XRD) at synchrotron facilities (SR) has played a central role for this purpose, along with nuclear magnetic resonance (NMR), mass spectroscopy, and especially more recently cryo electron microscopy (EM). Spectroscopic approaches for transition metal containing metalloenzymes are also an ideal target for X-ray spectroscopic methods, especially X-ray absorption and emission measurements to observe transitions around the transition metal K-edge region and have been often combined with theoretical approaches to gain a deeper understanding of the details of the electronic structure of the active sites. Structural biology has seen important advances in two new methods, cryo-EM and X-ray free electron lasers (XFELs). The former enables the visualization of large complexes at the atomic scale, at resolutions that approach or are equal to those accessible by X-ray crystallography from ensemble measurements of single molecules without the need of crystals (Kuhlbrandt, 2014; Saibil, 2022). The latter, while requiring crystals, can give a view of macromolecules and their time-dependent structural changes under physiological conditions (Barends, Stauch, Cherezov, & Schlichting, 2022; Chapman et al., 2011; Neutze, 2014). In addition, the femtosecond pulses from XFELs allow scientists to obtain X-ray spectroscopic information either using separate specialized experiments or in a combined approach, where both the spectroscopic and diffraction signal is collected at the same time from the same sample volume. In this chapter, we describe the X-ray crystallography and X-ray spectroscopy methods at XFELs relevant to the structural biology for mechanistic studies of metalloenzymes.

Over the decades, synchrotron X-ray crystallography has been a standard tool for structural biology. The introduction of cryopreservation methods in the 1980s was a major advance in the field. By reducing the rate of secondary radiation damage caused by the diffusion of radicals, it improved the useful X-ray exposure lifetime of crystals, and enabled the collection of high-resolution data. Cryo-trapping is likewise useful for other synchrotron-based structural techniques like X-ray Absorption Spectroscopy (XAS) for the same reasons (Yano et al., 2005). However, cryo-trapping methods can only reveal intermediates within the time resolution of the freezing process (usually several milliseconds, although highly dependent on the method used) and there has been an increasing desire to study systems under functional conditions (Keedy et al., 2015; Thompson et al., 2019) in order to understand processes that control the directionality of reactions. XFELs have opened up the possibility of such approaches, by enabling room temperature studies, through the ability to collect data before the onset of radiation damage. Also, method development for XFEL experiments has recently led to a technology transfer to synchrotron sources, enabling some of these experiments (for longer time scales and less radiation-sensitive samples) at synchrotron sources and hence allowing for wider access and utilization of such approaches (see for example Schulz et al., 2022).

The X-ray pulses generated by an XFEL are extremely intense with short pulse widths, containing as many photons (10^{12-13}) in one pulse (typically <50 fs, 10^4-fold shorter than pulses created by X-ray storage ring facilities) as synchrotrons generate in one second, focused into a μm-size spot. These beam properties, together with pulse-by-pulse "diffract and replace" data collection, make it possible to overcome radiation damage caused by the diffusion of hydroxyl radicals (~ps) and even by the diffusion of solvated electrons (sub ps); the XFEL pulses are generally shorter (fs) than the time scale of these secondary damage processes, and therefore enable X-ray diffraction data collection at room temperature without resorting to cryo-techniques (Chapman et al., 2011; Kern et al., 2013; Spence, 2018). This is also true for X-ray spectroscopy data collection, which probes electronic state information and structure with high resolution in the local vicinity of a given element (Bergmann et al., 2021). By combining diffraction methods and X-ray spectroscopy, scientists can examine the dynamics of the local chemical environment of the catalytic metal center of metalloenzymes in the context of the atomic structure of the entire enzyme (Fig. 1).

2. Sample preparation

Protein crystallization can be considered as a 'trial and error' procedure because it depends on the interplay of multiple parameters such as protein and precipitant concentrations, pH, temperature, ionic strength, etc. Particularly, the three-dimensional structure with appropriate degrees of flexibility and dynamics of proteins are required to maintain their functionalities. But some of these characteristics—especially those that increase heterogeneity—impede the crystal formation procedure as the protein molecules in a crystal lattice are held by weak and non-covalent

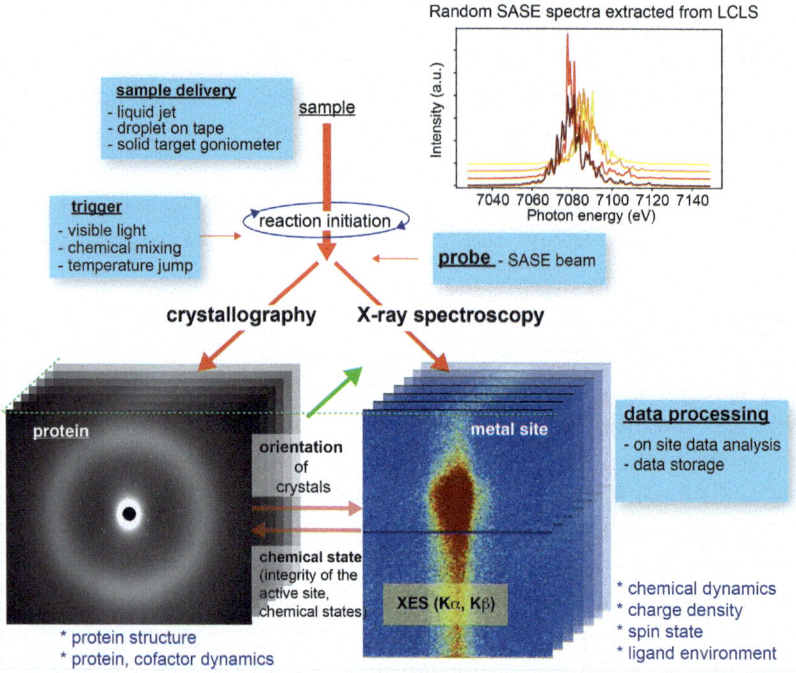

Fig. 1 Overview of the shot-by-shot data collection of X-ray crystallography and X-ray spectroscopy of biological systems at XFELs. Incoming self-amplified spontaneous emission (SASE) pulses at XFELs have a bandwidth of ~0.2% of the central energy.

intermolecular interactions that include ionic, dipole, van der Waals interactions and hydrogen bonds. Crystallizability of a protein is determined in part, by its surface characteristics meaning the distribution of charged and polar amino acids on the solvent-exposed protein surface. That is why certain proteins cannot be crystallized due to their surface characteristics that are incompatible with the crystal lattice (Bijelic & Rompel, 2018; Rupp, 2013). As a general workflow, determining optimal crystallization conditions through several trials would be essential steps to obtain protein crystals. However, before starting to find suitable crystallization conditions, purity and homogeneity of a protein solution play a prominent role for crystal formation and growth.

XFEL studies are usually conducted using micron size crystals (between 1 and ~50 μm in the longest dimension). As this size range is not commonly used for synchrotron-based experiments, some modifications to traditional crystallization protocols are necessary. Here, we describe sample preparation steps for microcrystal generation based on two protein examples. Their crystallization conditions are well-established and the proteins selected are photosystem II (PSII) and isopenicillin N synthase (IPNS).

2.1 Crystallization kinetics and phase diagram

To obtain single crystals, a highly purified protein solution needs to be brought to a supersaturated stage, which is required to overcome the nucleation free energy barrier (Fig. 2). A spontaneous appearance of a new phase, the so-called phase separation, can only happen when a system is in a non-equilibrium condition. During the phase separation, protein molecules are transitioned from the soluble state into the solid-phase to form critical nuclei. The protein concentration in solution decreases as the nuclei grow beyond the critical size, moving to the so-called growth phase, where the ordered aggregation of the molecules takes place. In this process, more and more protein molecules attach to the nuclei surface to form a single crystal. In the last step, a crystal stops growing when the protein molecules in solution decrease significantly (Bijelic & Rompel, 2018; Erdemir, Lee, & Myerson, 2019).

From thermodynamic perspectives, the two events of nucleation and crystal growth occur spontaneously when the Gibbs free energy (ΔG) value is negative. Crystal formation and growth lead to entropy reduction of the protein molecules as their movements are restricted upon incorporation into a crystal lattice, thus becoming a more ordered system. However, at the same time structured water molecules from the protein's surface can

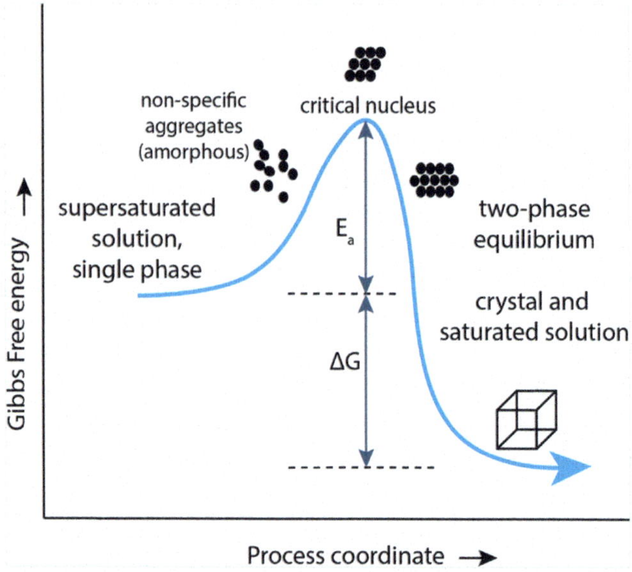

Fig. 2 Schematic representation of the activation energy barrier E_a and the driving force for crystallization ΔG during the crystallization process. The peak position represents a kinetic energy barrier that needs to be overcome to form a critical nucleus. Exceeding the critical nucleus size allows formation of a single crystal, inducing the desired phase separation. Small black circles represent protein molecules, and the square indicates a single crystal.

either be released in the solvent or additional water molecules are trapped during crystal formation. According to the calculations demonstrated in the previous literature (Vekilov, 2004; Vekilov, 2003; Vekilov and Chernov, 2002), around 5–30 water or solvent molecules are released in the solvent from each protein molecule bound to the crystal lattice. In the disordered bulk, the released water molecules have more translational and rotational degrees of freedom, leading to an increase in entropy of the water or solvent molecules that predominates the comparably small entropy loss of protein molecules attached to a crystal. In most cases, the release of water molecules around the protein molecules (upon attachment of the protein molecules to the crystal) is the main thermodynamic driving force for crystallization (Derewenda & Vekilov, 2006; Vekilov, 2019).

Protein crystal formation requires promoting a protein solution from the undersaturated state to the supersaturated states. This is illustrated by the phase diagram in Fig. 3. The solubility line divides the diagram into two zones: an undersaturated and supersaturated zone. This line represents

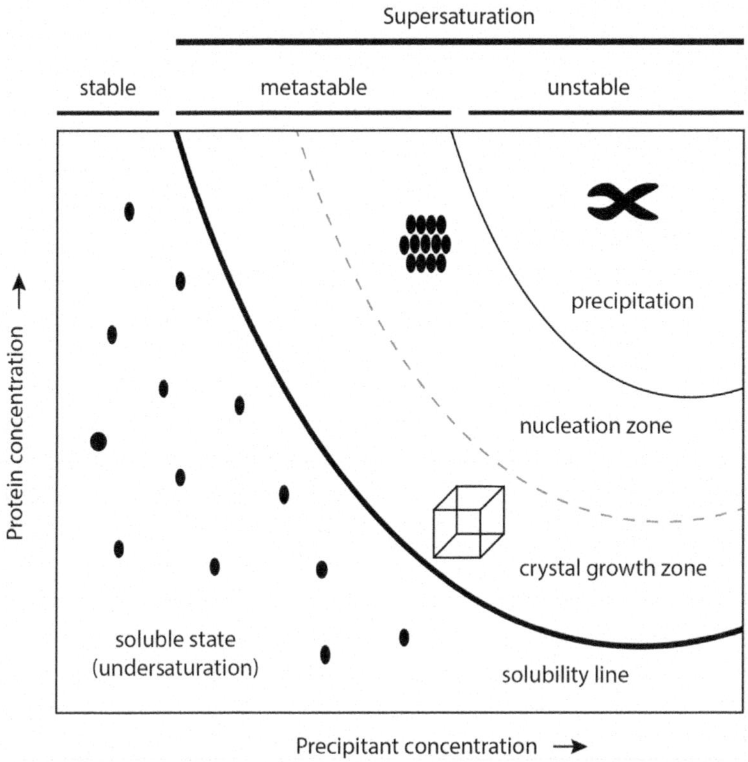

Fig. 3 Crystallization phase diagram, which is divided by a solubility line into two regions: an undersaturated and supersaturated zone. In the soluble state, neither crystal formation nor crystal growth occurs. Protein molecules represented as black circles stay stable. When reaching supersaturation, three events can happen: precipitation, nucleation and crystal growth. Under high supersaturated conditions, only amorphous solids precipitate (precipitation zone). With decreasing saturation, spontaneous crystal formation takes place (nucleation zone). Further moving towards even lower supersaturation conditions, crystal growth or metastable zone is reached. Black circles are demonstrated as protein molecules. Single crystal is represented as a square.

the equilibrium between the existence of the solid phase and the free-molecule phase. Protein crystallization is not possible in an undersaturated zone as protein molecules stay in solution within this zone. The supersaturated region is subdivided, depending on the states of supersaturations, into three zones: metastable zone, nucleation zone and precipitation zone. The metastable zone is a region for optimal crystal growth, but no nucleation takes place within this region. If supersaturation proceeds, spontaneous crystal formation occurs within the nucleation zone. Here, both nucleation and crystal growth are possible. Increasing the conditions towards an even higher supersaturation, the precipitation zone is reached where proteins precipitate amorphously (Bijelic & Rompel, 2018).

2.2 Crystallization techniques

Different crystallization techniques (Shoeman, Hartmann, & Schlichting, 2023; Schulz et al., 2022; Beale et al., 2019; García-Ruiz, 2003; Giegé, 2013; Rhodes, 2006) have been proposed to obtain single crystals. Here, we describe three most commonly used crystallization techniques: **Vapor diffusion, batch crystallization and crystal seeding.**

The **vapor diffusion** technique can be performed in two setups, the hanging or sitting drop format. Considering the hanging drop setup as an example, a small droplet, typically a few microliters between 0.1 and 5 μL, of sample mixed with crystallization buffer is placed on a siliconized glass cover slide. Crystallization buffer containing precipitant reagent is located in the reservoir. The initial reagent concentration is usually higher in the reservoir well than in the drop. The concentration gradient is needed to trigger the equilibrium process between droplet and reservoir. During this process wherein the drop and well are in a sealed environment, water from the droplet will evaporate towards the reservoir. As a result, the volume of the drop is decreased, the protein and precipitant concentration in the drop increase slowly until the solubility limit of the protein is exceeded to reach supersaturation. This is the requirement to initiate the crystallization process. Crystal formation and growth will occur under proper crystallization conditions.

Batch crystallization is based on the direct mixing of the protein and precipitant. There is no equilibrium with a reservoir needed. Through this method the protein is brought into the nucleation zone. When starting from a known crystallization condition, crystals can appear within a few minutes or hours.

Crystal seeding is a powerful tool to grow a large quantity of uniform crystals. A major advantage of using seeding techniques is to control the crystallization process based on separating the nucleation and growth events. When using the microseeding method, submicroscopic seeds (crushed crystals) are brought to a crystallization agent fulfilling the optimal conditions for the metastable zone where crystals from seeds can grow but no spontaneous nucleation can occur.

2.3 PSII purification and crystallization

PSII thylakoid membranes are isolated from cells of *Thermosynechococcus vestitus* (*T. vestitus*, previously named *T. elongatus*) and prepared according to (Kern et al., 2005). Particularly at the initial steps of isolation of PSII thylakoids, washing with high salt can remove a large amount of phycobilisomes. However, isolated thylakoids can still contain further impurities such as remaining phycobilisomes and other soluble proteins. Since high degrees of purity and homogeneity of protein samples are prerequisite to obtain high-resolution crystal structures as mentioned above, further purification steps of PSII thylakoids are essential. Isolated PSII thylakoid membranes are solubilized by using octaethylene glycol monododecyl ether ($C_{12}E_8$) detergent. The solubilization step is important in order to extract crude PSII particles from the membrane when using appropriate concentrations of detergent. The crude PSII extract after solubilization is further purified through two steps of weak and one step of strong anion exchange chromatography in order to obtain pure PSII dimers [dPSII, MW = ~750 kDa (Zouni et al., 2005)]. Detailed information about the purification steps of PSII dimers using $C_{12}E_8$ detergent is given in Hellmich et al. (2014). The purification effect and the purity of PSII dimers are demonstrated by the Blue native PAGE analysis shown in Fig. 4. Crude

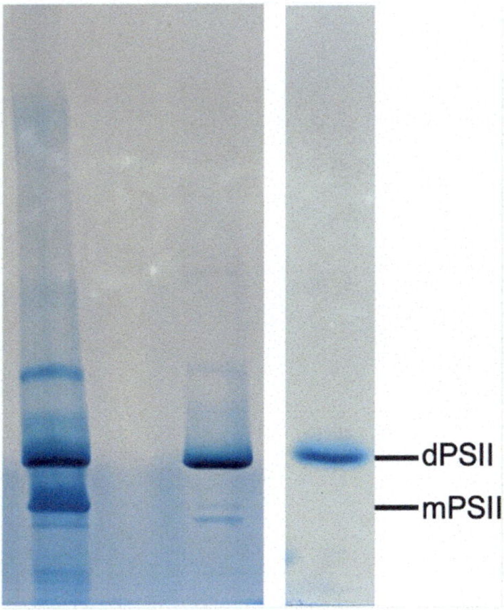

Fig. 4 Blue native PAGE analysis of crude PSII particles (lane 1), purified PSII dimers (lane 2) and PSII dimers after recrystallization (lane 3).

PSII particles are shown in lane 1. After column chromatography, PSII dimer in lane 2 shows a weak band in position equivalent to that of PSII monomer [mPSII, MW = ~350 kDa (Zouni et al., 2005)]. When detecting small amounts of PSII monomers present in the PSII dimers after column chromatography, recrystallization can be used as an additional purification step in order to obtain highly purified and monodisperse PSII dimers. The result after recrystallization is shown in lane 3.

2.3.1 Crystallization in microbatch plates

Crystals were obtained using the microbatch method in 96-well IMPACT Plates (Greiner Bio-One GmbH). Highly purified PSII protein at a final concentration of 2 mM Chl (chlorophyll) is mixed 1:1 with the crystallization buffer that contains 0.1 M Tris (pH 7.5), 0.1 M $(NH_4)_2SO_4$, 14–19% polyethylene glycol monomethyl ether 5000 (PEG 5000 MME). Crystals grown at 20 °C having a size between 100 and 200 μm could be observed after 1 or 2 days (Hellmich et al., 2014).

2.3.2 Microseeding procedure

For making seed stock solution, a few microcrystals were selected from a microbatch plate. On average between 5 and 10 crystals of a size between 100 and 200 μm (longest axis) shown in Fig. 5 were transferred to 50 μL of 15% PEG 5000 buffer (0.1 M Tris pH 7.5, 0.1 M $(NH_4)_2SO_4$, 15% PEG 5000 MME (w/v), 0.013% $C_{12}E_8$). A seed bead tool kit (Hampton Research) was

Fig. 5 Microbatch crystals were obtained from the purified PSII dimers. The scale bar represents 100 μm.

used to crush these crystals by vortexing for 5 min. The mixture was centrifuged at 13,000 r.p.m. for 1 min and vortexed again for another 5 min. After vortexing, the crushed crystal suspension was mixed with 450 μL of buffer 15% PEG 5000 buffer with detergent to prepare the first seed stock.

Microseeding crystals of PSII can be made by mixing the PSII protein sample at 2 mM Chl concentration and the seed stock solution in a ratio of 1:3 (for example: 10 μL of PSII protein and 30 μL of seed stock solution). The final PEG concentration is now 11.25%. The mixed solution is sequentially placed in an incubator at a constant temperature of 20 °C. Under this condition, a clear microcrystal pellet usually appears after 1 h of growth time. The size of microseeding crystals is on average between 20 and 50 μm (Fig. 6). [Microseeding protocol modified from Ibrahim et al. (2015).]

2.3.3 Post-crystallization treatment

According to the post-crystallization protocol in Hellmich et al. (2014), microseeding crystals are treated by stepwise transfer to a buffer containing the final PEG concentration of 35% (w/v), 0.1 M Tris (pH 7.5), 0.1 M $(NH_4)_2SO_4$ in steps of 5% PEG for 5–10 min. This crystal treatment is conducted at room temperature. Stepwise buffer exchange eventually results in the reduction of detergent and solvent content. Crystals are thus transformed from type II to type I showing a clear resolution improvement (Hellmich et al., 2014).

Fig. 6 Microseeding crystals of PSII dimers. The scale bar represents 20 μm.

After 30 min incubation in the Tris buffer at pH 7.5, the crystal suspension is transferred to a buffer that contains PEG concentration of 35% (w/v), 0.1 M MES (pH 6.5), 0.1 M $(NH_4)Cl$. The buffer exchange to MES at pH 6.5 is essential as keeping PSII samples too long in buffers at higher pH leads to reduction of the protein activity, meaning a reduction of the rate of light-dependent generation of O_2. Previous studies showed that nearly full activity can be restored in PSII samples when exchanging the sample back into the buffer at pH 6.5 (Young et al., 2016) within about 5 h after initial exposure of the sample to high pH. The final Chl concentration of PSII microcrystal suspension typically used for the XFEL experiments is between 0.5 and 1.2 mM. The number of crystals was determined by an automated cell counter (Invitrogen Countess II). 10 μL of microcrystal suspension was added to a cell-counting chamber slide and about 2×10^6 crystals/mL were estimated when the Chl concentration is approximately 0.5 mM and the average crystal size is about 10 μm.

2.4 IPNS

Highly purified (>95% by SDS–PAGE) recombinant IPNS was produced and purified as described (Roach et al., 1996) and stored in 25 mM Tris, pH 8.0. All protein handling and crystallization were conducted under anaerobic conditions (<2 ppm O_2) and a N_2 atmosphere inside an anaerobic chamber (Belle Technologies, UK) and at room temperature. IPNS crystals were obtained through the hanging drop method by using 24-well VDX plates (Hampton Research, US). To obtain the IPNS crystallization solution, 4 μL of freshly prepared 100 mM $FeSO_4$ were mixed with 80 μL of 50–52 mg/mL IPNS, followed by 4 times addition of 5 μL of ACV (L-δ-(α-aminoadipoyl)-L-cysteinyl-D-valine) (2.0 mg in 20 μL 25 mM Tris, pH 8.5). To screen the optimum conditions for macrocrystal generation, different conditions for the reservoir solution were generated by using 0.1 M Tris at different pH values between 8.1 and 8.7 and changing the salt concentration of Li_2SO_4 between 1.5 and 2.0 M. The crystallization droplet contained 3 μL of the reservoir solution mixed with 3 μL of protein solution and were equilibrated against 500 μL of the reservoir solution. Needle, plate- and hexagonal-shaped crystals appeared within 2 days at room temperature, and a few 100–150 μm long needle-shaped crystals (3–5 crystals harvested into 50 μL reservoir solution) obtained from these first round of crystallization setups were used to prepare seeds using the Seed Bead Kit as described by the manufacturer (Hampton Research, USA).

The batch method was applied to obtain microcrystals. In this procedure, a 96-chimney well plate (Corning, USA) was used. For each well, 6.5 μL of the IPNS–ACV–$FeSO_4$ mixture prepared as described above was mixed with 90 μL of 1.7 M Li_2SO_4 and 0.1 M Tris pH 8.5. To the mixture, 1 μL of the seed stock was added. The plate was sealed (Polyolefin StarSeal, Starlab, UK) and the mixed solution in the plate was kept under agitation using a microplate shaker (SciQuip, UK) at 700 r.p.m., until a crystal size between 40 and 60 μm (longest dimension) had been reached (Fig. 7). Microcrystal growth usually took between 16 and 30 h, depending on the temperature and the humidity of the anaerobic chamber. A final concentration of $\sim 2 \times 10^7$ crystals/mL was typically achieved. Detailed procedure of IPNS microcrystal preparation is described in Kamps (2020) and Rabe et al. (2020).

3. Sample delivery system

The pulse-by-pulse "diffract and replace" data collection method as described in the Introduction section using fs XFEL beams is known as

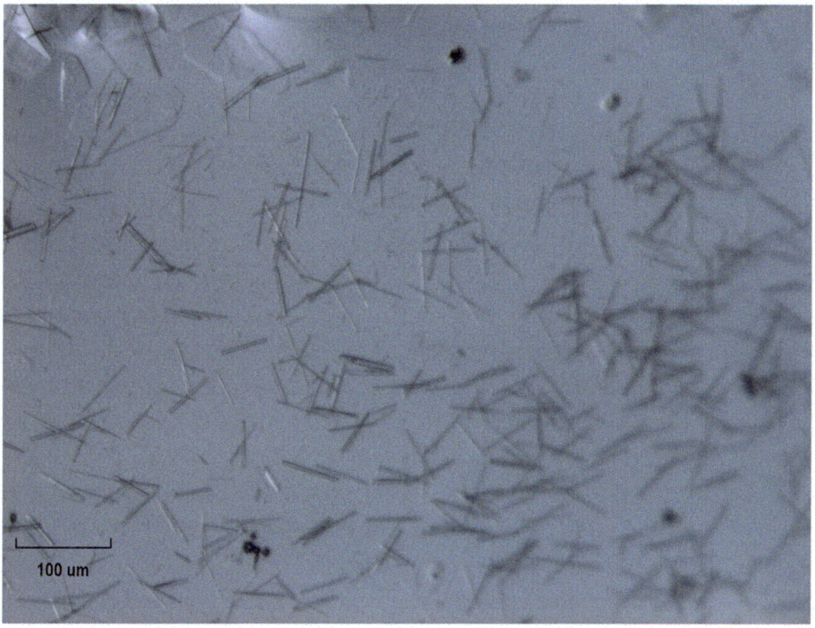

Fig. 7 Optical microscope image of IPNS • Fe • ACV microcrystals.

serial femtosecond crystallography (SFX) and requires a fresh batch of microcrystal sample to be delivered for (ideally) each pulse of XFEL beam to ensure efficient data collection. To conduct an SFX experiment that maximizes the beam usage and minimizes the sample waste, a sample delivery system that accounts for controlled infusion and ejection of the microcrystal sample is required. A variety of sample delivery methods have been developed to satisfy this need including on-demand ejection of crystal suspension droplets (Cheng, 2020). This droplet-on-demand approach is versatile in accommodating different experimental conditions, and has been integrated with other sample delivery methods such as micromesh and polyamide conveyor belt systems (Fuller et al., 2017; Roessler et al., 2013; Soares et al., 2011). With proper design and installation of the sample delivery system, it has been demonstrated that both the XRD and X-ray emission spectroscopic (XES) data can be collected simultaneously. Combined with a flexible reaction triggering module, time-resolved crystallographic and spectroscopic studies on protein dynamics that probes on the structural and electronic state of protein intermediates is made possible. In this section, the operation of the Drop-on-Tape (DOT) sample delivery system (Fuller et al., 2017), one of such methods that enable simultaneous time-resolved XRD and XES measurements on metalloenzymes at room temperature, is detailed.

Full design and composition of the DOT system are described in Fuller et al. (2017), and the core components are detailed in Fig. 8A. An image of the actual deployment of the DOT system, including the peripherals that are described in later sections, is shown in Fig. 9A. Briefly, the core of the system is composed of a conveyor belt and an acoustic droplet ejector (ADE) (Fig. 8A, a–c). ADE produces a droplet of discrete volume using an ultrasonic transducer (Fig. 8A, a) (Hadimioglu, Stearns, & Ellson, 2016). The ejected droplets are deposited onto a continuously moving polyimide conveyor belt, which transports a queue of droplets through the reaction triggering module and to the XFEL interaction point (Fig. 8A, i). The droplet on tape is a hemisphere with ~200–300 μm in diameter and 0.8–6 nL in volume. The ejection rate is variable, with an upward rate of 120 Hz. The mechanism of droplet ejection by an acoustic transducer does not require a confined sample reservoir. As such, relatively large microcrystals with a longest dimension in the range of ~100 μm can be delivered with ease using this system. The speed of the conveyor belt is a key variable that controls the timing of reaction triggering and is detailed in the sections below. In standard operation, the conveyor belt is run at ~200–300 mm/s.

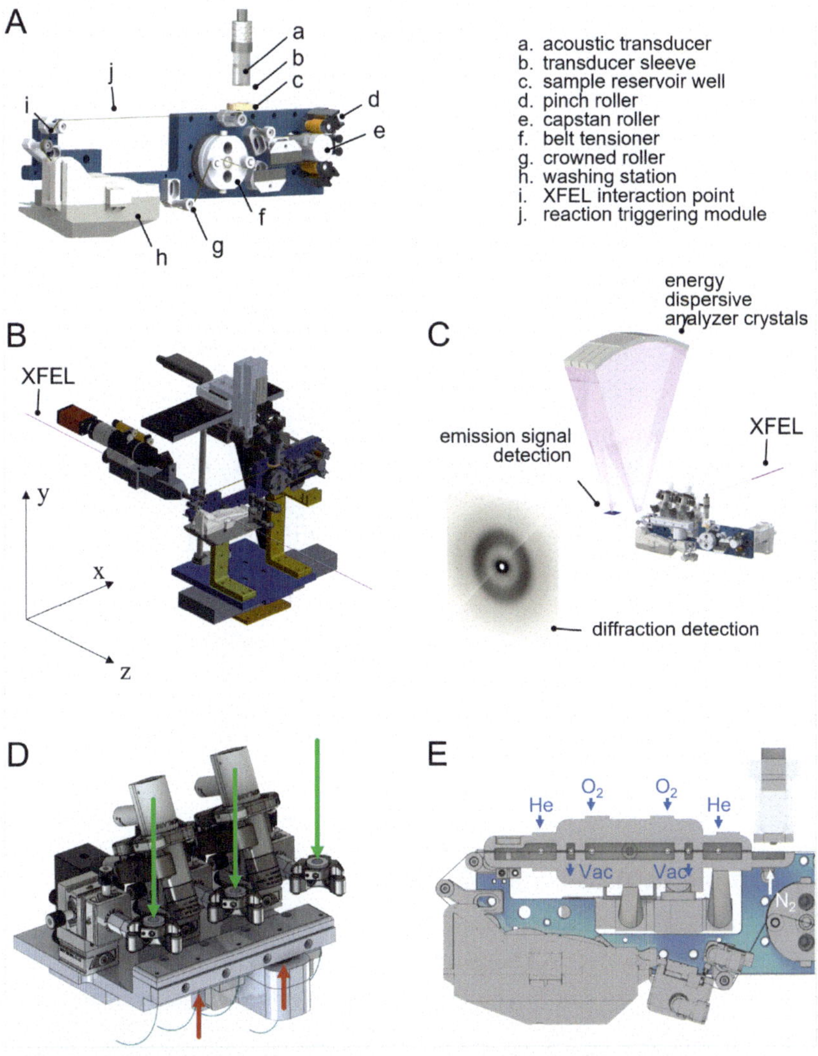

Fig. 8 Overview of the DOT sample delivery system and deployment configurations. (A) Core components of the DOT system. Different reaction triggering modules are inserted at position *j*. (B) A typical mounting scheme of the DOT system with a directional reference nomenclature used in this chapter. The XFEL beam is parallel to the *z*-axis. The core components shown in (A) are mounted onto a translational stage for movement in the *x*-direction using vertical brackets. The ADE setup is mounted on multi-axis stages to enable positional adjustments in *y* and *z*. A beam enclosure to provide a camera view to the interaction point in-line with the XFEL beam is also shown. (C) Relative positioning of the DOT setup with respect to the X-ray emission spectrometer and X-ray diffraction detector. Distance between the interaction point on the DOT system and the XRD detector is not drawn to scale. (D) Photoactivation

(Continued)

3.1 Sample handling and loading

To reduce the interaction of the XFEL beam with the atmospheric gases, and to shield O_2-sensitive samples from exposure to atmospheric oxygen in the chemical activation experiment, the main components of the DOT system, which includes the mounted core components and motorized stages and cameras for visualization (Fig. 8B), are enclosed in a large gas-tight chamber that is flushed with He. The sample is continuously infused into this chamber and to the reservoir pocket of the ADE (Fig. 8A,C) via a capillary feed line of 200 or 250 μm inner diameter connected to a 1-mL gas-tight syringe. In the case of PSII, which is light sensitive, the crystal suspension is loaded into a syringe (Hamilton Model 1001 PTFE Luer Lock, Hamilton Company, Reno, NV, USA) at ~0.5–1.2 mM Chl concentration and dark-adapted for 1 h before transporting to the experimental hutch. An aluminum foil is applied to the syringe to shield the sample from light exposure. The DOT chamber is covered by a regular or matte black aluminum foil, and the experimental hutch is kept in dark when possible. For IPNS and other O_2 sensitive samples, the suspension is loaded into a gastight syringe with a sample lock (Hamilton Model 1001 Sample Lock termination, Hamilton Company, Reno, NV, USA), and is transported to the experimental hutch in gas-tight container with an oxygen absorber (AnaeroPack-Anaero System, Mitsubishi Gas Chemical America, Inc., New York, NY, USA).

3.2 Sample infusion and ejection

The sample infusion rate is controlled by a syringe pump with a remote pump head (Legato 130, KD Scientific, Hamilton, MA, USA). The remote head of the syringe pump is mounted on a syringe shaker that rotates the pump head and syringe between 0° and 180° on an axis perpendicular to the long axis of the syringe to keep the microcrystals suspended in the syringe for the duration of the experiment. The rotation rate and duration of stops at each end of the rotation cycle need to be optimized for each

Fig. 8—Cont'd reaction module. A droplet travels from right to left in this diagram. A total of three photoactivation is possible using a fiber-guided laser aligned to the 1st, 3rd, and 5th apertures. A droplet position on the reaction module is tracked using a near infrared light probe aligned at the 2nd and 4th apertures. (E) A schematic of the O_2-activation module. A droplet travels from right to left in this diagram. The central chamber is saturated with O_2, while other chambers are used to create a differential pump.

crystal sample, depending on crystal size/shape and buffer density. The infusion rate is dependent on the droplet ejection rate and size, which in turn is determined by the rate of detection and frequency of XFEL pulses. At 30 Hz ejection rate, an infusion rate of 7–9 μL/min is sufficient to yield stable ejection of droplets with 4–5 nL volume. When using the above-described conditions for PSII and IPNS between 8 and 20 PSII or around 60–80 IPNS microcrystals are contained in a single droplet. The gas tight syringe pump sitting on the syringe shaker can be enclosed in different ways depending on the sample needs. For temperature sensitive samples the shaker can be mounted inside a refrigerator with a feedthrough for the sample capillary (care has to be taken that the syringes used for sample delivery are compatible with the temperature setting of the refrigerator. Gas-tight Hamilton syringes for example are not sealing well at temperatures below 12 °C). For light sensitive samples the shaker can be placed inside a light tight box or enclosure. The capillary line feeding out of the enclosure and to the DOT chamber is shielded from light by an opaque capillary tubing sleeve (PEEK Capillary Tubing Sleeve, IDEX Health & Science, LLC, Oak Harbor, WA, USA). For O_2 sensitive samples, a top loading enclosure filled with Ar gas is used.

Sample droplets are generated using the ADE as mentioned above. The main components of the ADE system are: a focused ultrasonic transducer, a housing for the transducer and the immersion water, and a sample reservoir pocket with a connection to the sample feed line (Fig. 8A). A sample droplet is produced by focusing sound energy on the interface between the liquid containing the microcrystal suspension and air to pinch off a small volume of sample into the air. The sample is introduced to the focal point of the ultrasonic transducer by infusing a set volume of sample at the specified infusion rate (from the syringe and via the capillary line) into the reservoir pocket (Fig. 8A,C). The sample reservoir pocket is cylindrical in shape and has a dimension of 2.5 mm in diameter and 2.0 mm in depth and has a capacity of 10 μL. In addition to the sample feed line, one to two additional feed lines are guided to the reservoir to infuse reservoir cleaning water/solution. Besides this reservoir with the standard dimensions, a smaller (1.55 mm diameter × 1.25 mm depth, 2.4 μL capacity) pocket has been employed to produce stable sample infusion and droplet ejection for challenging samples.

The droplet ejection is managed by sending a 15 MHz tone burst to the ultrasonic transducer. Here, a 15-MHz focused immersion transducer is employed (Olympus V319-SU, 1.00-in spherical focus). A waveform of a tone burst is generated using a function generator (Agilent 33600A). Timing and repetition rate of the waveform are controlled by a low-jitter digital delay generator (DG645, Stanford Research Systems). The repetition rate is adjusted to match the frequency of the XFEL pulse. This waveform is amplified by a 150 W 250 MHz or 500 W 250 MHz RF amplifier (150A250 or 500A250C, AR RF/Microwave Instrumentation) to a driving signal of ~40 V. The amplified tone burst is sent to a 20-dB bidirectional coupler (ZFBDC20-61HP+, Mini-Circuits, Brooklyn, NY, USA) that transmits the tone burst to the ultrasonic transducer. Along with the tone burst for droplet ejection, a probing signal is sent to the transducer. The probing signal is reflected upon interacting with the liquid–air surface, and is redirected as a return signal by the coupler. The reflected probe signal functions as an "echo" that indicates the fluid level in the sample reservoir. The filling of the reservoir changes the distance between the transducer focal point and the liquid–air interface, and the maximum "echo" signal indicates that the sample is filled to the focal point. Using this "echo" signal as a probe, the ejection rate (outflow rate) is adjusted by varying the burst cycle and amplitude sent out by the function generator. Once the "echo" is stabilized at the maximum signal amplitude, the infusion and ejection rates are balanced.

3.3 Tape drive control

The droplets ejected from the ADE setup are deposited onto a seamless Kapton polyimide tape (BPB, Inc.) that is 31 in. in length, 1.5 mm in width, and 25 μm in thickness. Two linear stages are used to control the positioning of the ADE setup with respect to the tape (Fig. 8B). A linear stage with 25-mm travel range (Newport, MFA-CC) controls the movement in the y-axis to adjust the vertical distance between the ADE setup and the surface of the tape. A second linear stage with 25-mm travel range (Newport, UMR8.25), motorized by a high-load actuator (Newport, LTA-HL), is used to adjust the position of the ejected droplet on the width of the tape in the z-axis. Unless deemed necessary, the ejected droplets are positioned at the center of the tape width. The seamless tape is passed through a series of low friction bearings with radially crowned aluminum caps, and tape transport components. The tape transport components include a tension arm (Fig. 8A,F), a capstan roller (Fig. 8A,E), and two pinch rollers to press the tape against the capstan (Fig. 8A,D). The shaft of the capstan is coupled to a stepper motor (Lexium M-Drive LMDCE) which serves as the only motorized drive for the tape control. The speed of the conveyor belt is adjusted via the control of this stepper motor. The set of rollers, the tape transport components, the motor drive, and the ADE setup are mounted onto two aluminum optical breadboards that are assembled using an aluminum angle bracket. This construction that carries the tape drive and ADE setup is then mounted onto a motorized linear stage with 100-mm travel range (Newport, UTS100CC) to enable adjustment of horizontal positioning of the DOT setup with respect to the XFEL beam in the *x*-axis. Details of the mechanical setup including construction drawings are described in Yano et al. (2017).

3.4 Delivering droplets to the XFEL beam

A brief procedure to bring the ejected droplets on the tape drive to interact with the XFEL beam is as follows. To facilitate each of these steps, a GigE camera with a zoom lens attachment is installed to provide a view of the X-ray interaction point nearly collinear with the direction of the XFEL beam with a sufficient magnification to visualize a ~200 μm droplet (Manta, Allied Vision, Stadtroda, Germany). The GigE camera is triggered to run at the same frequency as the XFEL beam, and its timing and exposure time is adjusted by the digital delay generator (DG645, Stanford Research Systems) and the camera's internal controls. First, to detect the position of the XFEL beam in the view of the collinear camera, a piece of reference scintillator (such as GaAs or YAG(Ce)) is brought into the predetermined XFEL/sample interaction point. This scintillator is mounted on a reference mount setup that is installed on the same 100-mm linear stage for *x*-axis movement as the tape drive setup at the same *z*-position as the droplet but at the offset *x*-position. The reference mount setup is brought into the XFEL interaction point using this linear stage. Using a long exposure time (to cover the period of the XFEL frequency) on the collinear camera, a beam position can be identified and marked on the camera view. In a sequential fashion, reduce the exposure time of the camera and adjust the timing of the camera acquisition such that the fluorescence from the XFEL beam on the scintillator is not blinking with 100 μs of exposure time or less. Once the position and timing of the XFEL beam is determined on the collinear camera, translate the DOT system in *x*-axis using the 100-mm linear stage to bring the tape to approximately 500 μm of the marked XFEL position. Generate a series of stable water (i.e., non-sample) droplets at the frequency that matches the XFEL. Further translate the DOT but in a small (~10 μm) increment in *x*-axis while scanning through the timing of droplet ejection until a droplet explosion is observed on the collinear camera

(Fig. 9C), or a water ring is detected on a diffraction detector. Through the course of an experiment, a buildup from the exploded sample droplet or infusion of inhomogeneous sample or air bubbles into the ADE setup may result in offset in timing and positioning. These slight offsets are corrected by adjusting the timing of droplet ejection at a 50 μs increment, and repositioning the *x*-axis linear stage that carries the DOT setup at a 5 μm increment.

3.5 Peripherals

The DOT system that includes the ADE setup, the tape drive setup, and the linear stage that carries these components are enclosed in a gas-tight housing. The housing is made of 0.5 in. acrylic with a cutout for the X-ray emission spectrometer on the top and for the X-ray diffraction detector on the downstream side. These cutouts are sealed by inflatable plastic bags that press against the analyzer crystals for the XES and a detector window for XRD (a protector surface in front of the actual detector elements) when inflated with He flow/circulation. The upstream panel includes a port for introducing the XFEL beam into the enclosure, and a cutout sealed by an anti-reflection coated Nd:YAG laser line window for introducing a free space excitation laser (described below). One of the side panels is fitted by a feedthrough panel to pass capillary feed lines as well as communication and signaling cables into the enclosure. To create a low-O_2/high-He environment inside the enclosure, a diaphragm pump (KNF Neuberger Inc., N1400) is used to purge the enclosure while flowing He at 50–90 SCFH to replenish the atmosphere. Depending on the flow rate of He, the enclosure can be purged to <2% O_2 in 10–90 min. Once the <2% level is reached, the helium flow is reduced to ~20 SCFH and the diaphragm pump is set to recirculate the helium within the enclosure. Alternatively, to further reduce the overall He consumption, the enclosure can be purged by flowing N_2. In this configuration, the flight tube collinear to the beam (depicted in Fig. 8B) and the X-ray emission shielding (Fig. 9A,G) are sealed and filled with He to secure the path of the XFEL beam, and the XES photons are in a low-scattering He environment. In the case of non-O_2 sensitive samples such as PSII, a configuration without a purged enclosure is also possible. In these cases, a beamstop or a collimator is mounted downstream of the X-ray interaction point (IP) to reduce the air scattering of the beam from reaching the XRD detector.

Upon interacting with the XFEL beam, the sample droplet is exploded, and the majority of the volume is removed from the tape. These sample residues, as well as any "missed" droplets, are a potential source for coating

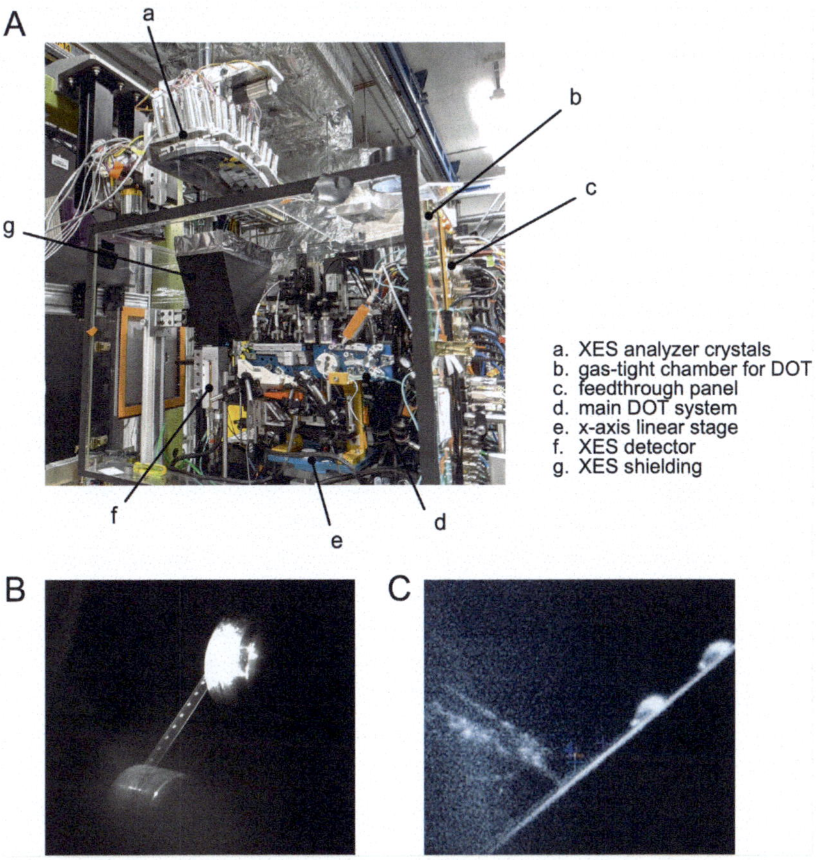

Fig. 9 (A) The DOT system deployed at the XFEL experimental hutch. In this configuration, the DOT system is enclosed in a gas-tight chamber which has a cutout for the XES spectrometer (a) and XRD detector (not shown). The cutouts are sealed by a thin inflatable plastic sheet for the XES spectrometer side, and by a window fitted by an X-ray transparent material for the XRD detector side. (B) A train of droplets ejected at a set frequency on a Kapton conveyor belt. The image is taken by a GigE camera triggered at the same frequency as the droplet ejection and XFEL beam, and shows a view onto the X-ray interaction point on the DOT. In this view, the tape is transporting droplets from top to bottom. (C) A droplet shot by a pulse of XFEL. The image is taken by a GigE camera with zoom lens attachment aligned to provide a collinear view as the incoming XFEL beam. In this view, the droplet is delivered from the top right corner of the image to the X-ray interaction point near the center of the image.

the tape surface. This has two negative effects: it reduces the hydrophobicity of the tape, leading to flattening of newly ejected droplets to a smear or "pancake" shape instead of a hemisphere, and if moisture collects on the tape, then this could lead to derailing the tape off the rollers. Additionally, coating of the tape by a sample residue is highly detrimental to the collection of XES as the accumulated residue could generate an XES signal that overlaps with a signal from the sample droplet. Therefore, proper cleaning and drying of the tape is essential to both the XRD and XES experiment. To remove the residue and "missed" droplets, the tape feeds into two cleaning stages past the XFEL/sample interaction point. In the first stage, a generic submerged water pump (TotalPond, 1200 gph waterfall pump) is used to spray water onto the tape via two 1-mm ID tubes with combined flow of 150 mL/min. In the second stage, four high pressure water jets with a combined flow rate of 50 mL/min are directed to the tape using a low pulsation HPLC (high-performance liquid chromatography) pump (Teledyne, LD-class HPLC). The outflow of the sprayed cleaning water is collected and removed from the enclosure by a peristaltic pump (MasterFlex, L/S driver fitted with Easy-Load II pump head). Cleaning water, which uses distilled water, is divided into two separate containers each with 14-gallon capacity. Initially, 12 gallons of distilled water is supplied to a tank for the second stage HPLC pump, and 2 gallons of distilled water is filled to the tank for the first stage submerged water pump. The outflow guided by the peristaltic pump is fed into the first stage tank. The cleaning water, therefore, is recirculated for the first stage but unidirectional for the second stage. Beyond the two cleaning stages, the tape is passed through a drying stage which injects compressed helium with two blowoff nozzles from the top and bottom of the tape.

To improve stability of droplet deposition on the tape the static charge of the tape needs to be minimized. For this purpose, an antistatic gun (Ptec IN3425 AC Ionizing Nozzle, Transforming Technologies, LLC., Toledo, OH, USA) is directed at the tape directly upstream of the sample deposition region and provided with a continuous flow of He gas.

3.6 Reaction triggering

Between the ADE setup and the XFEL/sample interaction point of the tape drive, a variable reaction triggering module can be installed to activate a reaction in the protein with controlled delay times (Fig. 8B,J). Here, two types of reaction triggering systems, the photoactivation module and the O_2-incubation module, are described.

3.6.1 Photoactivation

For samples that require multiple, precisely timed actinic flashes to advance the reaction cycle (such as PSII), a laser excitation system is installed in the module space (Fig. 8D). In this system, fiber-coupled lasers are guided with a specific spacing on the tape to enable up to three events of photoactivation. In order to align the droplet both in space and in timing to multiple excitations, two fiber-guided IR light sources (>800 nm) are positioned equidistant between the first/second excitation sources and second/third excitation sources. The droplets passing through the IR lights scatter the light, and the intensity and timing of the scattering are probed by photodiodes. These scattered lights serve as gates that indicate the positioning and timing of the droplets on the tape. In total, the laser excitation system holds three fiber-coupled lasers for the actinic source and two fiber-guided IR lights that function as the IR gates. The output of all five sources of light emitted through the tape are monitored remotely via an oscilloscope.

The timing of the excitation laser on the photoactivation module needs to be synchronized in relation to the timing of droplet ejection as well as the tape speed. The timing of the droplet ejection (deposition delay) itself is synchronized to the timing of the XFEL beam in a manner described above. To synchronize the excitation lasers on the droplet for a given deposition delay, the tape speed is first adjusted to create a temporal overlap on the IR gate signals on the oscilloscope. The overlap indicates that a droplet is present on each of the IR gates at the same time (the separation between two droplets is equal to the distance between two gates). Because the two gates are separated by the same distance as the space between two adjacent excitation laser sources, and a separation between a gate and excitation laser is half of the separation between two gates (or two laser sources), a combination of the tape speed that results in a droplet being present at all three laser source points can be calculated. By adjusting the delay of the laser flashes such that the timing overlaps with the timing of the IR gates and using the tape speed calculated by the spacing between the gates and laser sources, spatial and temporal overlap between droplets, excitation laser flashes, and XFEL beam can be achieved. A series of combinations and graphical representation of the tape speed calculation are detailed in Fuller et al. (2017). Briefly, for a standard operation of the PSII experiments, a droplet ejection rate is set at 20 Hz and the tape speed of 300 mm/s is applied.

For time-resolved photoactivated measurements, a free space laser source is introduced near the XFEL/sample interaction point, separate from the laser excitation system described above. Depending on the laser source, the delay between photoactivation and XFEL interaction can range between fs–μs timescale or ns-ms timescale. The free space laser is introduced into the enclosure and DOT via a laser line window installed on the upstream panel of the enclosure, and is guided onto the tape by the kinematic mirror mount with piezoelectric controls. The spot size on a tape is typically set to 400 μm in diameter, which is sufficient to cover a ~250 μm droplet. To properly guide the free space laser onto the tape and ensure a sufficient overlap with a droplet, another GigE camera is installed to provide a view of the surface of the tape around the interaction point. The camera, like the collinear one, has delay and repetition rate controlled by a digital delay generator to match the frequency and timing of the XFEL/sample interaction. Defining the timing of the XFEL/sample interaction as $t=0$, to measure the XRD and XES data of a photoactivated sample of, for example, 500 μs after photoactivation, the following steps are followed. First a good temporal and spatial overlap between free space laser, and XFEL beam, and a sample droplet is confirmed at $t=0$. Second, the delay on the side-viewing GigE camera and the free space laser is decreased by 500 μs. The camera shows the position of the droplet and free space laser 500 μs prior to the arrival of the XFEL beam. Using the piezo-controlled kinematic mirror mount, the laser spot is adjusted to overlay the spot onto a droplet. Lastly, the delay on the camera is reset to 0 to confirm that a temporal and special overlap between a droplet and XFEL beam (but not the free space laser which delay stays at −500 μs) is unaffected by this procedure.

3.6.2 O_2-activation

Many of the catalytic reactions of metalloenzymes are activated by the presence and uptake of O_2 in the surrounding atmosphere. To investigate the protein dynamics of such O_2-activated reactions, a reaction module can be replaced by an O_2-incubation platform (Fig. 8E). The platform is composed of an orifice with a series of chambers for the tape to pass. The central chamber extends for ~60 mm in length and is sandwiched by a pair of two 5 mm chambers, which are flanked by two chambers with 25/39 mm in length. An isolated region filled with O_2 gas is created in the central chamber by introducing a total flow of 1–2 SCFH into the chamber. The O_2 content in the chamber is monitored by an oxygen meter directly attached to this chamber. To limit O_2 from infiltrating the rest of the enclosure, the other chambers are used as a differential pump:

the 5 mm chambers on either side of the O_2 chamber are connected to a vacuum line which creates negative pressure, while the outermost chambers have helium flow to create positive pressure. Using the O_2-incubation platform, the main chamber can be saturated with O_2 (>90%) while retaining a low level of O_2 in the rest of the DOT enclosure (<2%). The exposure time is mainly controlled by adjusting the tape speed. The length of the incubation chamber (60 mm) and the distance from the end of the chamber to the IP (90 mm) can be used to calculate the O_2 incubation time as well as the total delay time between start of O_2 incubation and probing the sample with X-rays as a function of the belt speed. At the standard belt speed of 200 mm/s, this results in an incubation time of 300 ms and a total delay between start of incubation and measurement of 750 ms. By varying the belt speed, these times can be changed over a long range from about 200 and 500 ms (belt speed of 300 mm/s) up to 6 and 15 s (belt speed of 10 mm/s), respectively. For very long incubation times, care has to be taken that the sample is not drying out during the passage to the IP. For this purpose, the relative humidity in the He enclosure of the system and the O_2 incubation gas can be regulated to ~20–30% by bubbling either or both gasses through a wash bottle before feeding them in the experimental enclosure. Depending on the buffer composition, it is still common to observe formation of dried out PEG or salt in the form of powder diffraction at these conditions, and this is often the limiting factor in determining the longest possible incubation time. Also, it is important to note that the spacing between individual droplets becomes very small for slow belt speeds (e.g., at 250 μm drop diameter, 10 mm/s belt speed and 30 Hz deposition frequency the gap between individual drops is only 80 μm) and the deposition frequency might need to be reduced to accommodate very long incubation times.

4. XES data collection

Tandem XES on the time-resolved SFX experiments on metalloenzymes using XFEL is highly advantageous due to XES's high element specificity and XFEL's peak brightness that enable shot-by-shot collection of the spectra. Given these advantages, XES has been applied concurrently in a series of experiments to provide in situ confirmation of the reaction kinetics, batch-to-batch sample integrity, and oxidation state of the active site. Using the DOT system, XES of two different transitions (Kα and

$K\beta_{1,3}$) of three different elements (Ni, Fe, and Mn) has been collected using microcrystals of metalloenzymes in time-resolved SFX experiments.

The energy of the incoming XFEL beam must be set above the absorption edge of the element to be detected. The energy-dispersive spectrometer is based on the von Hamos geometry, and the assembly and performance of such a spectrometer is detailed in Alonso-Mori et al. (2012). Briefly, the spectrometer consists of a series of cylindrically bent analyzer crystals that are mounted on motorized stages to control pitch, yaw, and linear translation. In the case of $K\alpha$ XES, 2–4 analyzer crystals are utilized. For the $K\beta_{1,3}$ signal which is approximately an order of magnitude smaller than the $K\alpha_1$ signal, 16 crystals assembled as a 4×4 array are used. The spectrometer is mounted on two motorized linear stages and one rotational stage to enable adjustment in position horizontally and vertically (both directions are normal to the axis of the incoming XFEL beam). The diffracted energies were collected by a position sensitive detector (SLAC National Accelerator Laboratory, ePix100). Placement of the spectrometer and position sensitive detector is dependent on the direction of polarization of the XFEL beam to avoid the largest elastic scattering signal. For the current polarization of the Linac Coherent Light Source (LCLS) facility, the spectrometer is mounted on a vertical plane intersecting the interaction point, while the detector is placed on the horizontal plane (Fig. 8C). Using multiple sets of analyzer crystals and position sensitive detectors, XES signals of multiple different metals can be collected simultaneously, along with the XRD data.

To align the spectrometer and detector, one of the two methods can be applied: (1) Use the DOT system to generate stable droplet of a solution sample of molecular standards (such as FeCl or MnCl) in ~1–5 mM concentration, or (2) Use a metal foil of < 100 μm thickness. The foil is to be placed at the XFEL/sample interaction point but tilted to ~45° around the horizontal axis to ensure that the emitted radiation is not blocked by the foil itself. For this option the foil can be mounted on the reference mount setup described in the previous section and moved into the beam interaction region using the *x*-axis motorized translation stage of the DOT system. Using method 1, an alignment can be made with the accurate/actual droplet position. This approach is also highly reproducible as the relative droplet position is controlled by motorized stages. A solution of concentrated molecular standards, however, is a potential source of contamination of the setup as the concentration of the metals in microcrystal protein samples are an order of magnitude smaller. A metal foil in method 2, on the other hand, does not come into physical contact with the tape drive setup and will not introduce contamination to the system. However, the relative position of the foil is not coupled to the droplet position, and a slight discrepancy may exist between both the metal foil and the actual droplet sample with respect to where the signal appears on the position sensitive detector. As this can also lead to slight distortion in the energy axis, care must be taken that the foil is located as close as possible in the same position along the X-ray beam as the actual sample. To ensure that data taken during different times of an experiment are comparable it is important to regularly check the alignment of the spectrometer (Fig. 10).

For the collection of XES, the number of shots to be averaged depends on the metal concentration, the type of spectra to be collected, and number of analyzer crystals that are employed. For a typical PSII sample described in this chapter, using a spectrometer with cylindrically bent 16 Si(440) crystals (R = 500 mm), approximately 25,000 shots or, when operated at 20 Hz, 20 min of data collection yields Mn K$\beta_{1,3}$ spectra with sufficient signal-to-noise ratio. For the IPNS crystals described above, data collection of the Fe Kα spectra using four cylindrically bent (R = 250 mm) Ge(440) crystals yielded high quality spectra in about 7 min collection time at 30 Hz

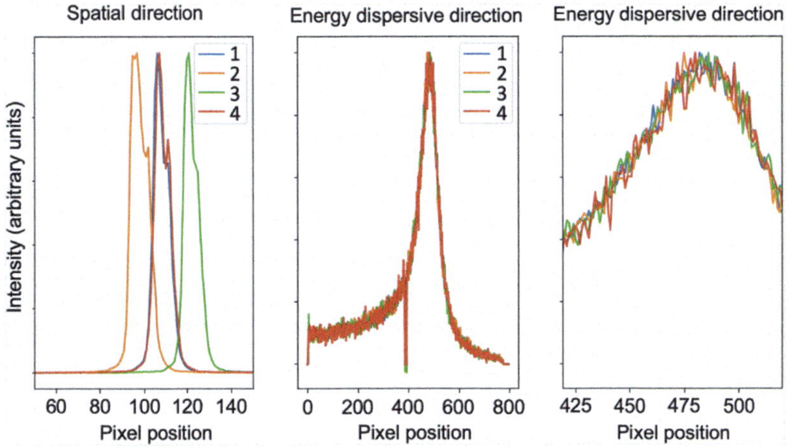

Fig. 10 Alignment of XES spectrometer, reproducibility over shifts. The Mn K$\beta_{1,3}$ spectra collected from a Mn foil for four consecutive days (1–4) of an experiment are plotted. While small changes in the position of the signal on the detector are visible (left panel, "Spatial direction") they perfectly align in the energy dispersive direction (middle and right panel). Thereby it was confirmed that the sample spectra from different days of the experiment can be directly compared with each other.

repetition rate. In most cases, however, as the XRD data collection requires a significantly higher number of shots or data collection time, data collection time for XES is usually not an issue.

Processing of the XES signals and examples of analysis of resulting spectra are outlined in Fig. 11. To process the XES signal collected on a position sensitive detector, pixels containing emission signal are defined as a region of interest (ROI) (Fig. 11A). Images from each shot are screened for "missed" shots by filtering by the relative signal intensity in the ROI. Areas surrounding the ROI are then used to extrapolate a background signal. The final XES spectrum is obtained by integrating over the spatial width of the signal in the ROI. To calculate a difference spectrum of different timepoints in reaction kinetics, the spectra at each timepoint are first normalized by the area under the curve. Several examples of these area-normalized spectra along with associated difference spectra are shown in Fig. 11C and E. In Fig. 11C, a clear shift can be observed in Mn $K\beta_{1,3}$ signal of PSII at two different reaction states (dark states and after excitation by two saturating flashes) in these area-normalized spectra. This shift is visualized in a difference spectrum, also included in Fig. 11C. Similarly, in Figs. 11E and F, a shift in Fe $K\alpha$ signal of IPNS at different reaction states are observable. Further details about the XES data processing can also be found in Fransson et al. (2018).

To quantitatively analyze the changes in XES at different reaction states beyond consideration of difference spectra, spectral parameters such as first moment and full width at half maximum can be calculated. By plotting these spectral parameters, change the oxidation state and/or ligand environment of the active site metal(s) can be tracked across different delay times following the reaction triggering. Examples of these kinetic trends are shown in Fig. 11D and G. As described in the next section, these trends determined via XES are to be analyzed concurrently with the information obtained from XRD to assist the interpretation of the structural changes in the protein dynamics.

5. XRD data collection

X-ray diffraction data are collected with a large position sensitive detector downstream of the interaction region. Detectors that have been used with the DOT setup include the Rayonix MX 340 (at LCSL/MFX), the Rayonix MX 300 (at the SPring-8 Angstrom Compact Free Electron

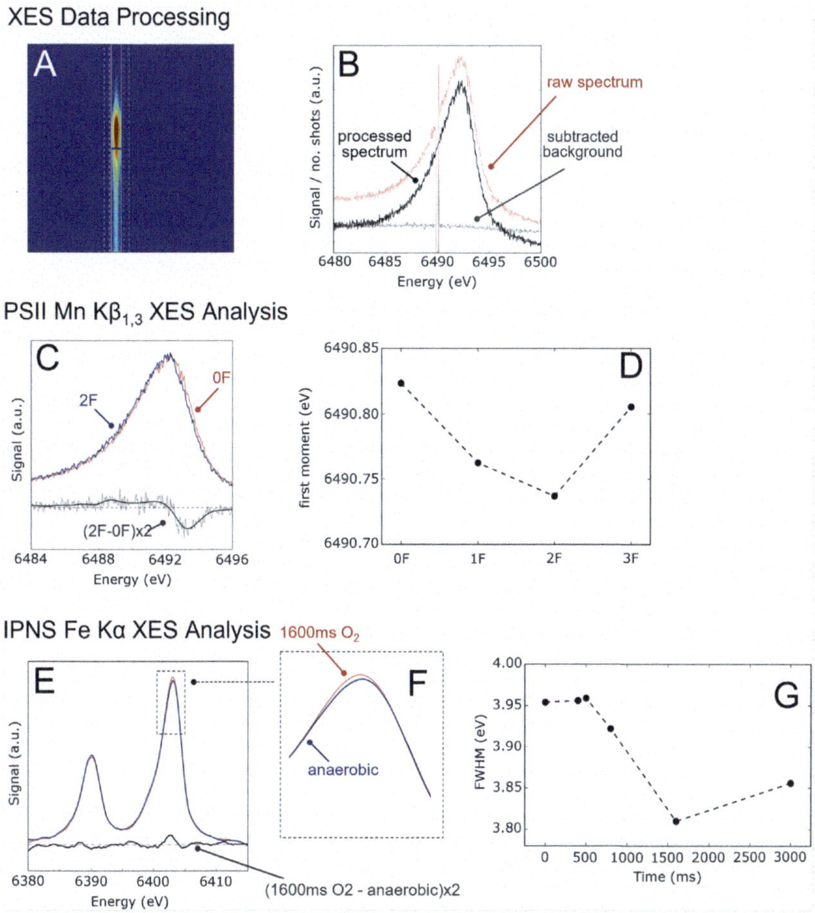

Fig. 11 Outline of the XES data processing, from the signals on the position sensitive detector to quantitative analysis of the spectra. (A) Signals collected on the position sensitive detector. The pixels containing the signals are cut out as ROI (defined by two solid lines). To correct for any drift in the background signal, two additional ROIs are defined outside the signal ROI (defined by four dashed lines). The intensities within the two background ROIs are extrapolated across the signal ROI to estimate the drift. (B) A raw spectrum calculated from the signal ROI (dotted line), extrapolated background signal (dashed line), and a processed spectrum where the background signal is subtracted from the raw spectrum. (C) Mn $K\beta_{1,3}$ signal of PSII, in the dark state ("0F") and after applying two actinic flashes ("2F"). A scaled DS shows a shift in the spectrum. (D) First moment of the Mn Kβ1,3 spectrum calculated for dark and different flash states of PSII. A shift in the first moment as the reaction cycle is advanced indicates a change in the Mn oxidation state through the reaction. (E) Fe Kα spectrum of IPNS, in the anaerobic state and after incubation with O_2 for 1600 ms. A scaled difference spectrum is also included. (F) A closer view of the shift in Fe $K\alpha_1$ peak. (G) Full-width half-maximum of Fe $K\alpha_1$ peak of IPNS across different O_2 incubation time.

Laser (SACLA) BL2/EH3), the Jungfrau 16 M (at SwissFEL/Bernina) or the ePix10K (at LCLS/MFX) detector. The choice of detector is dependent on the availability at the facility but also the needs of the experiment. For large unit cell samples, the spacing between individual Bragg spots for the long cell axis can become the limiting factor, making a detector with a large number of pixels necessary in order to cover diffraction out to the required resolution. Usually, a minimum separation of 4–5 pixels between two neighboring Bragg spots is required to properly integrate the intensities for these spots. The minimum detector distance achievable with the DOT setup depends on the detector type but for most cases 125 mm minimum distance can be achieved. This translates for example to a resolution of 1.45 Å for the inscribed circle when using an X-ray energy of 9.5 keV (1.3 Å) with the Rayonix MX 340 detector or to a resolution of 1.4 Å when using an X-ray energy of 10.5 keV (1.18 Å) and the Rayonix MX300.

As the precise distance between the detector and the X-ray interaction point with the sample is not known a priori this needs to be calibrated. For a first order estimation a diffraction pattern from a silver behenate powder is recorded. For this purpose, a thin pellet of silver(I) behenate (AgBeh, Alfa Aesar) is pressed using a pellet press. A small wedge of this pellet is then glued to the reference arm of the DOT setup. By changing the horizontal position perpendicular to the beam direction this reference arm is brought into the beam IP. Using a transmission of ~0.1–1% and rastering the AgBeh constantly through the beam by 10 μm translation steps every 50–100 shots during the measurement, a powder diffraction pattern (Huang, Toraya, Blanton, & Wu, 1993) can be recorded in about 1000 X-ray shots (16 s at 30 Hz). The known spacing of the diffraction rings for AgBeh (e.g. $d_{0\ 0\ 8}$ = 4.194 Å, $d_{0\ 0\ 10}$ = 3.35 Å, $d_{0\ 0\ 12}$ = 2.796 Å) is then used to calibrate the sample detector distance. For this, a tool in the cctbx.xfel data processing GUI can be used (Brewster et al., 2019).

While the data collection can be directly monitored visually via the online and side view cameras of the DOT system, allowing to fine tune droplet positioning and timing manually to optimize droplet hit rate, these do not give a complete picture of the success of the experiment. Using the *cctbx.xfel* GUI it is possible to track the progress of the experiment by plotting several statistics, including droplet hit rate, crystal indexing rate and diffraction quality (Fig. 12). Following the output on the GUI enables scientists to detect problems in the sample delivery. For example, Fig. 12A shows unstable droplet ejection as evidenced by the highly variable droplet

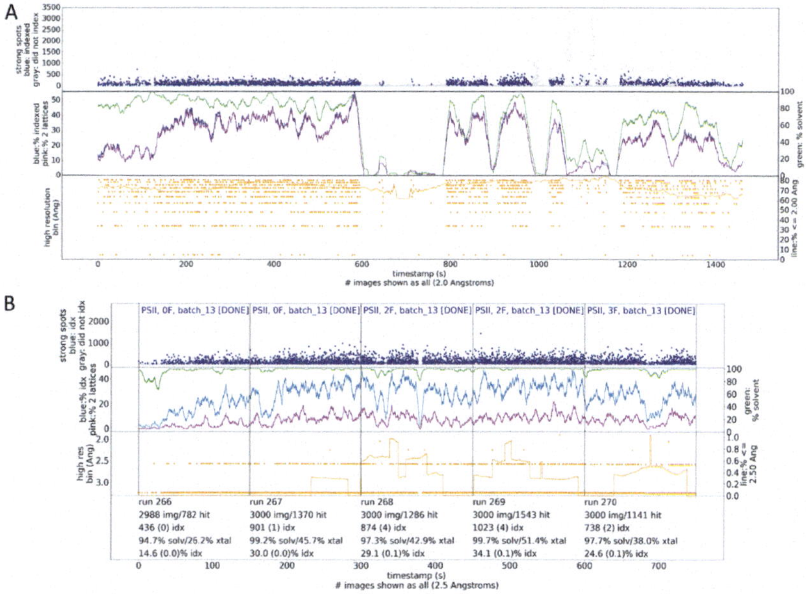

Fig. 12 Examples for direct feedback for diffraction experiments using the cctbx.xfel GUI. (A) Runs of a sample that exhibits several problems in the injection behavior and (B) sample that exhibits stable injection condition. The middle row shows the droplet hit rate (green line) and the indexing rate (blue line) as well as the percentage of double hits (purple). The top row shows the number of diffraction spots identified on each indexed image (blue dots) and on each non-indexed image (gray dots). The bottom row shows the maximum resolution quality detected (yellow dots) and the percentage of diffraction patterns that are better than 2.0 Å (A) and 2.5 Å (B) (yellow line). Below the graphs details about indexed images and hit rates are given for each experimental run.

hit rate (green trace in the middle row). In addition, many of the observed diffraction patterns cannot be properly indexed (gray dots in the top row), leading to an indexing rate of only a few % (blue trace in the middle row). On the positive side it is visible that nearly all indexed patterns show diffraction to better than 2.5 Å (yellow line in the bottom row) and often high-quality diffraction to 1.6 Å or better is observed (yellow dots in the bottom row). In contrast, the example in Fig. 12B shows a very stable droplet hit rate of nearly 100% and an indexing rate of 25–30% of all shots. But the diffraction quality of the indexed patterns is weaker with only very few crystals diffracting to better than 2.5 Å. In the first case, one should consider changing the droplet ejection parameters, maybe the buffer conditions, and the crystal concentration to achieve more stable operation.

In the second case, these conditions are all well-tuned but the diffraction quality of the crystals is suboptimal and consequently, it would be worth changing to a different protein batch in such a situation.

The direct processing of the diffraction and spectroscopic data also provides fast feedback on the unit cell and the catalytic activity of the measured sample. Fig. 13A shows the observed unit cell distribution for two different batches of photosystem II crystals, generated with the *cctbx.xfel* GUI (Brewster et al., 2019). The small difference in the distribution for the *c*-axis length indicates problems with the post crystallization treatment for one batch compared to the other. Consequently, this feedback can be used to optimize the treatment procedure on the fly to achieve a very uniform distribution over many different protein batches,

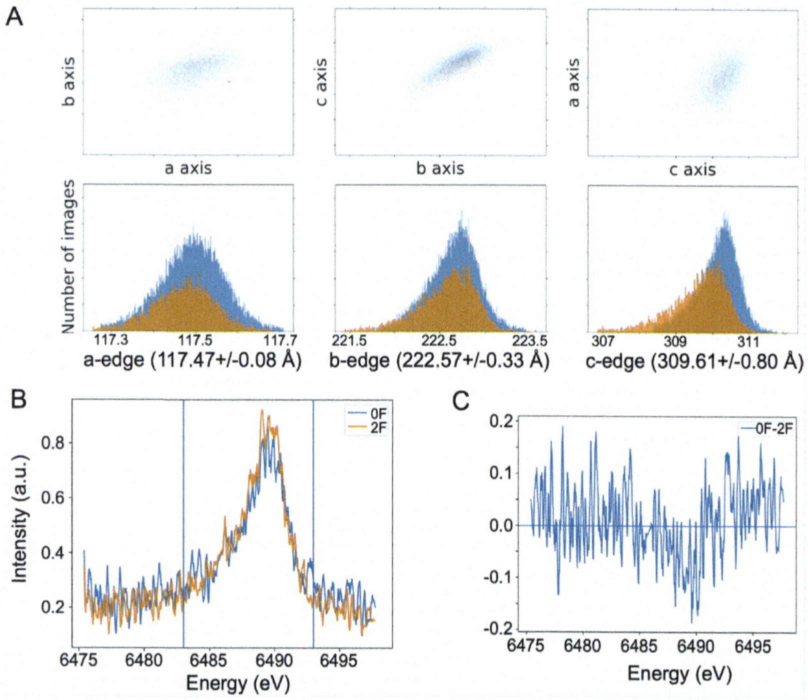

Fig. 13 Additional feedback during the XES/SFX experiment. (A) Unit cell distribution plot for two different batches of PSII crystals (orange and blue) generated by the *cctbx.xfel* GUI. A difference especially for the *c* axis length is visible between the two batches. (B) Mn Kβ XES signal recorded from PSII single crystals in the dark (0F, blue trace) and doubly illuminated (2F, orange trace) state and (C) difference spectrum between the 0F and 2F XES data both showing the expected difference due to light induced oxidation of the Mn cluster in PSII.

which is necessary to obtain highly isomorphous data. Fig. 13B,C shows the results of the XES processing during the experiment for one PSII protein batch. Even after only recording for a few minutes for each condition, it is possible to see the difference in the signal between the dark state ("0F") and the double illuminated state ("2F"). The observed shift between the spectra for the two illumination conditions is indicative for the sample turning over properly and serves as confirmation that the illumination conditions are set correctly and that the sample is intact (see also Fransson et al., 2018 for details).

While collecting data it is important to keep track of the number of indexed images for one condition (e.g., one illumination state) as a first indicator for the data set quality. To obtain a more complete picture of the dataset quality and estimate how much more data collection is necessary, repeated merging of the indexed data including subsequent experimental runs is very helpful. The *xfel.merge* program of the *cctbx* package (Brewster et al., 2018; Sauter, 2015) provides an informative output log with information regarding the percentage of indexed images that diffract to a certain resolution bin as well as merging statistics, including the multiplicity and the half data set correlation coefficient ($CC_{1/2}$) for each resolution bin. Example outputs are shown in Fig. 14. From our experience, the number of accepted lattices needs to be at least 2000–3000 in the highest resolution

A

Image Statistics

Bin	Resolution Range	Lattices	Accepted (%)
1	-1.0000 - 4.0716	16102	99.99
2	4.0716 - 3.2317	16076	99.83
3	3.2317 - 2.8231	16050	99.66
4	2.8231 - 2.5650	16013	99.43
5	2.5650 - 2.3811	15870	98.55
6	2.3811 - 2.2407	15552	96.57
7	2.2407 - 2.1285	15439	95.87
8	2.1285 - 2.0358	15021	93.27
9	2.0358 - 1.9574	14285	88.70
10	1.9574 - 1.8899	13776	85.54
11	1.8899 - 1.8308	12069	74.94
12	1.8308 - 1.7784	11530	71.60
13	1.7784 - 1.7316	10187	63.26
14	1.7316 - 1.6894	8925	55.42
15	1.6894 - 1.6510	8097	50.28
16	1.6510 - 1.6158	6612	41.06
17	1.6158 - 1.5835	5940	36.89
18	1.5835 - 1.5536	4917	30.53
19	1.5536 - 1.5259	3939	24.46
20	1.5259 - 1.5000	3274	20.33
All		16104	

B

Image Statistics

Bin	Resolution Range	Lattices	Accepted (%)
1	-1.0000 - 5.1574	1541	100.00
2	5.1574 - 4.0934	1540	99.94
3	4.0934 - 3.5759	1539	99.87
4	3.5759 - 3.2490	1506	97.73
5	3.2490 - 3.0161	1421	92.21
6	3.0161 - 2.8382	1200	77.87
7	2.8382 - 2.6961	912	59.18
8	2.6961 - 2.5787	785	50.94
9	2.5787 - 2.4794	675	43.80
10	2.4794 - 2.3938	512	33.23
11	2.3938 - 2.3190	448	29.07
12	2.3190 - 2.2527	387	25.11
13	2.2527 - 2.1934	282	18.30
14	2.1934 - 2.1399	231	14.99
15	2.1399 - 2.0912	193	12.52
16	2.0912 - 2.0467	124	8.05
17	2.0467 - 2.0058	94	6.10
18	2.0058 - 1.9679	73	4.74
19	1.9679 - 1.9328	39	2.53
20	1.9328 - 1.9000	21	1.36
All		1541	

C

Image Statistics

Bin	Resolution Range	Lattices	Accepted (%)
1	-1.0000 - 5.1574	19306	100.00
2	5.1574 - 4.0934	19303	99.98
3	4.0934 - 3.5759	19288	99.91
4	3.5759 - 3.2490	19082	98.84
5	3.2490 - 3.0161	18386	95.23
6	3.0161 - 2.8382	16506	85.50
7	2.8382 - 2.6961	13484	69.84
8	2.6961 - 2.5787	11770	60.97
9	2.5787 - 2.4794	10334	53.53
10	2.4794 - 2.3938	8097	41.94
11	2.3938 - 2.3190	7083	36.69
12	2.3190 - 2.2527	6230	32.27
13	2.2527 - 2.1934	4800	24.86
14	2.1934 - 2.1399	4149	21.49
15	2.1399 - 2.0912	3440	17.82
16	2.0912 - 2.0467	2356	12.20
17	2.0467 - 2.0058	1823	9.44
18	2.0058 - 1.9679	1395	7.23
19	1.9679 - 1.9328	768	3.98
20	1.9328 - 1.9000	524	2.71
All		19306	

Fig. 14 Data statistics generated by *cctbx.xfel*. Acceptance statistics per resolution bin for data collected from IPNS (A) and PSII (B,C). Based on a rule of thumb of ~3000 indexed images needed in the highest resolution bin one would estimate a useable resolution of 1.53 Å for the IPNS data set, no useable dataset after 10 min of PSII data collection and about 2.1 Å after 75 min of data collection for PSII. It can be seen that the acceptance statistics obtained even after a short collection time (10 min, B) can be used as a good estimate for the data quality and the time needed to obtain a complete data set to the desired resolution (~2.1 Å, 75 min, C).

bin to result in acceptable statistics for the resulting data set up to that resolution. Monitoring the acceptance rate at the desired resolution bin together with the indexing rate thus allows one to predict how much more time one needs to spend on collecting at a certain condition (see Fig. 14B,C).

Two other criteria used by us to decide about the usable resolution cut off for the collected data are the $CC_{1/2}$ value and the multiplicity of observations (Fig. 15). The $CC_{1/2}$ value should show a monotonous decrease from low to high resolution and should not exhibit jumps or sudden changes from resolution bin to resolution bin. While a low $CC_{1/2}$ could be indicative of weak data, sudden changes in the $CC_{1/2}$ values can also be caused by outlier images (e.g. a strong background peak not properly masked out) and excluding such outlier images can help improve

A

Bin	Resolution Range	Completeness	%	<asu multi>	<obs multi>	n_meas	asu_m_meas	Merged <I>	Merged <I/sig(I)>	CC int	N int
1	-1.0000 - 4.0716	[2796/2817]	99.25	293.07	295.27	825568	2796	177884	14.867	99.0%	2796
2	4.0716 - 3.2317	[2691/2691]	100.00	188.38	188.38	506931	2691	139082	11.818	98.6%	2691
3	3.2317 - 2.8231	[2644/2644]	100.00	156.82	156.82	414633	2644	72584	8.936	98.3%	2644
4	2.8231 - 2.5650	[2633/2633]	100.00	136.93	136.93	360537	2633	37549	7.503	98.2%	2633
5	2.5650 - 2.3811	[2622/2622]	100.00	122.19	122.19	320388	2622	25976	6.700	97.3%	2622
6	2.3811 - 2.2407	[2622/2622]	100.00	110.44	110.44	289577	2622	22324	5.933	97.3%	2622
7	2.2407 - 2.1285	[2615/2615]	100.00	102.64	102.64	268410	2615	17660	5.255	97.0%	2615
8	2.1285 - 2.0358	[2581/2581]	100.00	91.11	91.11	235158	2581	12893	4.441	96.9%	2581
9	2.0358 - 1.9574	[2589/2589]	100.00	82.29	82.29	213044	2589	9209	3.651	96.1%	2589
10	1.9574 - 1.8899	[2562/2562]	100.00	73.56	73.56	188458	2562	6789	3.030	95.6%	2562
11	1.8899 - 1.8308	[2612/2612]	100.00	61.24	61.24	159953	2612	5033	2.435	93.9%	2612
12	1.8308 - 1.7784	[2582/2582]	100.00	54.55	54.55	140837	2582	3889	2.035	91.8%	2582
13	1.7784 - 1.7316	[2575/2575]	100.00	44.71	44.71	115140	2575	3216	1.685	91.0%	2575
14	1.7316 - 1.6894	[2567/2567]	100.00	38.12	38.12	97856	2567	2559	1.396	83.1%	2565
15	1.6894 - 1.6510	[2591/2591]	100.00	31.91	31.91	82673	2591	2408	1.267	80.6%	2590
16	1.6510 - 1.6158	[2572/2572]	100.00	25.38	25.38	65288	2572	2017	1.034	72.3%	2563
17	1.6158 - 1.5835	[2570/2570]	100.00	21.17	21.17	54401	2570	1809	0.874	66.6%	2557
18	1.5835 - 1.5536	[2590/2591]	99.96	16.24	16.24	42071	2590	1655	0.755	50.8%	2560
19	1.5536 - 1.5259	[2553/2555]	99.92	12.67	12.68	32374	2553	1513	0.620	45.2%	2442
20	1.5259 - 1.5000	[2551/2573]	99.14	8.63	8.70	22198	2551	1442	0.472	28.4%	2121
All		[52118/52164]	99.91	85.03	85.10	4435495	52118	28314	4.313	99.2%	51522

B

Bin	Resolution Range	Completeness	%	<asu multi>	<obs multi>	n_meas	asu_m_meas	Merged <I>	Merged <I/sig(I)>	CC int	N int
1	-1.0000 - 5.1574	[32708/32857]	99.55	271.92	273.16	8934459	32708	3253670	17.166	99.7%	32708
2	5.1574 - 4.0934	[32085/32085]	100.00	171.29	171.29	5495728	32085	2319555	12.000	99.1%	32085
3	4.0934 - 3.5759	[31869/31869]	100.00	141.58	141.58	4512033	31869	1163158	7.649	98.1%	31869
4	3.5759 - 3.2490	[31788/31788]	100.00	119.15	119.15	3787558	31788	571541	4.739	95.4%	31788
5	3.2490 - 3.0161	[31697/31697]	100.00	95.02	95.02	3011712	31697	294067	2.927	89.3%	31697
6	3.0161 - 2.8382	[31708/31708]	100.00	71.24	71.24	2258846	31708	190364	2.071	81.4%	31708
7	2.8382 - 2.6961	[31623/31623]	100.00	52.76	52.76	1668387	31623	137412	1.551	69.4%	31623
8	2.6961 - 2.5787	[31607/31607]	100.00	41.88	41.88	1323582	31607	104786	1.191	53.0%	31604
9	2.5787 - 2.4794	[31563/31563]	100.00	32.87	32.87	1037539	31563	88101	0.962	42.9%	31554
10	2.4794 - 2.3938	[31507/31507]	100.00	25.42	25.42	800988	31507	77642	0.794	29.3%	31433
11	2.3938 - 2.3190	[31536/31537]	100.00	20.77	20.78	655163	31536	71181	0.683	19.2%	31352
12	2.3190 - 2.2527	[31500/31509]	99.97	16.44	16.44	517914	31500	66387	0.585	13.3%	30991
13	2.2527 - 2.1934	[31503/31538]	99.89	12.70	12.72	400629	31503	61106	0.487	9.8%	30063
14	2.1934 - 2.1399	[31376/31477]	99.68	9.99	10.02	314504	31376	55538	0.403	8.8%	27887
15	2.1399 - 2.0912	[30951/31506]	98.24	7.19	7.32	226566	30951	52786	0.326	6.2%	22683
16	2.0912 - 2.0467	[29148/31401]	92.83	4.76	5.12	149350	29148	49681	0.260	3.1%	14444
17	2.0467 - 2.0058	[25469/31441]	81.01	3.19	3.94	100313	25469	44705	0.207	3.4%	7913
18	2.0058 - 1.9679	[17901/31425]	56.96	1.75	3.08	55054	17901	43853	0.183	5.4%	2630
19	1.9679 - 1.9328	[9914/31442]	31.53	0.80	2.54	25224	9914	43433	0.167	3.1%	527
20	1.9328 - 1.9000	[4838/31410]	15.40	0.35	2.29	11085	4838	40558	0.142	-3.1%	86
All		[562291/632990]	88.83	55.75	62.76	35286634	562291	494331	3.091	99.0%	486645

Fig. 15 Merging statistics for IPNS (A) and PSII data (B), showing the multiplicity ("asu multi"), $CC_{1/2}$ ("CC int") and $I/\sigma(I)$ ("Merged < I/sig(I) >") values for each resolution bin. Indicators used to estimate the resolution cut-off for each data set are highlighted in green.

the $CC_{1/2}$ statistics. When examining for example the statistics for the IPNS dataset shown in Fig. 15A, the $CC_{1/2}$ decreases continuously up to the highest resolution bin, indicating that the data is self-consistent up to 1.5 Å resolution. For the PSII data in Fig. 15B, the $CC_{1/2}$ number shows a monotonous decrease up to bin 16 and then starts to fluctuate, hence bin 16 should be used as the cut off for the merging based on $CC_{1/2}$. When considering the multiplicity (how many times each HKL value was observed in all indexed images included in the merging) as computed by the *xfel.merge* program a value of ~10 was used as an (somewhat arbitrary) cut off in our past work. Following this the IPNS data shown in Fig. 15A should be usable up to 1.53 Å resolution while the PSII data (Fig. 15B) can be used up to 2.14 Å resolution. It should be noted that multiplicity can be computed in different ways and the number obtained for the same data can be very different, depending on what reflections are included or excluded for the computation of the multiplicity (see also discussion in the SI material for Ibrahim et al., 2020). In addition, one should monitor the $I/\sigma(I)$ value to ensure that there is still a measurable signal in the last bin of the final merged data set. Again, it is difficult to compare these numbers directly to numbers from synchrotron rotation crystallography as the error estimates are not as well established for SFX data and hence the $\sigma(I)$ computation is highly dependent on the chosen error model. For a detailed discussion about the effect of the error model on SFX data processing see (Brewster, Bhowmick, et al., 2019; Hattne et al., 2014). Due to this uncertainty, we generally accept lower $I/\sigma(I)$ values (e.g., down to 0.3 for the highest resolution bin) compared to the standard $I/\sigma(I)$ cut off of ~1 conventionally used for rotation crystallography work. It should be noted that also for synchrotron work, recent developments indicate that including all data, even when having very weak $I/\sigma(I)$ values is beneficial for the quality of the electron density maps and one should utilize an iterative refinement process to decide up to what resolution bin should be included in the final data (Hattne et al., 2014; Karplus & Diederichs, 2015).

Depending on the unit cell dimensions, diffraction quality of the crystals, and the fraction that yield lattices to fill the desired high-resolution shell, data collection can take anywhere from 10 min to 6 h for a complete diffraction data set. For example, using 30 Hz repetition rate (dictated by 4×4 detector binning, 177 μm pixels) we were able to collect a 1.55 Å data set from IPNS microcrystals in about 15 min. The sample consumption rate was 6 μL/min, so we needed a total amount of 90 uL crystal suspension for this dataset. For soluble methane monooxygenase a collection time of 45 min at 20 Hz repetition rate (dictated by 3 × 3 detector binning, 133 μm pixels) resulted in a dataset of 1.9 Å resolution using a total of 360 μL of crystal suspension (8 μL/min consumption rate). For the weaker diffracting crystals of PSII a collection time of 150 min at 20 Hz repetition rate gave data out to 2.15 Å resolution, using a flow rate of 5 μL/min and a total sample volume of 750 μL.

Further processing of the merged diffraction data can be done using standard crystallography software like phenix (Liebschner et al., 2019) or the ccp4 suite (Agirre et al., 2023) and visualization and manual model building can be performed in PyMOL (Schrödinger, 2015) and Coot (Emsley, Lohkamp, Scott, & Cowtan, 2010).

When interpreting the XRD data it is very helpful to include the information obtained from the XES part of the experiment. In the case of IPNS the observed changes in the Fe Kα emission signal measured from the crystals after different O_2 incubation times was in good agreement with observations of changes around the Fe atom in the active site of IPNS at these time points (Fig. 16A) (Rabe et al., 2021). For PSII we were able to correlate the kinetics of the change in the Mn oxidation state in the S_2–S_3 transition observed via the Mn $K\beta_{1,3}$ XES with the appearance of electron density for an additional oxygen atom in the cluster (Ibrahim et al., 2020) (Fig. 16B), allowing us to conclude that the oxidation and oxygen insertion process happen concomitantly.

6. Summary and conclusions

Conducting combined diffraction and emission measurements on metalloenzymes utilizing the fs pulses from XFEL sources allows to obtain information at room temperature that is not affected by the usual problems of radiation damage one encounters with synchrotron measurements. Hence this approach opens the door to conduct time-resolved studies to probe details of the reaction mechanisms of many radiation-sensitive metalloenzymes. We described protocols to generate suitable crystalline samples for these kinds of experiments using seeding and batch methods. Conducting these experiments with the drop-on-tape setup described here is very efficient in terms of sample consumption compared to commonly used liquid jet approaches. It enables scientists to collect not only diffraction but also X-ray emission data concomitantly from the same sample at the same time. This allows the experimenter to confirm if the sample is in

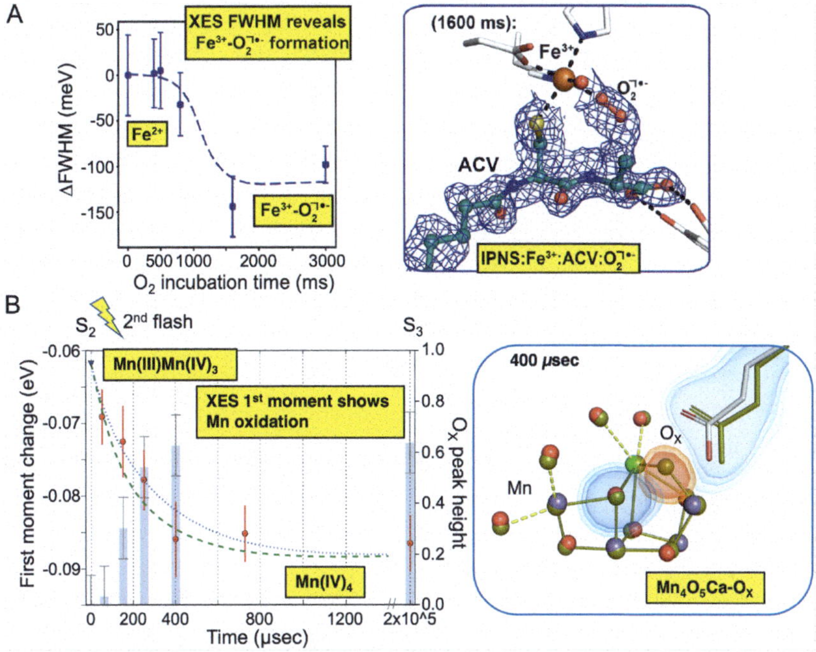

Fig. 16 Changes in XES and XRD for different time points along the reaction process in IPNS (A) and PSII (B). (A) The full width half max (FWHM) of the Fe Kα signal from IPNS microcrystals is plotted as a function of O_2 incubation time (left) showing a clear change in Fe oxidation state between 800 and 1600 ms (left). Accordingly, a density feature is observed in the 1600 ms O_2 incubation XRD data (right), that is modeled as an O_2^- species bound to the Fe^{3+} at the active site of IPNS. (B) The change in the first moment for the Mn Kβ XES is plotted as a function of time after the 2nd light flash given to PSII microcrystals (left). The change in the first moment is indicative of an oxidation of the Mn cluster from the Mn(III)Mn(IV)$_3$ state to the Mn(IV)$_4$ state. Concomitantly, an additional density is observed in the region of the Mn cluster starting at around 200 μs and the buildup of that density correlates well with the extent of Mn oxidation (right). This density was modeled as an additional bridging oxygen (O_X) in the Mn cluster. *(A) Adapted from Rabe et al. (2021). (B) Adapted from Ibrahim et al. (2020).*

the expected starting state with respect to oxidation state and gives feedback about changes in the electronic structure e.g. due to reaction triggering under the experimental conditions. The described approach for XES data collection can be adapted to different transition metals and also to measuring signals from two different transition metals at the same time (Fuller et al., 2017; Gul et al., 2015), e.g. for metalloenzymes that have a heteronuclear active site. The different reaction triggering options discussed (laser illumination and in-situ gas incubation) target various enzyme systems and can be modified to match the needs of the system under study (for an expansion using chemical activation via in-situ mixing see for example the drop-on-drop mixing approach described in Butryn et al., 2021). Fast feedback during the experiment for both the emission and diffraction data can be used to optimize the use of limited experimental time and the quality of the data. Our approach has been applied to a number of metalloenzymes, including not only IPNS and PSII but also soluble methane monooxygenase (Srinivas et al., 2020), ribonucleotide reductase R2b (John et al., 2022) and methyl coenzyme M reductase (Ohmer et al., 2022). We hope that other users can apply this approach to their systems and help to broaden the nascent field of room temperature time resolved structural studies on metalloenzymes.

Acknowledgments

We thank Isabel Bogacz, Philipp Simon and Patrick Rabe for discussion of different aspects of this manuscript and providing critical feedback. This work was supported by the Director, Office of Science, Office of Basic Energy Sciences (OBES), Division of Chemical Sciences, Geosciences, and Biosciences (CSGB) of the Department of Energy (DOE) (J.Y., J.K.) for X-ray spectroscopy and crystallography data collection and analysis, and methods development for photosynthetic systems, by the National Institutes of Health (NIH) Grants GM110501 (J.Y.) and GM126289 (J.K.) for instrumentation development for XFEL experiments. This research used resources of NERSC, a User Facility supported by the Office of Science, DOE, under Contract No. DE-AC02–05CH11231. XFEL data was collected at LCLS/SLAC, Stanford, funded by DOE OBES under Contract No. DE-AC02–76SF00515, and structural biology work at the LCLS is supported by NIH grant P41GM139687 and the Rayonix detector at the MFX instrument at LCLS was funded by a NIH grant S10 OD023453.

References

Agirre, J., Atanasova, M., Bagdonas, H., Ballard, C. B., Baslé, A., Beilsten-Edmands, J., ... Yamashita, K. (2023). The CCP4 suite: Integrative software for macromolecular crystallography. *Acta Crystallographica, Section D. Structural Biology, 79*(6), 449–461. https://doi.org/10.1107/S2059798323003595.

Alonso-Mori, R., Kern, J., Sokaras, D., Weng, T. C., Nordlund, D., Tran, R., ... Bergmann, U. (2012). A multi-crystal wavelength dispersive x-ray spectrometer. *The Review of Scientific Instruments, 83*(7), 073114. https://doi.org/10.1063/1.4737630.

Barends, T. R. M., Stauch, B., Cherezov, V., & Schlichting, I. (2022). Serial femtosecond crystallography. *Nature Reviews Methods Primers, 2*(1), 59. https://doi.org/10.1038/s43586-022-00141-7.

Beale, J. H., et al. (2019). Successful sample preparation for serial crystallography experiments. *Journal of Applied Crystallography, 52*, 1385–1396. https://doi.org/10.1107/S1600576719013517.

Bergmann, U., Kern, J., Schoenlein, R. W., Wernet, P., Yachandra, V. K., & Yano, J. (2021). Using X-ray free-electron lasers for spectroscopy of molecular catalysts and metalloenzymes. *Nature Reviews Physics, 3*(4), 264–282. https://doi.org/10.1038/s42254-021-00289-3.

Bijelic, A., & Rompel, A. (2018). Polyoxometalates: More than a phasing tool in protein crystallography. *ChemTexts, 4*(3), 10. https://doi.org/10.1007/s40828-018-0064-1.

Brewster, A. S., Bhowmick, A., Bolotovsky, R., Mendez, D., Zwart, P. H., & Sauter, N. K. (2019). SAD phasing of XFEL data depends critically on the error model. *Acta Crystallographica, Section D. Structural Biology, 75*(Pt 11), 959–968. https://doi.org/10.1107/S2059798319012877.

Brewster, A. S., Waterman, D. G., Parkhurst, J. M., Gildea, R. J., Young, I. D., O'Riordan, L. J., ... Sauter, N. K. (2018). Improving signal strength in serial crystallography with DIALS geometry refinement. *Acta Crystallographica, Section D. Structural Biology, 74*(Pt 9), 877–894. https://doi.org/10.1107/S2059798318009191.

Brewster, A. S., Young, I. D., Lyubimov, A., Bhowmick, A., & Sauter, N. K. (2019). Processing serial crystallpgrahpic data from XFELs or synchrotrons using the cctbx.xfel GUI. *Computational Crystallography Newsletter, 10*, 22–39.

Butryn, A., Simon, P. S., Aller, P., Hinchliffe, P., Massad, R. N., Leen, G., ... Orville, A. M. (2021). An on-demand, drop-on-drop method for studying enzyme catalysis by serial crystallography. *Nature Communications, 12*(1), 4461. https://doi.org/10.1038/s41467-021-24757-7.

Chapman, H. N., Fromme, P., Barty, A., White, T. A., Kirian, R. A., Aquila, A., ... Spence, J. C. H. (2011). Femtosecond X-ray protein nanocrystallography. *Nature, 470*(7332), 73–77. https://doi.org/10.1038/Nature09750.

Cheng, R. K. (2020). Towards an optimal sample delivery method for serial crystallography at XFEL. *Crystals, 10*(3), 215. https://doi.org/10.3390/cryst10030215.

Derewenda, Z. S., & Vekilov, P. G. (2006). Entropy and surface engineering in protein crystallization. *Acta Crystallographica. Section D, Biological Crystallography, 62*(1), 116–124. https://doi.org/10.1107/S0907444905035237.

Emsley, P., Lohkamp, B., Scott, W. G., & Cowtan, K. (2010). Features and development of Coot. *Acta Crystallographica. Section D, Biological Crystallography, 66*(Pt 4), 486–501. https://doi.org/10.1107/S0907444910007493.

Erdemir, D., Lee, A. Y., & Myerson, A. S. (2019). Crystal nucleation. In A. Y. Lee, A. S. Myerson, & D. Erdemir (Eds.). *Handbook of industrial crystallization* (pp. 76–114)(3rd ed.,). Cambridge University Press. ⟨https://doi.org/10.1017/9781139026949.003⟩.

Fransson, T., Chatterjee, R., Fuller, F. D., Gul, S., Weninger, C., Sokaras, D., ... Yano, J. (2018). X-ray emission spectroscopy as an in situ diagnostic tool for X-ray crystallography of metalloproteins using an X-ray free-electron laser. *Biochemistry, 57*(31), 4629–4637. https://doi.org/10.1021/acs.biochem.8b00325.

Fuller, F. D., Gul, S., Chatterjee, R., Burgie, E. S., Young, I. D., Lebrette, H., ... Yano, J. (2017). Drop-on-demand sample delivery for studying biocatalysts in action at X-ray free-electron lasers. *Nature Methods, 14*(4), 443–449. https://doi.org/10.1038/nmeth.4195.

García-Ruiz, J. M. (2003). Counterdiffusion methods for macromolecular crystallization. *Methods in Enzymology,* 130–154. https://doi.org/10.1016/s0076-6879(03)68008-0.

Giegé, R. (2013). A historical perspective on protein crystallization from 1840 to the present day. *The FEBS Journal, 280*, 6456–6497. https://doi.org/10.1111/febs.12580.

Gul, S., Ng, J. W., Alonso-Mori, R., Kern, J., Sokaras, D., Anzenberg, E., ... Yano, J. (2015). Simultaneous detection of electronic structure changes from two elements of a bifunctional catalyst using wavelength-dispersive X-ray emission spectroscopy and in situ electrochemistry. *Physical Chemistry Chemical Physics: PCCP, 17*(14), 8901–8912. https://doi.org/10.1039/c5cp01023c.

Hadimioglu, B., Stearns, R., & Ellson, R. (2016). Moving liquids with sound: The physics of acoustic droplet ejection for robust laboratory automation in life sciences. *Journal of Laboratory Automation, 21*(1), 4–18. https://doi.org/10.1177/2211068215615096.

Hattne, J., Echols, N., Tran, R., Kern, J., Gildea, R. J., Brewster, A. S., ... Sauter, N. K. (2014). Accurate macromolecular structures using minimal measurements from X-ray free-electron lasers. *Nature Methods, 11*(5), 545–548. https://doi.org/10.1038/nmeth.2887.

Hellmich, J., Bommer, M., Burkhardt, A., Ibrahim, M., Kern, J., Meents, A., ... Zouni, A. (2014). Native-like photosystem II superstructure at 2.44 Å resolution through detergent extraction from the protein crystal. *Structure (London, England: 1993), 22*(11), 1607–1615. https://doi.org/10.1016/j.str.2014.09.007.

Huang, T. C., Toraya, H., Blanton, T. N., & Wu, Y. (1993). X-ray powder diffraction analysis of silver behenate, a possible low-angle diffraction standard. *Journal of Applied Crystallography, 26*, 180–184. https://doi.org/10.1107/S0021889892009762.

Ibrahim, M., Chatterjee, R., Hellmich, J., Tran, R., Bommer, M., Yachandra, V. K., ... Zouni, A. (2015). Improvements in serial femtosecond crystallography of photosystem II by optimizing crystal uniformity using microseeding procedures. *Structural Dynamics, 2*(4), 041705. https://doi.org/10.1063/1.4919741.

Ibrahim, M., Fransson, T., Chatterjee, R., Cheah, M. H., Hussein, R., Lassalle, L., ... Yano, J. (2020). Untangling the sequence of events during the $S_2 \rightarrow S_3$ transition in photosystem II and implications for the water oxidation mechanism. In: *Proceedings of the National Academy of Sciences of the United States of America, 117*(23), 12624–12635. https://doi.org/10.1073/pnas.2000529117.

John, J., Aurelius, O., Srinivas, V., Saura, P., Kim, I.-S., Bhowmick, A., ... Högbom, M. (2022). Redox-controlled reorganization and flavin strain within the ribonucleotide reductase R2b–NrdI complex monitored by serial femtosecond crystallography. *ELife, 11*, e79226. https://doi.org/10.7554/eLife.79226.

Kamps, J. (2020). *Biophysical studies on enzymes related to antibiotics.* ⟨http://purl.org/dc/dcmitype/Text⟩, University of Oxford. ⟨https://ora.ox.ac.uk/objects/uuid:c6b3089c-ff7c-408d-a8a3-c5e7a673746e⟩.

Karplus, P. A., & Diederichs, K. (2015). Assessing and maximizing data quality in macromolecular crystallography. *Current Opinion in Structural Biology, 34*, 60–68. https://doi.org/10.1016/j.sbi.2015.07.003.

Keedy, D. A., Kenner, L. R., Warkentin, M., Woldeyes, R. A., Hopkins, J. B., Thompson, M. C., ... Fraser, J. S. (2015). Mapping the conformational landscape of a dynamic enzyme by multitemperature and XFEL crystallography. *Elife,* 4. https://doi.org/10.7554/eLife.07574.

Kern, J., Alonso-Mori, R., Tran, R., Hattne, J., Gildea, R. J., Echols, N., ... Yano, J. (2013). Simultaneous femtosecond X-ray spectroscopy and diffraction of photosystem II at room temperature. *Science (New York, N. Y.), 340*(6131), 491–495. https://doi.org/10.1126/science.1234273.

Kern, J., Loll, B., Luneberg, C., DiFiore, D., Biesiadka, J., Irrgang, K. D., ... Zouni, A. (2005). Purification, characterisation and crystallisation of photosystem II from Thermosynechococcus elongatus cultivated in a new type of photobioreactor. *Biochimica et Biophysica Acta, 1706*(1–2), 147–157. https://doi.org/10.1016/j.bbabio.2004.10.007.

Kuhlbrandt, W. (2014). Biochemistry. The resolution revolution. *Science (New York, N. Y.), 343*(6178), 1443–1444. https://doi.org/10.1126/science.1251652.

Liebschner, D., Afonine, P. V., Baker, M. L., Bunkoczi, G., Chen, V. B., Croll, T. I., ... Adams, P. D. (2019). Macromolecular structure determination using X-rays, neutrons and electrons: Recent developments in Phenix. *Acta Crystallographica. Section D, Structural Biology, 75*(Pt 10), 861–877. https://doi.org/10.1107/S2059798319011471.

Neutze, R. (2014). Opportunities and challenges for time-resolved studies of protein structural dynamics at X-ray free-electron lasers. *Philosophical Transactions of the Royal Society of London. Series B, Biological Sciences, 369*(1647), 20130318. https://doi.org/10.1098/rstb.2013.0318.

Ohmer, C. J., Dasgupta, M., Patwardhan, A., Bogacz, I., Kaminsky, C., Doyle, M. D., ... Ragsdale, S. W. (2022). XFEL serial crystallography reveals the room temperature structure of methyl-coenzyme M reductase. *Journal of Inorganic Biochemistry, 230*, 111768. https://doi.org/10.1016/j.jinorgbio.2022.111768.

Rabe, P., Beale, J. H., Butryn, A., Aller, P., Dirr, A., Lang, P. A., ... Owen, R. L. (2020). Anaerobic fixed-target serial crystallography. *IUCrJ,* 7(5), 901–912. https://doi.org/10.1107/s2052252520010374.

Rabe, P., Kamps, J., Sutherlin, K. D., Linyard, J. D. S., Aller, P., Pham, C. C., ... Schofield, C. J. (2021). X-ray free-electron laser studies reveal correlated motion during isopenicillin N synthase catalysis. *Science Advances,* 7(34), eabh0250. https://doi.org/10.1126/sciadv.abh0250.

Rhodes, G. (2006). *Crystallography made crystal clear: A guide for users of macromolecular models.* Burlington: Elsevier.

Roach, P. L., Clifton, I. J., Hensgens, C. M. H., Shibata, N., Long, A. J., Strange, R. W., ... Hajdu, J. (1996). Anaerobic crystallisation of an isopenicillin N synthase Fe(II) substrate complex demonstrated by X-ray studies. *European Journal of Biochemistry, 242*(3), 736–740. https://doi.org/10.1111/j.1432-1033.1996.0736r.x.

Roessler, C. G., Kuczewski, A., Stearns, R., Ellson, R., Olechno, J., Orville, A. M., ... Heroux, A. (2013). Acoustic methods for high-throughput protein crystal mounting at next-generation macromolecular crystallographic beamlines. *Journal of Synchrotron Radiation, 20*, 805–808. https://doi.org/10.1107/S0909049513020372.

Rupp, B. (2013). Macromolecular crystallography: Overview. In G. C. K. Roberts (Ed.). *Encyclopedia of biophysics* (pp. 1346–1353) Springer. ⟨https://doi.org/10.1007/978-3-642-16712-6_655⟩.

Saibil, H. R. (2022). Cryo-EM in molecular and cellular biology. *Molecular Cell, 82*(2), 274–284. https://doi.org/10.1016/j.molcel.2021.12.016.

Sauter, N. K. (2015). XFEL diffraction: Developing processing methods to optimize data quality. *Journal of Synchrotron Radiation, 22*(2), 239–248. https://doi.org/10.1107/S1600577514028203.

Schrödinger LLC (2015). *The PyMOL molecular graphics system, version 2.0.*

Schulz, E. C., Yorke, B. A., Pearson, A. R., & Mehrabi, P. (2022). Best practices for time-resolved serial synchrotron crystallography. *Acta Crystallographica. Section D, Structural Biology, 78*, 14–29. https://doi.org/10.1107/S2059798321011621.

Shoeman, R. L., Hartmann, E., & Schlichting, I. (2023). Growing and making nano- and microcrystals. *Nature Protocols, 18*, 854–882. https://doi.org/10.1038/s41596-022-00777-5.

Soares, A. S., Engel, M. A., Stearns, R., Datwani, S., Olechno, J., Ellson, R., ... Orville, A. M. (2011). Acoustically mounted microcrystals yield high-resolution X-ray structures. *Biochemistry, 50*(21), 4399–4401. https://doi.org/10.1021/bi200549x.

Spence, J. C. H. (2018). X-ray lasers for structure and dynamics in biology. *IUCrJ, 5*(Pt 3), 236–237. https://doi.org/10.1107/S2052252518005365.

Srinivas, V., Banerjee, R., Lebrette, H., Jones, J. C., Aurelius, O., Kim, I. S., ... Hogbom, M. (2020). High-resolution XFEL structure of the soluble methane monooxygenase hydroxylase complex with its regulatory component at ambient temperature in two oxidation states. *Journal of the American Chemical Society, 142*(33), 14249–14266. https://doi.org/10.1021/jacs.0c05613.

Thompson, M. C., Barad, B. A., Wolff, A. M., Sun Cho, H., Schotte, F., Schwarz, D. M. C., ... Fraser, J. S. (2019). Temperature-jump solution X-ray scattering reveals distinct motions in a dynamic enzyme. *Nature Chemistry, 11*(11), 1058–1066. https://doi.org/10.1038/s41557-019-0329-3.

Vekilov, P. G. (2019). Crystallization of proteins. In A. Y. Lee, A. S. Myerson, & D. Erdemir (Eds.). *Handbook of industrial crystallization* (pp. 414–459)(3rd ed.,). Cambridge University Press. ⟨https://doi.org/10.1017/9781139026949.014⟩.

Vekilov, P. G. (2004). In X. Y. Lui (Ed.). *Nanoscale structure and assembly at solid–fluid interfaces, Vol. II: Assembly in hybrid and biological systems* (pp. 145–200) Dordrecht: Kluewer Academic Publishers.

Vekilov, P. G. (2003). Solvent entropy effects in the formation of protein solid phases. *Methods in Enzymology, 368*, 84–105. https://doi.org/10.1016/s0076-6879(03)68006-7.

Vekilov, P. G., & Chernov, A. A. (2002). In F. Spaepen (Ed.). *Solid state physics* (pp. 1–147) New York: Academic Press.

Yano, J., Fuller, F., Gul, S., Chatterjee, R., Kern, J., & Yachandra, V. (2017). *Droplet on tape: Protocol.* February 28. ⟨https://doi.org/10.1038/protex.2017.017⟩.

Yano, J., Kern, J., Irrgang, K.-D., Latimer, M. J., Bergmann, U., Glatzel, P., ... Yachandra, V. K. (2005). X-ray damage to the Mn4Ca complex in photosystem II crystals: A case study for metallo-protein X-ray crystallography. *Proceedings of the National Academy of Sciences of the United States of America, 102*, 12047–12052.

Young, I. D., Ibrahim, M., Chatterjee, R., Gul, S., Fuller, F., Koroidov, S., ... Yano, J. (2016). Structure of photosystem II and substrate binding at room temperature. *Nature, 540*(7633), 453–457. https://doi.org/10.1038/nature20161. Epub 2016 Nov 21. PMID: 27871088; PMCID: PMC5201176.

Zouni, A., Kern, J., Frank, J., Hellweg, T., Behlke, J., Saenger, W., & Irrgang, K.-D. (2005). Size determination of cyanobacterial and higher plant photosystem II by gel permeation chromatography, light scattering, and ultracentrifugation. *Biochemistry, 2005*(44), 4572–4581.

CHAPTER ELEVEN

Exploring the structural dynamics of proteins by pressure perturbation using macromolecular crystallography

Nathalie Colloc'h[a], Anne-Claire Dhaussy[b], and Eric Girard[c,*]
[a]Imagerie et stratégies thérapeutiques pour les cancers et tissus cérébraux (ISTCT), CNRS Université de Caen Normandie, Centre Cyceron, Caen, France
[b]Normandie Université, CRISMAT UMR, Caen, France
[c]Univ. Grenoble Alpes, CEA, CNRS, IBS, Grenoble, France
*Corresponding author: e-mail address: eric.girard@ibs.fr

Contents

Abstract

High pressure is a convenient thermodynamic parameter to probe the dynamics of proteins as it is intimately related to volume which is essential for protein function. To be biologically active, a protein fluctuates between different substates. Pressure

ISSN 0076-6879, https://doi.org/10.1016/bs.mie.2023.06.007

perturbation can promote some hidden substates by modifying the population between them. High pressure macromolecular crystallography (HPMX) is a perfect tool to capture and to characterize such substates at a molecular level providing new insights on protein dynamics. The present chapter describes the use of the diamond anvil cell to perform HPMX experiments. It also provides tips on sample preparation and optimal data collection as well as on efficient analysis of the resulting high-pressure structures.

1. Introduction

One may wonder, why put protein crystals under high hydrostatic pressure? High hydrostatic pressure is in fact a wonderful tool to obtain structural information with a great level of precision on high-energy substates, which correspond to physiologically relevant conformers involved in functional reactions, interactions with partners, initiation of the unfolding process, etc. (Frauenfelder et al., 1990). Indeed, to be able to interact with ligands, partner proteins, and other binding partners, proteins need to access an excited state, which is low-populated at ambient pressure. The effect of high pressure is remarkably subtle—in fact, proteins compress with an overall reduction in volume of about 1% per 200 MPa. Interestingly, the effects are non-isotropic and thus, high pressure provides a way to subtly perturb structure in a controlled manner in relation to functional properties (Barstow, Ando, Kim, & Gruner, 2008), and to access functionally relevant high-energy substates of the native state. In solution, high pressure can induce unfolding transitions (which are typically preceded by dissociation of subunits for oligomeric proteins) (Winter, 2019).

Pressure allows one to explore the conformational landscape of proteins and other macromolecules. In solution, proteins are in equilibrium between different conformers: a dominant conformer corresponding to the ground state with the lowest internal free energy and multiple conformers at higher energy corresponding to transient intermediates (Frauenfelder, Sligar, & Wolynes, 1991). These latter conformers are less populated at ambient pressure in the absence of any partners and are thus difficult to characterize. The differences in free energy between the conformer in the ground state and the other conformers are small, on the order of −10 kJ/mol. Pressure favors the transition from the ground state toward conformers with smaller partial molar volumes, which correspond to higher energy substates. The equilibrium between different conformers which coexist in solution is thus modified by pressure to promote those with smaller partial volumes.

Pressure thus increases the population of transient high-energy substates, allowing for their characterization (Akasaka, 2006; Fourme, Girard, & Akasaka, 2012; Williamson & Kitahara, 2019; Winter, 2019). Nuclear magnetic resonance (NMR) spectroscopy studies under pressure have demonstrated that these high-energy substates stabilized by pressure are functionally relevant and correspond for example, to an open conformation of ubiquitin that is able to bind to specific enzymes for its activation (Kitahara, Yokoyama, & Akasaka, 2005) or a mechano-sensitive switch in a domain of talin toward a conformer able to bind vinculin (Baxter, Zacharchenko, Barsukov, & Williamson, 2017).

Proteins generally crystallize in their predominant conformer corresponding to their ground state with the lowest internal free energy. Moreover, the folded state of the protein is generally stabilized by the crystal lattice. Putting protein crystals under pressure pushes the equilibrium toward high-energy conformers corresponding to transient intermediates. One may wonder then how a protein crystal can withstand pressure without cracking. However, the first study in 1987 demonstrated that lysozyme crystals can withstand 100 MPa without modification of the crystal intrinsic quality, and later, it was shown that they can withstand even higher pressures up to 920 MPa (Kundrot & Richards, 1987; Yamada, Nagae, & Watanabe, 2015). This is because protein crystals are full of water molecules and other small molecules (between 30% and 70% of solvent), and hydrostatic pressure can be transmitted by the solvent (e.g., mother liquor) and enable structural modifications as long as they have no effect on the crystal packing. It happens that high pressure can modify the space group of the crystal, either to a more-ordered one as is the case for crystals of Cowpea mosaic virus (CpMV) and dihydrofolate reductase (DHFR) with the M20 loop in an open conformation (Girard et al., 2005; Nagae, Yamada, & Watanabe, 2018) or toward a less-ordered one as it is the case in lysozyme crystals (Yamada et al., 2015).

Since protein crystals are highly hydrated, the pressure-induced structural modifications could be in part due to increased contacts between symmetrical proteins, but these effects are small compared to the promotion of high-energy substates. In many examples which will be described later, crystallographic structures at high pressure indeed correspond to substates involved in functional reactions and/or in interactions with partners. Regions that are rigid under pressure are likely to be rigid at ambient pressure, thus less likely to be involved in transitions between conformational states. On the other hand, regions that are very compressible and undergo pressure-induced conformational changes are likely to be functionally relevant. The dynamics of proteins thus is efficiently mapped out using high hydrostatic pressure macromolecular crystallography (HPMX) (Collins, Kim, & Gruner, 2011; Dhaussy & Girard, 2015; Fourme, Girard, et al., 2012; Fourme, Girard, Kahn, Dhaussy, & Ascone, 2009; Kurpiewska & Lewiński, 2010).

2. Diamond anvil cell

The common method to obtain a crystallographic structure in macromolecular crystallography (MX) is to achieve the largest sampling of reciprocal space by rotating the crystal in the beam. This has a direct consequence for high-pressure measurements (i.e., in HPMX) in the sample cell design. Other parameters to consider include the highest pressure desirable for HPMX and the experimental conditions needed for optimal signal-to-noise in the diffraction signal. Prior work suggested that the latter can be achieved by using a tightly collimated X-ray beam at ultra-short wavelength (24–41 keV) (Fourme, Honkimäki, et al., 2012), and beyond pressures of 2.5 GPa, crystals often do not maintain diffraction quality, possibly due to the formation of Ice VI or other phenomena causing disorder within the crystal lattice. Diamond Anvil Cell (DAC) technology is ideally suited for achieving pressures below this upper limit and for use with high-energy X-rays.

The basic principle of the DAC is very simple and can be explained as follows. A sample is confined within a metal gasket that is sandwiched between the flat parallel faces of two opposed diamond anvils (see Fig. 1). A pressure medium within this space transmits hydrostatic pressure within the crystal when an external force is applied on the upper diamond. The majority of the DACs used for X-ray diffraction studies operate in transmission mode, in the sense that the X-ray beam goes through the diamonds (Liu, 2016; Shen & Mao, 2016) (Fig. 1). The collimated or focused beam enters the sample cavity through one of the anvils and the diffracted beam goes through the opposite one. In our past experiments, we used DACs of the piston-cylinder type with a membrane used to apply the force on the piston (Chervin, Canny, Besson, & Pruzan, 1995; Letoullec, Pinceaux, & Loubeyre, 1988). Such DACs can be purchased from Almax-easyLab (http://www.almax-easylab.com) or from BETSA (https://www.betsa.fr). In our experience (Girard, Fourme, et al., 2010), using a piston-based

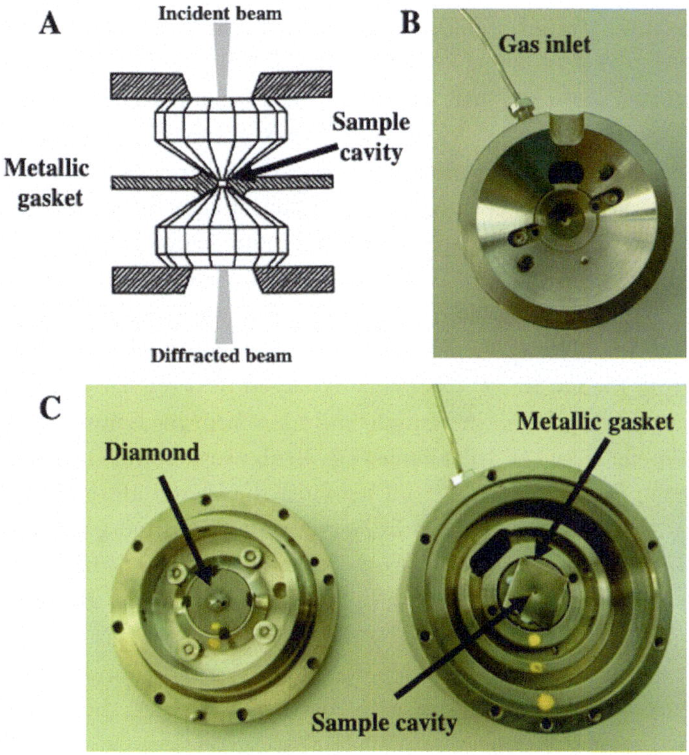

Fig. 1 Diamond anvil cell. (A) Principle. (B and C) Pneumatic DAC developed for optimal HPMX experiments and manufactured by BETSA (https://www.betsa.fr) (closed and opened respectively).

DAC without a membrane (i.e., one that is driven directly by gas) has provided greater pressure control in the 0–2.5 GPa range, especially when pressure increments need to be small to induce the desired substate. The angular rotation range available for data collection is limited by the geometry of the DAC, specifically the opening angle achievable by the exit X-ray window. To overcome this limitation, DACs specialized for HPMX feature a wide aperture angle (close to 100°) (Fig. 1) while remaining compatible with the 2.5 GPa upper limit (Girard, Fourme, et al., 2010).

DACs provide a powerful way to apply a hydrostatic pressure to the sample in situ. Measurements are performed at room temperature. To apply hydrostatic pressure, a medium is used to transmit the pressure uniformly to the sample, and the pressure is maintained at a constant value during the entire data collection so that the sample remains in equilibrium with pressure. Typical transmitting media used in non-biological high-pressure experiments include helium, argon, silicon oil or a 4:1 methanol–ethanol mixture (Chen et al., 2021; Couzinet, Dahan, Hamel, & Chervin, 2003; Dewaele & Loubeyre, 2007; Klotz, Chervin, Munsch, & Marchand, 2009; Klotz, Paumier, Le March, & Munsch, 2009; Takemura, 2021). NaCl crystals have been also found to be an excellent choice (Celeste, Borondics, & Capitani, 2019). Unfortunately, none are compatible with biological samples, which require hydration, and in the case of macromolecular crystals, we have to use the so-called mother liquor (or a slightly modified composition of it) that is used for crystallization as the transmitting medium.

As mentioned above, the sample and transmitting medium is contained within a metallic gasket. In practice, a gasket is prepared first by introducing the metallic foil in the virgin cell between the diamonds, and pressure is applied to produce an indentation in the foil. Then the gasket is finalized by drilling a hole by electro-erosion at the center of the indentation insuring a perfect alignment of the gasket with respect to the pair of diamonds. The cavity size is linked to the upper limit of high pressure to be reached. Sample cavity diameter should be kept around one half of the diamond culet size, and similarly, the cavity thickness should be half of the cavity diameter. For our HPMX experiments up to 2.5 GPa, we typically use a hole diameter of 400 μm since the diamond culets present in our DAC are 800 μm. Inconel, stainless steel and Waspaloy (a Ni-based alloy) are classically used as the gasket metal foil. The thickness of the foil may vary from 250 μm (Inconel) to 200 μm (stainless steel) leading to a thickness of 200–180 μm respectively after indentation. For our HPMX experiments at low pressure (<0.7 GPa), we prefer soft gaskets made of copper for improved pressure stability (see (Girard et al., 2007) for more details), and for higher pressure (0.7–2.5 GPa), several tests have been carried out (Girard, Fourme, et al., 2010). Based on our experience, we have found that Inconel alloys provide better and reliable results, even at very low pressure. Two types of Inconel have been tested. We used Inconel X750 up to 1.2 GPa. Between 1.2 GPa and 1.8 GPa, we switched to Inconel 718, which is more tough, ensuring better performance under mechanical stress. To determine the hardness of materials like metals, the Brinell test is commonly used and referenced (ASTM E10–18 database). The higher the value, the harder and more resistant the material is to mechanical stress. Metal foils can be annealed or full hard. Annealed metal foil is significantly less resistant to mechanical stress than hard metal foils.

For instance, the Hardness Brinell (HB) for annealed Inconel X750 is 150 (250 for Inconel 718). For all our experiments, we used full hard metal foils for which the HB have been estimated to 280 (Inconel X750) and 410 (Inconel 718). Stainless steel may be considered an alternative material as the resistance to stress is excellent (HB: 190–440), but it requires more time to machine the cavity. We also would like to point out that the gasket, in addition to providing the pressure medium containment, also acts as a supporting ring between the diamonds to prevent failure of the anvils due to the concentration of stress at the edge of the anvil faces.

3. Sample managing

Sample crystallization is performed classically by vapor diffusion methods. As several crystals are often required to perform an extensive high-pressure study, where data collection at several pressures may be required, reproducible samples are necessary. Moreover, sample loading within the DAC has to be repeated during an HPMX experiment, so that greaseless screw-in crystallization plates, such as those provided by the NeXtal supplier, can facilitate sample transfer as these plates are easy to open and close.

If several crystallization conditions are available for a given protein, one would select the one leading to a good compromise between the following:

i. diffraction quality, in particular the attainable resolution
ii. crystal symmetry, as high symmetry space group will be more favorable since the DAC geometry limits the accessible rotation range (90° in the case of our DAC)
iii. crystallization solution components. For example, high salt concentrations may complicate sample loading within the DAC, and some crystallization mixtures may crystallize under pressure, damaging the protein crystals. If the latter occurs, addition of molecules usually exploited for crystal cryo-preservation in conventional MX can solve the issue.
iv. crystal morphology. Some morphologies, such as platelet-shaped crystals, will lead to preferred orientation of the crystal with respect to the diamond culets. As a consequence and depending on the space group, this may prevent acquisition of a complete data set.

Issues related to points ii and iv can be overcome by changing the crystal orientation within the DAC cavity in order to increase coverage of reciprocal space and obtain a complete data set. Orientation changes can be achieved by exploiting the natural crystal morphology to select complementary positions. If the crystal morphology does not allow favorable changes in orientation, modification of crystal orientation with respect to the diamond culets can be done with tiny X-ray-transparent splinters, such as small pieces of a broken diamond anvil (Girard et al., 2007), or with cigarette-filter fibers (Nagae et al., 2012).

DAC-based HPMX experiments are performed at room temperature ensuring that the physico-chemical state of the sample is perfectly characterized, particularly in term of pressure equilibria. As a consequence, precautions have to be taken for crystal transportation to the synchrotron sources as for any diffraction experiments expected to be done at room temperature. Transportation of crystallization plates should be done taking precautions to limit vibrations, the effects of which can be mitigated with the use of sitting drop settings. The temperature has an influence on the crystallization diagram of a protein, in particular on the solubility curve which defines the protein equilibrium state between the solution state and the crystalline state. Thus, the temperature has to be controlled during transportation to prevent the crystals from getting damaged. With this respect, reusable temperature-controlled shipping container can be considered. This topic is also covered in Chapter 1 of this volume (Pei, Bharr, Wang, Ando, & Meisburger, submitted).

Another alternative is the old fashion way of putting crystals in one or more capillaries filled with the mother liquor. In that case, vibration during transportation is less of a concern, but unless the crystals are grown inside the capillaries, there are risks associated with the physically transferring of crystals from crystallization drops to the capillaries.

4. DAC-based experiment

4.1 Pressure generation

Several systems have been developed to generate the pressure in the DAC cavity (Bassett, 2009; Girard, Fourme, et al., 2010; Letoullec et al., 1988; Letoullec, Pinceaux, & Loubeyre, 1990; Merrill & Bassett, 1974). One widely used method is based on pressure-driving screws (where an external force is applied by turning screws that bring the two opposing anvils closer together). This method, which is typically used to study materials under geologically relevant pressures, cannot be easily used for biological HPMX experiments due to the difficulty in controlling the applied pressure below 2.5 GPa and in preventing uncontrolled pressure jumps. Indeed, in some cases, a pressure jump of dozens of MPa could induce large modifications on protein conformations.

In our experiments, we adopted the Pneumatic Drive System (PDS) to generate the pressure inside the cavity. The PDS is a portable unit containing a bottle of compressed gas (e.g. nitrogen or helium), a manometer, two valves and a pressure sensor. It is used to generate and control the external force applied to the anvil by manually varying the gas pressure inflated in the membrane or directly in the piston cylinder. The PDS is connected to a gas bottle which can deliver up to 20 MPa allowing many cycles before reloading. Pressures of a few MPa delivered by the PDS are amplified to hundreds of MPa in the sample cavity thanks to the surface reduction between the diamond support and diamond culet. An Automatic Pneumatic Drive System (APDS) has been designed more recently by BETSA. In this improved system, the external pressure is regulated automatically by the use of a pressure controller system. Such a feature allows a more gradual increase in the cavity pressure by keeping the speed and duration of steps constant.

4.2 Pressure measurement

In earlier years of use, the pressure inside the DAC was estimated by force over area estimation, or by internal markers such as NaCl or silver crystals, for which pressure-induced unit cell changes can be tracked by diffraction (Decker & Worlton, 1972; Dewaele et al., 2012; Dorogokupets & Dewaele, 2007; Syassen & Holzapfel, 1978). NaCl crystals are incompatible with most of biological samples, while silver is a relatively heavy metal and therefore a strong x-ray absorber/scatterer, which reduces the signal to noise during the data collection. Therefore, we use the fluorescence line of ruby excited by a laser as marker of the pressure. It was shown previously by Forman et al. (Forman, Piermarini, Barnett, & Block, 1972) that the ruby fluorescence lines shift linearly to higher wavelength with hydrostatic pressure (and that they broaden if the ruby experiences non-hydrostatic stresses). For HPMX experiments, a tiny 5–15 μm ruby sphere is placed in the pressure medium along with the sample, and we follow the R1 line of ruby (394.2 nm) (Shen et al., 2020) to determine in situ the pressure in the cavity. The pressure measurement remains linear until the gasket indentation pressure is reached. In our case, this occurs at around 3 MPa of external force on Inconel and 6 MPa with stainless steel. It is worth noting that this value is directly dependent on the type of cell. Once the indentation pressure is reached, the pressure increase is exponential and incrementation steps should be made smaller and smaller. If the pressure doesn't increase as much as expected, it means that the maximum mechanical stress the gasket can handle has been reached and has caused an increase of the sample cavity diameter. Eventually, the risk is to break the diamond anvils.

Pressure measurement precision with ruby can be considered to be around 25 MPa. Greater precision can be achieved if precautions are taken. First, the apparatus used for ruby fluorescence measurement should be properly calibrated. Second, each ruby ball possesses its own response to laser excitation, and thus, the corresponding R1 line wavelength value has to be measured. This sets the reference wavelength value for ambient pressure that will then be used for pressure determination. The attained precision in pressure measurement is illustrated by our work on urate oxidase (UOX) from *Aspergillus flavus*, in which three independent HPMX experiments were performed at three different beamlines and synchrotrons leading to reproducible values of the unit-cell volume as a function of pressure (see Fig. S3 in Prangé et al., 2022).

4.3 Incident beam energy

Setting the beam energy to short wavelength, typically in the range 20–30 keV, both limits the absorption by the two diamonds and the effects of the DAC geometry, in particular on data resolution, by closing the diffraction cone. Moreover, such beam conditions have proven to be a good compromise to mitigate sample radiation damage (Dickerson & Garman, 2019; Fourme, Honkimäki, et al., 2012; Storm, Axford, & Owen, 2021). Such beam conditions were initially seen as non-optimal for MX experiments mainly related to the fact that detector efficiencies were not optimized. New detectors equipped with CdTe sensors have changed the game and have allowed us to revisit the interest in short-wavelength X-rays for efficient MX data collection. The incident beam energy should be set with precision as for any diffraction experiment.

4.4 Sample loading

Prior to sample loading, the gasket is seated on the lower diamond keeping the same orientation as it had for the indentation step. A ruby sphere is then placed on the culet of the second diamond (upper anvil) under a microscope. Subsequently, the sample cavity is filled with a large excess of mother liquor (2–3 μL), and the sample crystal is rapidly transferred using a loop from the crystallization drop to the drop of mother liquor to avoid sample damage. The crystal is then gently guided into the cavity. Finally, the DAC is sealed by bringing the upper anvil onto the gasket ensuring a hermetic environment for the sample.

4.5 Data collection

In our developments of DAC-based HPMX over the last 20 years, experiments were performed at a minimum of 5 European synchrotrons (European Synchrotron Radiation Facility (ESRF), Soleil synchrotron, Diamond Light Source (DLS), Swiss Light Source (SLS) and Elettra synchrotron) using several beamlines fully dedicated to high-pressure studies (ID27 and ID15B at ESRF) or not (CRISTAL at Soleil, XRD1 at Elettra and I19 at DLS) as well as MX beamlines equipped with the G-rob robotic arm (BM30A-FIP at ESRF and PXIII at SLS). Experiments begin by calibrating the sample-to-detector distance with precision using standard materials, such as LaB_6, that possess referenced powder diffraction spectra.

In conventional X-ray diffraction experiments, the sample centering method is often based on optical observations, which are used for positioning the crystal at the center of rotation and requires a 180° sample rotation to be performed. Optical techniques are not suited for DACs for two main reasons: first, the cell rotation is limited particularly because of the connecting system between the cell and the gas bottle supply, and second, the high refractive index of diamonds modifies the optical path during the cell rotation and complicates proper alignment of the sample.

For these reasons, X-ray transmission through the DAC has been used to center the sample. The method consists of scanning the sample holder in the two directions perpendicular to the beam axis for two symmetric angles. The coordinate of the centroid is calculated for each scan. Then a correction is applied to position the sample on the rotation axis (Budzianowski & Katrusiak, 2004; Kunz et al., 2005; Smith & Desgreniers, 2009).

DAC preparation, sample loading, DAC centering and pressure ramping may take up to 1 h. Thus, to get the best data and to save time, our initial strategy was to set the beam size to typically 30–50 μm and to translate the sample several times during the experiment in order to irradiate fresh zones within the crystal (see Fig. S1 in Prangé et al., 2022). Pixel array detectors (PADs) based on hybrid photon counting technology have revolutionized the way MX experiments are now performed with, among all the improvements, shutterless data collection. Such improvements have also benefited HPMX by allowing, for example, data collection without the need for sample translation, in particular with CdTe PADs as mentioned above. Such data collection strategies associated with the use of short-wavelength incident X-rays greatly reduce the apparent radiation damage (Prangé et al., 2022).

A typical DAC-based experiment starts with the determination of the evolution of the crystal cell parameters with pressure. This step allows one (i) to evaluate the pressure range accessible to the sample under consideration, (ii) to observe how the crystal diffraction quality changes with pressure and (iii) to determine the crystal unit-cell compressibility curve. This is performed by collecting diffraction over a 10° total rotation (e.g., by collecting 10 frames at 1° rotation) and by processing the corresponding diffraction frames in order to get precise values of the unit-cell parameters. At each pressure value, the sample is translated to irradiate a fresh zone and to get the corresponding unit-cell parameters. Eventually, the procedure is repeated on a new sample.

The second important point to be considered is a reference data set, i.e. at ambient pressure, collected in the same experimental conditions, in particular with a sample within the DAC. This highly limits bias in structure comparisons (see Section 7).

Finally, as illustrated by our work on the oncogenic protein Ras (Girard et al., 2022), night periods are also exploited for long sample incubation under pressure that may lead to *in-crystallo* protein conformational transitions. The incubation pressure can be identified from the unit-cell compressibility curve determined in the first step.

5. Data processing

Diffraction frames are integrated using conventional MX software, such as *XDS* (Kabsch, 2010), with standard settings except for the crystal-to-detector distance. To ensure that pressure-dependent unit cell parameters are correctly determined, integration should be performed with the crystal-to-detector distance parameter fixed to the value experimentally determined using a reference compound as described in Section 4.5. Integrated intensities are scaled and merged using software such as *SCALA/AIMLESS* and *TRUNCATE* from the CCP4 suite of programs

(Winn et al., 2011). For merging software, batch mode analysis is preferred. This is because diffraction frames may be contaminated by diamond diffraction spots, which are far more intense compared to protein diffraction, or by powder diffraction rings that can arise when the X-ray beam interacts with the gasket especially when the DAC samples large rotation angles. Batch mode analysis thus allows for outlier rejection of such contaminated diffraction frames as they may degrade the final data set quality.

6. Structure determination and refinement

Structure determination and refinement are performed using conventional protein crystallography methods and programs. Usually, the structures are solved by molecular replacement (MR), with software like *MOLREP* or *PHASER*, starting from the closest structural model extracted from the Protein Data Bank (PDB) as the reference structure at ambient pressure. For the high-pressure structures, the search model can be either the same PDB file, or the refined ambient-pressure structure, depending on the user. To minimize any bias due to the starting model, MR can also been performed with a poly-alanine chain followed by automatic side chain reconstruction using *BUCCANEER*. This procedure was done for example in (Colloc'h et al., 2017). All these programs are part of the CCP4 package (Winn et al., 2011). Another way to minimize any bias is to remove the areas known to be involved in large conformational changes from the MR model. This has been done, for example, for Ras protein, using an MR model without the switch I and switch II, two regions known to exhibit large conformational changes upon transition between the guanosine-5′-triphosphate (GTP) bound state to the guanosine-5′-diphosphate (GDP) bound state (Girard et al., 2022).

Refinement of the MR structures is then carried out using conventional software such as *REFMAC* (Murshudov et al., 2011) or *PHENIX* (Liebschner et al., 2019). The graphic program *COOT* (Emsley, Lohkamp, Scott, & Cowtan, 2010) is used in most cases to visualize the electron density maps and for manual rebuilding during refinement steps. It is crucial that the structures at different pressures be refined under similar conditions with the same software and procedure to avoid any refinement bias, as previously explained in (Girard et al., 2005; Urayama, Phillips, & Gruner, 2002).

7. Data analysis

7.1 Comparisons with a reference structure

To minimize potential experimental bias and to provide accurate comparisons between structures, it is crucial to compare high-pressure structures with a reference structure whose crystallographic data were collected in an apparatus as close as possible as the one used for the high-pressure structures. If possible, the optimal reference structure is the ambient-pressure structure coming from a crystal of the same batch, whose data are collected on the same beamline or laboratory diffractometer, with the same pressure cell. As much as possible, all the parameters should be identical except the pressure. This careful procedure has been applied in many studies for example in (Collins, Hummer, Quillin, Matthews, & Gruner, 2005; Colloc'h et al., 2017; Girard et al., 2022; Girard, Marchal, et al., 2010; Kundrot & Richards, 1987; Kurpiewska et al., 2018, 2019; Prangé et al., 2022). If such a reference structure is not available, it is better to compare with a conventional ambient-pressure structure whose data have been collected at room temperature, not in cryo conditions (100 K), to avoid mixing both pressure and temperature effects. This has been done for example in (Girard et al., 2005; Hamajima et al., 2016; Nagae et al., 2012, 2018; Prangé et al., 2019; Yamada et al., 2015). Lastly, if there is no reference structure available at room temperature, the comparison can still be done with a structure whose data were collected at 100 K, but the comparison should be interpreted with care, as for example in (Ascone, Savino, Kahn, & Fourme, 2010).

7.2 Geometrical comparison

To compare the high-pressure structures to their reference at ambient pressure, or to compare different high-pressure structures with each other, root-mean-square deviations (r.m.s.d.) between Cα atoms, backbone atoms, or all atoms, are usually calculated, using programs from the CCP4 package (Winn et al., 2011) or with *SSM* (Schneider, 2002). The average r.m.s.d. between Cα or backbone atoms are usually quite low, estimated to be significant above 0.1 Å by (Urayama et al., 2002). As a general tendency, the mean r.m.s.d. values are always smaller for β-strands than for α-helices and loops, these latter being more compressible than the β-strands (Collins et al., 2011; Fourme et al., 2009; Fourme, Girard, et al., 2012). The very first structural study of a protein under high hydrostatic pressure, that of hen egg-white lysozyme (HEWL) at 100 MPa, revealed that the β-strand domain is almost incompressible while the helical domain and the interdomain region are more compressible (Kundrot & Richards, 1987). Structural alignments can be performed on one domain to highlight displacement of domains toward or away from each other, as was the case for the T4 lysozyme mutant L99A at 200 MPa pressure, where the structural alignment of the C-terminal domain revealed a structural rearrangement of subdomains relative to each other (Collins, Quillin, Hummer, Matthews, & Gruner, 2007). To better analyze collective displacements, the r.m.s.d. can be computed on the center-of-mass of residues averaged over a window of 5 residues, from residues $n - 2$ to residue $n + 2$; this was done in the analysis of sperm whale myoglobin at 150 MPa, revealing that six regions underwent large collective displacement (Urayama et al., 2002).

Areas shifted by pressure by more than 1 Å revealed conformational modifications corresponding to transitions between structural high-energy substates. It highlighted for example the intrinsic flexibility of the loop between helix F and G in neuroglobin (Ngb) which was displaced by almost 2 Å (Colloc'h et al., 2017) or the hinge movement in dihydroorotase (DHOase) where a loose proline was displaced by more than 4 Å (Prangé et al., 2019).

The transition between structural states induced by pressure was studied in detail for the Ras protein, where the r.m.s.d. analyses were done in an inventive way to reveal subtle differences and similarities between states. The r.m.s.d. values between Ras structures at two different pressures were compared to the r.m.s.d. between Ras in two different already characterized structural states. This comparison revealed for example that the structural shift in the central segment was similar in the transition from 200 MPa and 500 MPa and in the transition between state 2(T) and state 1(T) (Fig. 2).

By making all possible comparisons between structures at different pressures and structures in different structural states, the segments of Ras which underwent structural modifications have been systematically mapped out. For example, the central segment was in a state similar to state 2(T) at ambient and 200 MPa pressure, to the state 1(T) at 500 MPa, to the state 2(T)*-RBD at 650 MPa and to the state 1(0) at 900 MPa (Girard et al., 2022).

However, r.m.s.d. analysis alone does not give any information on the direction of the shifts. This can be done with the visualization of the displacement vectors, using the *PYMOL* script modevector.py (https://pymolwiki.org/index.php/Modevectors), to highlight for example the closure of the entrance to the active site and the opening of a groove in

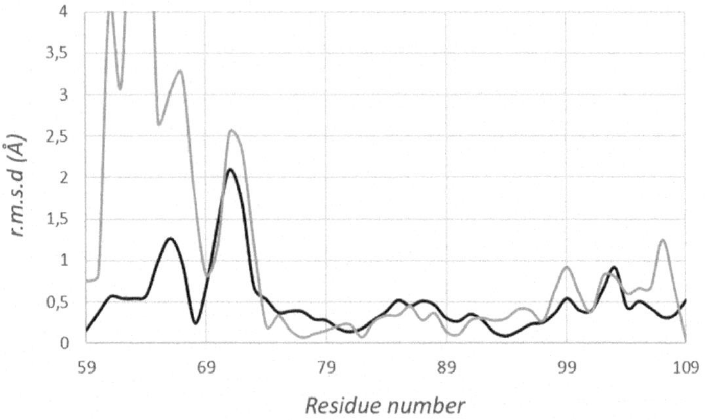

Fig. 2 Cα r.m.s. deviations in the central segment (59–109) between Ras(wt) structures at 200 MPa and 500 MPa (black line) and between 3tgp and 5b30 (gray line), two structures representatives of state 2(T) state and state 1(T) respectively. *Reprinted from Girard et al. (2022) with permission from the Royal Society of Chemistry.*

isopropylmalate dehydrogenase (IPMDH) at 580 MPa (Nagae et al., 2012) (Fig. 3), or domains displacement in HEWL at 920 MPa (Yamada et al., 2015).

The direction of the displacements can also be estimated by tracking whether the centroids of groups of residues (e.g., the whole protein or protein assembly or a domain within a protein) move toward the center of mass or the bulk solvent, corresponding to compression and expansion, respectively. This analysis highlighted for example a displacement of domain 1 toward the center of mass in HEWL at 100 MPa (Kundrot & Richards, 1987), a radial movement of 1.4 Å toward the center of the capsid in the Cowpea mosaic virus (CpMV) under 330 MPa (Girard et al., 2005), a shift of almost 2 Å toward the bulk of the FG loop in Ngb at 310 MPa (Colloc'h et al., 2017), a displacement of three helices toward the cavity in T4 lysozyme L99A at 200 MPa (Collins et al., 2005, 2007) or a shift of loop A toward the center while the two other loops B and C were shifted in the opposite direction in DHOase at 120 MPa (Prangé et al., 2019).

To visualize subdomains moving toward or away from each other, or correlated motions between different segments, r.m.s.d. analysis can be combined with a difference distance matrix (DDM) analysis, which can be performed through software like *ESCET* (Schneider, 2000), *CMView* (Vehlow et al., 2011), or *DynDom* (Hayward & Berendsen, 1998). For example, DDM analysis revealed in urate oxidase (UOX) at 150 MPa a correlated displacement of a loop and a helix, both containing residues

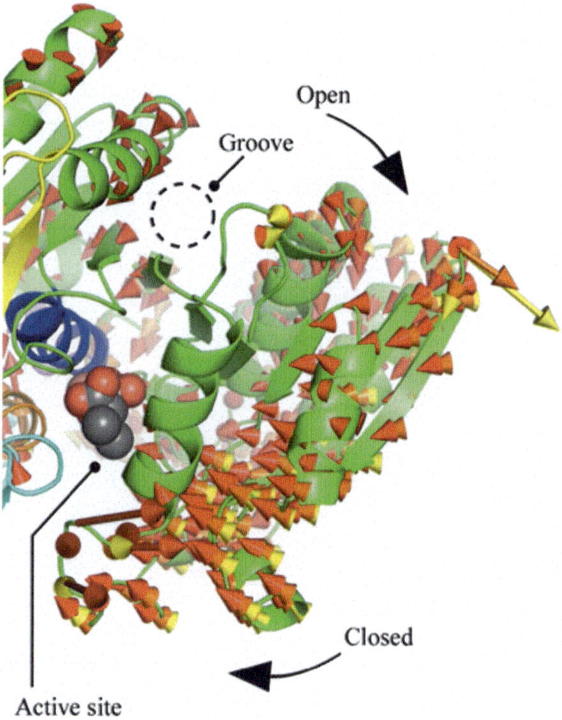

Fig. 3 The vectors between the Cα positions of the ambient and the high-pressure structures are represented as arrows (yellow, 340–0.1 MPa: orange, 410–0.1 MPa; red, 580–0.1 MPa). *Reprinted from open access Nagae et al. (2012) with permission from IUCr Journals*

lining the hydrophobic cavity and the active-site pocket (Girard, Marchal, et al., 2010), or in HEWL at 100 MPa and at 920 MPa the correlated displacement of the β-strand domain toward the helical domain to close the active site (Kundrot & Richards, 1987; Yamada et al., 2015) (Fig. 4).

DDM analysis also revealed the correlated displacement of 6 segments getting closer and 5 segments moving away in β-lactoglubulin (BLG) in complex with dodecane at 430 MPa, which could reflect the first step of the dimer dissociation by pressure (Kurpiewska et al., 2018). Moreover, DDM analysis also highlighted the largest displacements in BLG in complex with myristic acid at 550 MPa, which occurred in the areas responsible for the allergen properties of BLG, the exposed epitopes being more buried in the high-pressure structure, and hence less accessible to antibodies (Kurpiewska et al., 2019).

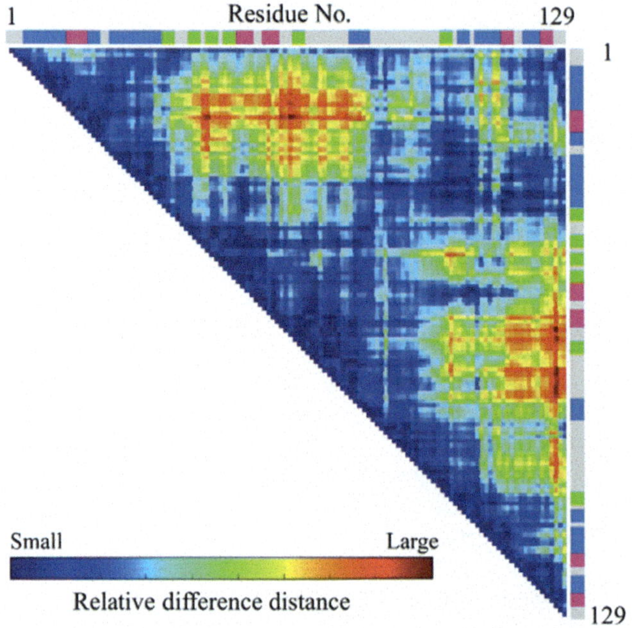

Fig. 4 Difference-distance matrix (DDM) plot for Cα atoms between the structures of HEWL at 0.1 and 890 MPa. The two color bars indicate the secondary structure with helices in blue, turns in red and sheets in green. *Reprinted from open access Yamada et al. (2015) with permission from IUCr Journals.*

7.3 Thermal B-factors analysis

The analysis of thermal B-factor differences between the high-pressure structures and the reference structure at ambient pressure ($B_{HP} - B_{AP}$) can be done only if the reference structure data set was collected in the same condition, at least in terms of temperature. To limit bias from the refinement procedure, one can also calculate the relative B-factor differences which has a distribution with a zero mean:

$$((B_{HP} - <B_{HP}>) - (B_{AP} - <B_{AP}>))/(<B_{HP}> - <B_{AP}>)$$

with <X> referring to the mean of X values.

The areas where B-factors are higher than average are the most pressure-sensitive, and conversely, the areas where B-factors are smaller than average are the least pressure-sensitive. The analysis of the B-factor trends allows one to map the flexible and rigid zones that are related to function.

The most flexible regions are usually those involved in conformational modifications between structural states or in response to interacting with partners. Conversely, the areas which are the least pressure-sensitive correspond to areas with a small number of accessible conformational states, and are thus less likely to be involved in protein function.

In many high-pressure structures, the B-factors decreased on average with increasing pressure, for example by 1 $Å^2$ in HEWL at 100 MPa (Kundrot & Richards, 1987) and by 9 $Å^2$ at 920 MPa (Yamada et al., 2015), by 9 $Å^2$ in CpMV at 330 MPa (Girard et al., 2005) or by 10 $Å^2$ in BLG at 550 MPa (Kurpiewska et al., 2019). However, in some cases, the B-factors increased on average with pressure, by 5 $Å^2$ in UOX at 150 MPa, 9 $Å^2$ at 210 MPa and 28 $Å^2$ at 310 MPa (Girard, Marchal, et al., 2010; Prangé et al., 2022).

The distribution of B-factor differences may be inhomogeneous giving information on the flexibility and rigidity of different areas. For example, in dihydrofolate reductase (DHFR), regardless of whether the M20 loop is in the open (DHFR-M20-open) or in closed (DHFR-M20-closed) conformations, the average B-factor was reduced by pressure, but increased in some loops, such as the FG loop and the M20 loop that are highly pressure-sensitive (Nagae et al., 2018).

In UOX at 150 MPa, the pressure sensitivity deduced from the B-factor analyses was higher for α-helices than for β-strands, as already shown from the r.m.s.d. analyses. In addition, the three loops lining the active site entrance were highly pressure-sensitive, explaining the loss of substrate affinity at high pressure (Girard, Marchal, et al., 2010). At higher pressure (210 MPa and 310 MPa), the least pressure-sensitive residues of UOX were located in the central tunnel and the β-barrel dimeric interface where the active site is located, highlighting the role of the ligand in strengthening the dimer against pressure effects. In contrast, the most pressure-sensitive residues were located at the face-to-face dimeric interface, suggesting a destabilization of this interface during the pressure-induced dissociation of the UOX tetramer. This hypothesis was confirmed with the structure of UOX under 200 MPa of argon, where the crystallographic symmetry was lost at this interface (Prangé et al., 2022).

In neuroglobin wild-type (Ngb-wt), the B-factors increased on average by 2 $Å^2$ at 270 MPa and 3 $Å^2$ at 310 MPa, while the inactive mutant Ngb-F106W which has a decreased affinity for CO, was more pressure-sensitive with B-factors increasing by 11 $Å^2$ at 280 MPa. The uneven distribution of B-factors in both structures was very informative. In Ngb-wt, the most pressure-sensitive residues were located close to the heme and in the FG

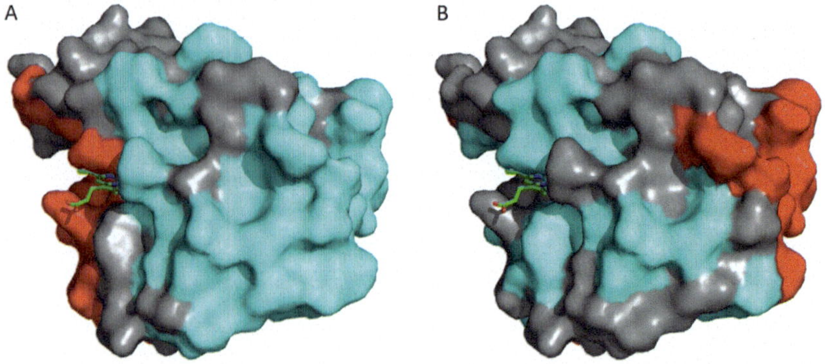

Fig. 5 (A) B-factor differences between Ngb-wt at 270 MPa and at 0.1 MPa (on average 2 Å^2). The molecular surface is colored according to the B-factor differences, in red if the differences are higher than 4 Å^2, in cyan if the differences have negative values, in gray otherwise. (B) B-factors differences between Ngb-F106W at 280 MPa and at 0.1 MPa (on average 10 Å^2). The molecular surface is colored according to the B-factor differences, in red if the differences are higher than 15 Å^2, in cyan if the differences are lower than 5 Å^2, in gray otherwise. The heme is shown in stick representation.

loop while the least sensitive residues were located at the back of the heme cavity. Interestingly, in the inactive mutant Ngb-F106W, the distribution was the opposite, with the least pressure-sensitive residues near the heme and the most pressure-sensitive residues at the back (Fig. 5). The intrinsic flexibility of the FG loop in front of the heme thus seemed necessary for Ngb activity (Colloc'h et al., 2017).

7.4 Molecular and cavities volumes

According to Le Chatelier's principle, conformers with a smaller partial molar volume (PMV) are favored by pressure. The PMV is the sum of the molecular volume (defined as the sum of the van der Waals volumes of the constituent atoms and the volume of void cavities) and the change in volume of the system due to hydration of the protein. The molecular volume can be calculated with *VOIDOO* (Kleywegt & Jones, 1994) or *PISA* (Krissinel, 2015) with a probe radius of 1.4 Å, with *ProteinVolume* (Chen & Makhatadze, 2015), or estimated through the use of Voronoi cells (Dupuis, Sadoc, Jullien, Angelov, & Mornon, 2005). In all the crystallographic structures under pressure, the molecular volume is normally decreased by the pressure, as expected. The compressibility of the molecular volume is usually in the range between 0.04 GPa^{-1} and 0.09 GPa^{-1}

(Dhaussy & Girard, 2015). For example, in CpMV, the volume of the capsid was reduced by 3.7% at 330 MPa.

The volumes of internal cavities are often computed with *CASTp* (Tian, Chen, Lei, Zhao, & Liang, 2018), with a probe radius between 0.9 Å and 1.4 Å, and also with the *HOLLOW* program (Ho & Gruswitz, 2008). The total volume of void cavities usually decrease with pressure, like in CpMV where the total volume of cavities was decreased by 40% at 330 MPa (Girard et al., 2005).

However, the volume of a few cavities may increase under pressure due to water penetration. This was the case for the T4 lysozyme L99A mutant, although the total cavity volume decreased by 3% at 200 MPa. Water molecules were observed to cooperatively penetrate into the large hydrophobic cavity of 160 $Å^3$ created by the mutation. This cavity was observed to be remarkably rigid and resistant to pressure (Collins et al., 2005, 2007). Importantly, this study established that volume reduction under pressure can be achieved by water molecules penetrating hydrophobic cavities and that this penetration is further stabilized by hydrogen-bonding between multiple water molecules.

In IPMDH, the total volume of internal cavities was compressed by 0.73 GPa^{-1}. However, the volume of the cavity located at the interface between subunits, first decreased at 160 MPa and then increased between 340 MPa and 650 MPa, because two water molecules penetrated into this interfacial cavity, with an occupancy increasing with pressure. This process could initiate the pressure-induced destabilization of the dimer (Nagae et al., 2012) (Fig. 6).

In HEWL, the compressibility of the cavity volume was 0.008 GPa^{-1} up to 710 MPa. All cavity volumes were decreased by pressure, except one located above Trp108, whose volume decreased with pressure up to 600 MPa, then increased at 710 MPa due to penetration of a water molecule, and decreased again at 800 MPa (Yamada et al., 2015).

In DHFR, all cavities were compressed, except one cavity in the DHFR-M20-open at 500 MPa where water penetrated. In DHFR-M20-closed, no water penetration was observed, even at 800 MPa, while in DHFR-M20-open at 500 MPa, water molecules penetrated in the active site, near the folate, some of them taking the place of the nicotinamide adenine dinucleotide phosphate ($NADP^+$) moiety that have flipped (Fig. 7) (Nagae et al., 2018).

In contrast, no water molecules could be visualized in the large hydrophobic cavity of UOX, whose volume was reduced by 16% at

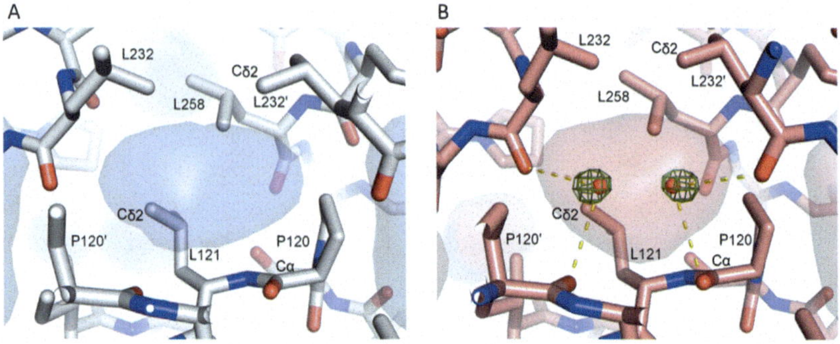

Fig. 6 Internal cavity located at the interface between subunits in IPMDH dimer at 0.1 MPa (A) and 580 MPa (B) where two water molecules penetrate. The difference electron-density map is shown as a green mesh contoured at 4σ. *Reprinted from open access Nagae et al. (2012) with permission from IUCr Journals.*

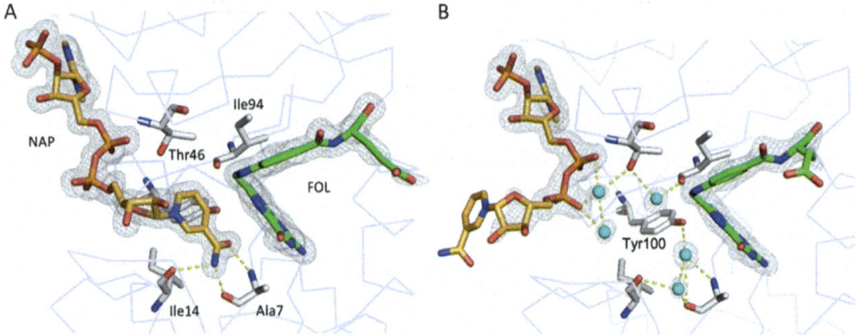

Fig. 7 DHFR with M20 loop in open conformation at 0.1 MPa (A) and 500 MPa (B). The electron density map covering $NADP^+$ (noted NAP), folate (FOL) and water molecules is shown at 1.0 σ. The structures have been determined at 1.7 Å and 1.9 Å resolution, respectively. *Reprinted from open access Nagae et al. (2018) with permission from IUCr Journals.*

150 MPa then 38% at 210 MPa (Girard, Marchal, et al., 2010; Prangé et al., 2022). The shrinkage of this large hydrophobic cavity seemed to be related to the pressure-induced enlargement of the nearby active site pocket. The cavity plays an allosteric role of a connecting vessel to allow the active site to adapt to large catalytic intermediates (Colloc'h & Prangé, 2014).

Similarly, no water molecule was visualized in the hydrophobic cavity located behind the heme in Ngb-wt or its F106W mutant, with a reduction in volume of 13% in Ngb-wt at 310 MPa and 8% in Ngb-F106W at 310 MPa (Colloc'h et al., 2017).

These different results confirm that void cavities inside proteins play a key role in the unfolding process. They either may constitute a path for solvent, which penetrates cavities and thereby induces unfolding. They may also constitute empty space allowing extensive conformational modifications, which induce unfolding.

7.5 Hydration with pressure

HMPX is a powerful method to detect the pressure-induced modifications in the hydration shell within or around proteins. The number of ordered water molecules is usually increased by pressure in almost all high-pressure structures when comparing structures at resolutions similar to or better than that at ambient pressure. Studies have shown that the number of ordered water molecules can increase by 4–120 per 100 MPa pressure. For example, the number of ordered water molecules increased from 108 to 198 in T4 lysozyme L99A at 200 MPa (Collins et al., 2005), from 99 to 195 in CpMV at 330 MPa (Girard et al., 2005), from 184 to 275 in IPMDH at 340 MPa (Nagae et al., 2012), from 100 to 171 in HEWL at 890 MPa (Yamada et al., 2015), from 87 to 134 in DHFR-M20-closed at 400 MPa, from 110 to 134 in DHFR-open-20 at 750 MPa (Nagae et al., 2018), from 103 to 163 in Ras wild-type at 650 MPa and from 82 to 210 in Ras D33K at 900 MPa (Fig. 8) (Girard et al., 2022). In DHOase at 120 MPa, the number of ordered water molecules increased from 33 to 174, and the expected catalytic water molecule linked to the zinc center in DHOase became visible (Prangé et al., 2019).

It has been suggested that the overall pressure-induced shortening of hydrogen-bond lengths between amino acids and water molecules could

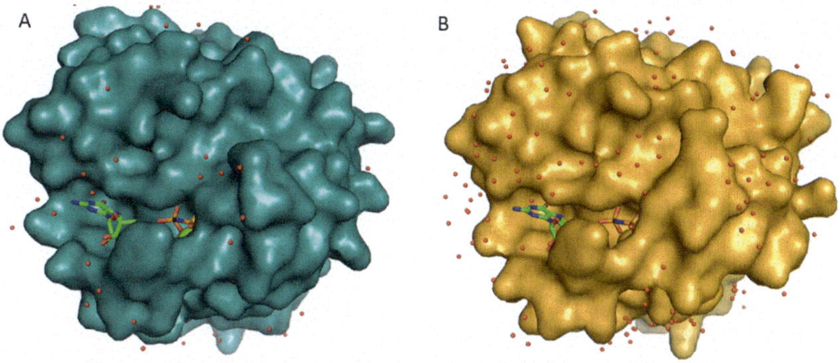

Fig. 8 Molecular surface of Ras-D33K (A) at 0.1 MPa (B) at 900 MPa. The water molecules are shown with red spheres and the GppNHp in stick representation.

explain the higher number of ordered water molecules in the crystallographic structures under pressure, as shown in the CpMV structure under high pressure (Girard et al., 2005). Pressure is thus a way to stabilize the water molecule network, increasing the stability of the protein-hydration shell system.

7.6 Intrinsic dynamics of proteins revealed by pressure

Pressure-induced structural modifications, revealed by r.m.s.d. and DDM analyses, the B-factor distribution, the reduction in the molecular volume, the reduction or expansion in several cavity volumes, and the hydration of certain cavities and binding sites, allow one to analyze the intrinsic dynamics of proteins by the mapping of the flexibility and rigidity of areas that are directly related to functional properties. The high-energy substates that can be characterized in great detail in high pressure crystallographic structures may correspond to functionally relevant conformers that are directly involved in protein activity.

In sperm whale myoglobin, the pressure-induced structural modifications were found to be similar to the pH-induced structural modifications confirming that the conformational substates characterized by pressure were functionally relevant (Urayama et al., 2002).

In the structures of HEWL at 710 MPa and above, the water molecules that ordered with pressure were located at the back of the active site. The catalytic Glu35 residue, which has an elevated pKa essential for the catalytic reaction, was flipped by pressure to an alternate position that stabilized the penetrating water molecules. This also highlighted the role of a Trp residue lining the cavity in the abnormal pKa of Glu35. This position of Glu involved in the HEWL enzymatic reaction showed that pressure stabilizes a biologically relevant high-energy conformer. This position was never observed before in the other crystallographic structures of HEWL (Yamada et al., 2015).

In Ngb, in addition to the B-factor analysis, the mechanical properties of residues were computed using *ProPHet* (Sacquin-Mora, 2014), highlighting that five hydrophobic residues lining a small cavity located just behind the heme underwent an important increase in their force constant. These five residues constituted a mechanical nucleus, around which Ngb hinges, between its high and low pressure-sensitive areas as revealed by the B-factor analysis. Moreover, in parallel to the reduction in the volume of the internal heme cavity, the tunnel linking the heme to the bulk disappeared with pressure. Altogether, these results suggested that NO gas could be hosted in the cavity to react with the oxygen bound to the heme, while no other gaseous molecules could migrate from the bulk, since the cavity was occluded. Pressure would thus have trapped a relevant conformer of Ngb involved in its NO detoxification activity (Colloc'h et al., 2017).

In DHFR with M20 loop in close conformation (DHFR-M20-close), the loop stayed in the closed conformation until 400 MPa and became more flexible at 650 MPa and above. Moreover, at 800 MPa, many loops became flexible, associated with a loss in occupancies for the $NADP^+$ cofactor and the substrate FOL. In DHFR with M20 in the open conformation (DHFR-M20-open), the loop was very stable at every pressure confirming that the M20 loop remains open in its high-energy substate. Moreover, the nicotinamide moiety of $NADP^+$ flipped out of the active site at 500 MPa, inducing a displacement of the nearby FG loop. The high-pressure structures of DHFR-M20-open at 500 MPa and above could represent a Michaelis complex-formation process. In the high-pressure structures of DHFR-M20-close, FOL and $NADP^+$ occupancies decreased with pressure without a flip of the $NADP^+$ moiety, suggesting that the product release would be faster when NADP+ has not flipped out of the active site (Nagae et al., 2018).

In IPMDH, the opening of the groove in the active site by pressure was accompanied by the entrance of several water molecules at 580 MPa. Both the opening of the groove and the water penetration in a non-polar environment could be considered as an initial step of pressure-induced unfolding of IPMDH (Nagae et al., 2012). Interestingly, in the piezophilic IPMDH, there is an alanine residue which lines the groove instead of a serine in the non-piezophilic IPMDH. In the mutant S266A of the non-piezophilic IPMDH at 580 MPa, there was almost no water penetration and only a slight enlargement of the groove, which could explain the pressure-tolerant activity of this mutant (Hamajima et al., 2016).

In BLG in complex with dodecane at 430 MPa, the volume of the ligand-binding pocket increased by 25%, corresponding to an increase in length, with a modification of the conformation of the ligand and of some residues at the entrance of the pocket. The two sides of the β-barrel fluctuated independently of each other. There was also an expansion of the uncharged regions by pressure, with more exposed hydrophobic patches, along with a decrease of interactions between the two monomers. These modifications suggested that the structural characteristics of the high-pressure structure of BLG reflected the first signs of pressure-induced dimer dissociation and could correspond to conformational modifications found in a partially unfolded state (Kurpiewska et al., 2018). At 550 MPa, the structure of BLG in complex with myristic acid revealed important structural shifts in the epitopes, explaining why pressure induces more allergenicity. The polar end of the myristic acid flipped toward the opening of the ligand-binding pocket, with two water molecules penetrating into the empty space between the two subunits. The compression of the ligand-binding pocket by pressure could prevent the ligand release (Kurpiewska et al., 2019).

In the non-productive DHOase, which is isolated out of its complex with aspartate transcarbamoylase (ATCase), some residues from loop A, which was not detectable in the electron density map at ambient pressure, became visible at 120 MPa. This loop, which occludes the catalytic site in isolated DHOase by blocking the access of the substrate, was shifted away under pressure, thus opening the active site. In parallel, the distance between the catalytic Asp305 residue and the zinc was significantly shortened by pressure. Both the opening of the active site and the reconnection of the catalytic Asp to the zinc in the high-pressure structure of DHOase could explain the restoration of its catalytic activity by pressure (Prangé et al., 2019).

In UOX at 210 MPa, the active site was enlarged, with a loss of hydrogen bonds between the catalytic water and an Asn residue which had a backward displacement. Moreover, a second inhibitor was localized in the tunnel connecting the bulk to the active site. This second inhibitor was loosely hydrogen-bonded to a lysine residue, which was among the least pressure-sensitive residues, suggesting that pressure stabilized this area. The high-pressure structures of UOX could therefore correspond to high-energy substates where the active site is enlarged to accommodate large catalytic intermediates with a backward displacement of Asn254 allowing the dioxygen to migrate from the nearby hydrophobic cavity to the active site. The allosteric role of this cavity — i.e., acting as a connecting vessel to provide flexibility to the active site — was confirmed with the structure of UOX under 200 MPa argon. The argon is bound in this allosteric cavity, preventing its contraction, and thus no enlargement nor structural modification in the neighboring active site could occur (Prangé et al., 2022).

The high-pressure structures of Ras-wt and the D33K mutant revealed that the allosteric switch between the two lobes of Ras, which mimics the allosteric effect of Ras membrane binding and which can be induced *in cristallo* upon the addition of an exogenous agent, can now be induced just using pressure. The role of two tyrosine residues, Tyr32 and Tyr71, which were described as an indicator of the conformational state of Ras (Parker, Meyer, Golemis, & Dunbrack, 2022), was confirmed in the high-pressure structures of Ras. Tyr32 flipped at 900 MPa in a position oriented toward the nucleotide, in agreement with a pressure-induced transition from a state 2(T) to a state 2(T)*-RBD of this segment. The Tyr71 residue oscillated between a buried and an exposed orientation upon increasing pressure, switching between a position where it interacts with the switch I and a position where it can interact with effectors and regulators (Girard et al., 2022).

All of these different examples illustrate the power of high-pressure macromolecular crystallography to finely decipher the intrinsic dynamics and plasticity of proteins and the subtle balance between rigidity and flexibility in the context of functional efficiency, and to explore structural and functional properties of transient high-energy states relevant to protein mechanisms, without the use of mutants nor exogenous agents.

8. Perspectives

HPMX could benefit from recent developments in data interpretation such as the use of Fo − Fo difference maps in serial crystallography. Such an approach would allow one to derive information about protein dynamics by optimally determining the proportion of each observed conformation. Accessibility of such features to the structural community have been facilitated through the development of software, such as *Xtrapol8* (De Zitter, Coquelle, Oeser, Barends, & Colletier, 2022). However, in HPMX, isomorphism may be a concern since unit-cell parameters vary under pressure.

In a recent review, Helliwell (2020) discussed that studies at physiologically relevant temperatures are required to understand biological systems. Recently, a crystallographic study of 3 model proteins demonstrated that is can be possible to measure diffraction at temperatures higher than room temperature by putting crystals in oil to prevent them from dehydration (Doukov, Herschlag, & Yabukarski, 2020). As mentioned previously, a DAC provides a hermetic environment for the sample and can thus be considered for MX at temperatures greater than 293 K. Typically, our DAC can be equipped with a Peltier system placed close to the gasket to access high temperature in a controlled way. It also opens the potential for crystallographic studies of the effect of both pressure and temperature on biological systems to better describe the conformational landscape of proteins and the associated energetics. This would also allow one to better evaluate the dynamic behavior of proteins isolated from organisms such as *Thermococcus piezophilus* or *Pyrococcus yayanosii CH1* (Birrien et al., 2011; Johler, Stephan, Althaus, Ehling-Schulz, & Grunert, 2016) which are both hyperthermophilic and obligate piezophilic species.

Acknowledgments

The authors acknowledge the different synchrotrons for beamtime provisions and greatly thanks all the staff members on the beamlines for their invaluable help. A special thought for Roger Fourme who initiated HPMX development in France as well as for Richard Kahn who was deeply involved in the early stages of the project. We also thank Thierry Prangé who participated to several HPMX campaigns as well as for his comments on the present contribution. IBS acknowledges integration into the Interdisciplinary Research Institute of Grenoble (IRIG, CEA).

References

Akasaka, K. (2006). Probing conformational fluctuation of proteins by pressure perturbation. *Chemical Reviews, 106*(5), 1814–1835. https://doi.org/10.1021/cr040440z

Ascone, I., Savino, C., Kahn, R., & Fourme, R. (2010). Flexibility of the Cu,Zn superoxide dismutase structure investigated at 0.57 GPa. *Acta Crystallographica. Section D, Biological Crystallography, 66*(Pt 6), 654–663. https://doi.org/10.1107/S0907444910012321

Barstow, B., Ando, N., Kim, C. U., & Gruner, S. M. (2008). Alteration of citrine structure by hydrostatic pressure explains the accompanying spectral shift. *Proceedings of the National Academy of Sciences of the United States of America, 105*(36), 13362–13366. https://doi.org/10.1073/pnas.0802252105

Bassett, W. A. (2009). Diamond anvil cell, 50th birthday. *High Pressure Research, 29*(2), 163–186. https://doi.org/10.1080/08957950802597239

Baxter, N. J., Zacharchenko, T., Barsukov, I. L., & Williamson, M. P. (2017). Pressure-dependent chemical shifts in the R3 domain of talin show that it is thermodynamically poised for binding to either vinculin or RIAM. *Structure (London, England: 1993), 25*(12), 1856–1866.e2. https://doi.org/10.1016/j.str.2017.10.008

Birrien, J.-L., Zeng, X., Jebbar, M., Cambon-Bonavita, M.-A., Quérellou, J., Oger, P., ... Prieur, D. (2011). Pyrococcus yayanosii sp. Nov., an obligate piezophilic hyperthermophilic archaeon isolated from a deep-sea hydrothermal vent. *International Journal of Systematic and Evolutionary Microbiology, 61*(12), 2827–2881. https://doi.org/10.1099/ijs.0.024653-0

Budzianowski, A., & Katrusiak, A. (2004). High-pressure crystallographic experiments with a CCD-detector. In A. Katrusiak, & P. McMillan (Eds.). *High-pressure crystallography* (pp. 101–112). Netherlands: Springer. https://doi.org/10.1007/978-1-4020-2102-2_7

Celeste, A., Borondics, F., & Capitani, F. (2019). Hydrostaticity of pressure-transmitting media for high pressure infrared spectroscopy. *High Pressure Research, 39*(4), 608–618. https://doi.org/10.1080/08957959.2019.1666844

Chen, C. R., & Makhatadze, G. I. (2015). ProteinVolume: Calculating molecular van der Waals and void volumes in proteins. *BMC Bioinformatics, 16*(1), https://doi.org/10.1186/s12859-015-0531-2

Chen, X., Lou, H., Zeng, Z., Cheng, B., Zhang, X., Liu, Y., ... Zeng, Q. (2021). Structural transitions of 4:1 methanol–ethanol mixture and silicone oil under high pressure. *Matter and Radiation at Extremes, 6*(3), 038402. https://doi.org/10.1063/5.0044893

Chervin, J. C., Canny, B., Besson, J. M., & Pruzan, Ph (1995). A diamond anvil cell for IR microspectroscopy. *Review of Scientific Instruments, 66*(3), 2595–2598. https://doi.org/10.1063/1.1145594

Collins, M. D., Hummer, G., Quillin, M. L., Matthews, B. W., & Gruner, S. M. (2005). Cooperative water filling of a nonpolar protein cavity observed by high-pressure crystallography and simulation. *Proceedings of the National Academy of Sciences of the United States of America, 102*(46), 16668–16671. https://doi.org/10.1073/pnas.0508224102

Collins, M. D., Kim, C. U., & Gruner, S. M. (2011). High-pressure protein crystallography and NMR to explore protein conformations. *Annual Review of Biophysics, 40*, 81–98. https://doi.org/10.1146/annurev-biophys-042910-155304

Collins, M. D., Quillin, M. L., Hummer, G., Matthews, B. W., & Gruner, S. M. (2007). Structural rigidity of a large cavity-containing protein revealed by high-pressure crystallography. *Journal of Molecular Biology, 367*(3), 752–763. https://doi.org/10.1016/j.jmb.2006.12.021

Colloc'h, N., & Prangé, T. (2014). Functional relevance of the internal hydrophobic cavity of urate oxidase. *FEBS Letters, 588*(9), 1715–1719. https://doi.org/10.1016/j.febslet.2014.03.017

Colloc'h, N., Sacquin-Mora, S., Avella, G., Dhaussy, A.-C., Prangé, T., Vallone, B., & Girard, E. (2017). Determinants of neuroglobin plasticity highlighted by joint coarse-grained simulations and high pressure crystallography. *Scientific Reports, 7*(1), 1858. https://doi.org/10.1038/s41598-017-02097-1

Couzinet, B., Dahan, N., Hamel, G., & Chervin, J.-C. (2003). Optically monitored high-pressure gas loading apparatus for diamond anvil cells. *High Pressure Research, 23*(4), 409–415. https://doi.org/10.1080/0895795031000095847

De Zitter, E., Coquelle, N., Oeser, P., Barends, T. R. M., & Colletier, J.-P. (2022). Xtrapol8 enables automatic elucidation of low-occupancy intermediate-states in crystallographic studies. *Communications Biology, 5*(1), 640. https://doi.org/10.1038/s42003-022-03575-7

Decker, D. L., & Worlton, T. G. (1972). Compression of NaCl and CsCl to 32 kbar. *Journal of Applied Physics, 43*(11), 4799–4800. https://doi.org/10.1063/1.1661013

Dewaele, A., & Loubeyre, P. (2007). Pressurizing conditions in helium-pressure-transmitting medium. *High Pressure Research, 27*(4), 419–429. https://doi.org/10.1080/08957950701659627

Dewaele, A., Belonoshko, A. B., Garbarino, G., Occelli, F., Bouvier, P., Hanfland, M., & Mezouar, M. (2012). High-pressure—High-temperature equation of state of KCl and KBr. *Physical Review B, 85*(21), 214105. https://doi.org/10.1103/PhysRevB.85.214105

Dhaussy, A.-C., & Girard, E. (2015). Functional sub-states by high-pressure macromolecular crystallography. *Sub-cellular Biochemistry, 72*, 215–235. https://doi.org/10.1007/978-94-017-9918-8_11

Dickerson, J. L., & Garman, E. F. (2019). The potential benefits of using higher X-ray energies for macromolecular crystallography. *Journal of Synchrotron Radiation, 26*(4), 922–930. https://doi.org/10.1107/S160057751900612X

Dorogokupets, P. I., & Dewaele, A. (2007). Equations of state of MgO, Au, Pt, NaCl-B1, and NaCl-B2: Internally consistent high-temperature pressure scales. *High Pressure Research, 27*(4), 431–446. https://doi.org/10.1080/08957950701659700

Doukov, T., Herschlag, D., & Yabukarski, F. (2020). Instrumentation and experimental procedures for robust collection of X-ray diffraction data from protein crystals across physiological temperatures. *Journal of Applied Crystallography, 53*(6), 6. https://doi.org/10.1107/S1600576720013503

Dupuis, F., Sadoc, J.-F., Jullien, R., Angelov, B., & Mornon, J.-P. (2005). Voro3D: 3D Voronoi tessellations applied to protein structures. *Bioinformatics (Oxford, England), 21*(8), 1715–1716. https://doi.org/10.1093/bioinformatics/bth365

Emsley, P., Lohkamp, B., Scott, W. G., & Cowtan, K. (2010). Features and development of Coot. *Acta Crystallographica. Section D, Biological Crystallography, 66*(Pt 4), 486–501. https://doi.org/10.1107/S0907444910007493

Forman, R. A., Piermarini, G. J., Barnett, J. D., & Block, S. (1972). Pressure measurement made by the utilization of ruby sharp-line luminescence. *Science (New York, N. Y.), 176*(4032), 284–285. https://doi.org/10.1126/science.176.4032.284

Fourme, R., Girard, E., & Akasaka, K. (2012). High-pressure macromolecular crystallography and NMR: Status, achievements and prospects. *Current Opinion in Structural Biology, 22*(5), 636–642. https://doi.org/10.1016/j.sbi.2012.07.007

Fourme, R., Girard, E., Kahn, R., Dhaussy, A.-C., & Ascone, I. (2009). Advances in high-pressure biophysics: Status and prospects of macromolecular crystallography. *Annual Review of Biophysics, 38*, 153–171. https://doi.org/10.1146/annurev.biophys.050708.133700

Fourme, R., Honkimäki, V., Girard, E., Medjoubi, K., Dhaussy, A.-C., & Kahn, R. (2012). Reduction of radiation damage and other benefits of short wavelengths for macromolecular crystallography data collection. *Journal of Applied Crystallography, 45*(4), 4. https://doi.org/10.1107/S0021889812019164

Frauenfelder, H., Alberding, N., Ansari, A., Braunstein, D., Cowen, B., Hong, M., ... Yue, K. (1990). Proteins and pressure. *The Journal of Physical Chemistry,* 1024–1037.

Frauenfelder, H., Sligar, S. G., & Wolynes, P. G. (1991). The energy landscapes and motions of proteins. *Science (New York, N. Y.), 254*(5038), 1598–1603.

Girard, E., Dhaussy, A.-C., Couzinet, B., Chervin, J.-C., Mezouar, M., Kahn, R., ... Fourme, R. (2007). Toward fully fledged high-pressure macromolecular crystallography. *Journal of Applied Crystallography, 40*(5), 912–918. https://doi.org/10.1107/S0021889807033833

Girard, E., Fourme, R., Ciurko, R., Joly, J., Bouis, F., Legrand, P., ... Kahn, R. (2010). Macromolecular crystallography at high pressure with pneumatic diamond anvil cells handled by a six-axis robotic arm. *Journal of Applied Crystallography, 43*(4), 762–768. https://doi.org/10.1107/S0021889810016146

Girard, E., Kahn, R., Mezouar, M., Dhaussy, A.-C., Lin, T., Johnson, J. E., & Fourme, R. (2005). The first crystal structure of a macromolecular assembly under high pressure: CpMV at 330 MPa. *Biophysical Journal, 88*(5), 3562–3571. https://doi.org/10.1529/biophysj.104.058636

Girard, E., Lopes, P., Spoerner, M., Dhaussy, A.-C., Prangé, T., Kalbitzer, H. R., & Colloc'h, N. (2022). Equilibria between conformational states of the Ras oncogene protein revealed by high pressure crystallography. *Chemical Science, 13*(7), 2001–2010. https://doi.org/10.1039/d1sc05488k

Girard, E., Marchal, S., Perez, J., Finet, S., Kahn, R., Fourme, R., ... Colloc'h, N. (2010). Structure-function perturbation and dissociation of tetrameric urate oxidase by high hydrostatic pressure. *Biophysical Journal, 98*(10), 2365–2373. https://doi.org/10.1016/j.bpj.2010.01.058

Hamajima, Y., Nagae, T., Watanabe, N., Ohmae, E., Kato-Yamada, Y., & Kato, C. (2016). Pressure adaptation of 3-isopropylmalate dehydrogenase from an extremely piezophilic bacterium is attributed to a single amino acid substitution. *Extremophiles: Life Under Extreme Conditions, 20*(2), 177–186. https://doi.org/10.1007/s00792-016-0811-4

Hayward, S., & Berendsen, H. J. (1998). Systematic analysis of domain motions in proteins from conformational change: New results on citrate synthase and T4 lysozyme. *Proteins, 30*(2), 144–154.

Helliwell, J. R. (2020). What is the structural chemistry of the living organism at its temperature and pressure? *Acta Crystallographica. Section D, 76*(2), 87–93. https://doi.org/10.1107/S2059798320000546

Ho, B. K., & Gruswitz, F. (2008). HOLLOW: Generating accurate representations of channel and interior surfaces in molecular structures. *BMC Structural Biology, 8*(1), 49. https://doi.org/10.1186/1472-6807-8-49

Johler, S., Stephan, R., Althaus, D., Ehling-Schulz, M., & Grunert, T. (2016). High-resolution subtyping of Staphylococcus aureus strains by means of Fourier-transform infrared spectroscopy. *Systematic and Applied Microbiology, 39*(3), 189–194. https://doi.org/10.1016/j.syapm.2016.03.003

Kabsch, W. (2010). XDS. *Acta Crystallographica. Section D, Biological Crystallography, 66*(Pt 2), 125–132. https://doi.org/10.1107/S0907444909047337

Kitahara, R., Yokoyama, S., & Akasaka, K. (2005). NMR snapshots of a fluctuating protein structure: Ubiquitin at 30 bar-3 kbar. *Journal of Molecular Biology, 347*(2), 277–285. https://doi.org/10.1016/j.jmb.2005.01.052

Kleywegt, G. J., & Jones, T. A. (1994). Detection, delineation, measurement and display of cavities in macromolecular structures. *Acta Crystallographica. Section D, Biological Crystallography, 50*(Pt 2), 178–185. https://doi.org/10.1107/S0907444993011333

Klotz, S., Chervin, J.-C., Munsch, P., & Marchand, G. L. (2009). Hydrostatic limits of 11 pressure transmitting media. *Journal of Physics D: Applied Physics, 42*(7), 075413. https://doi.org/10.1088/0022-3727/42/7/075413

Klotz, S., Paumier, L., Le March, G., & Munsch, P. (2009). The effect of temperature on the hydrostatic limit of 4:1 methanol–ethanol under pressure. *High Pressure Research, 29*(4), 649–652. https://doi.org/10.1080/08957950903418194

Krissinel, E. (2015). Stock-based detection of protein oligomeric states in jsPISA. *Nucleic Acids Research, 43*(W1), W314–W319. https://doi.org/10.1093/nar/gkv314

Kundrot, C. E., & Richards, F. M. (1987). Crystal structure of hen egg-white lysozyme at a hydrostatic pressure of 1000 atmospheres. *Journal of Molecular Biology, 193*(1), 157–170. https://doi.org/10.1016/0022-2836(87)90634-6

Kunz, M., MacDowell, A. A., Caldwell, W. A., Cambie, D., Celestre, R. S., Domning, E. E., ... Clark, S. M. (2005). A beamline for high-pressure studies at the Advanced Light Source with a superconducting bending magnet as the source. *Journal of Synchrotron Radiation, 12*(Pt 5), 650–658. https://doi.org/10.1107/S0909049505020959

Kurpiewska, K., & Lewiński, K. (2010). High pressure macromolecular crystallography for structural biology: A review. *Central European Journal of Biology, 5*(5), 531–542. https://doi.org/10.2478/s11535-010-0044-y

Kurpiewska, K., Biela, A., Loch, J. I., Lipowska, J., Siuda, M., & Lewiński, K. (2019). Towards understanding the effect of high pressure on food protein allergenicity: β-lactoglobulin structural studies. *Food Chemistry, 270*, 315–321. https://doi.org/10.1016/j.foodchem.2018.07.104

Kurpiewska, K., Biela, A., Loch, J. I., Świątek, S., Jachimska, B., & Lewiński, K. (2018). Investigation of high pressure effect on the structure and adsorption of β-lactoglobulin. *Colloids and Surfaces. B, Biointerfaces, 161*, 387–393. https://doi.org/10.1016/j.colsurfb.2017.10.069

Letoullec, R., Pinceaux, J. P., & Loubeyre, P. (1988). The membrane diamond anvil cell: A new device for generating continuous pressure and temperature variations. *High Pressure Research, 1*(1), 77–90. https://doi.org/10.1080/08957958808202482

Letoullec, R., Pinceaux, J. P., & Loubeyre, P. (1990). The membrane diamond anvil cell. *High Pressure Research, 5*(1–6), 871–873. https://doi.org/10.1080/08957959008246284

Liebschner, D., Afonine, P. V., Baker, M. L., Bunkóczi, G., Chen, V. B., Croll, T. I., ... Adams, P. D. (2019). Macromolecular structure determination using X-rays, neutrons and electrons: Recent developments in Phenix. *Acta Crystallographica, Section D: Structural Biology, 75*(10), 10. https://doi.org/10.1107/S2059798319011471

Liu, J. (2016). High pressure x-ray diffraction techniques with synchrotron radiation. *Chinese Physics B, 25*(7), 076106. https://doi.org/10.1088/1674-1056/25/7/076106

Merrill, L., & Bassett, W. A. (1974). Miniature diamond anvil pressure cell for single crystal x-ray diffraction studies. *Review of Scientific Instruments, 45*(2), 290–294. https://doi.org/10.1063/1.1686607

Murshudov, G. N., Skubák, P., Lebedev, A. A., Pannu, N. S., Steiner, R. A., Nicholls, R. A., ... Vagin, A. A. (2011). REFMAC5 for the refinement of macromolecular crystal structures. *Acta Crystallographica. Section D, Biological Crystallography, 67*(Pt 4), 355–367. https://doi.org/10.1107/S0907444911001314

Nagae, T., Kawamura, T., Chavas, L. M. G., Niwa, K., Hasegawa, M., Kato, C., & Watanabe, N. (2012). High-pressure-induced water penetration into 3-isopropylmalate dehydrogenase. *Acta Crystallographica. Section D, Biological Crystallography, 68*(Pt 3), 300–309. https://doi.org/10.1107/S0907444912001862

Nagae, T., Yamada, H., & Watanabe, N. (2018). High-pressure protein crystal structure analysis of Escherichia coli dihydrofolate reductase complexed with folate and NADP. *Acta Crystallographica. Section D, Structural Biology, 74*(Pt 9), 895–905. https://doi.org/10.1107/S2059798318009397

Parker, M. I., Meyer, J. E., Golemis, E. A., & Dunbrack, R. L. (2022). Delineating the RAS Conformational Landscape. *Cancer Research, 82*(13), 2485–2498. https://doi.org/10.1158/0008-5472.CAN-22-0804

Pei, X., Bharr, N., Wang, H., Ando, N., & Meisburger, S.P. (submitted). Introduction to diffuse scattering and data collection.

Prangé, T., Carpentier, P., Dhaussy, A. C., van der Linden, P., Girard, E., & Colloc'h, N. (2022). Comparative study of the effects of high hydrostatic pressure per se and high argon pressure on urate oxidase ligand stabilization. *Acta Crystallographica. Section D, Structural Biology, 78*(Pt 2), 162–173. https://doi.org/10.1107/S2059798321012134

Prangé, T., Girard, E., Fourme, R., Dhaussy, A.-C., Edwards, B., Vaishnav, A., ... Evans, D. R. (2019). Pressure-induced activation of latent dihydroorotase from Aquifex aeolicus as revealed by high pressure protein crystallography. *The FEBS Journal, 286*(6), 1204–1213. https://doi.org/10.1111/febs.14758

Sacquin-Mora, S. (2014). Motions and mechanics: Investigating conformational transitions in multi-domain proteins with coarse-grain simulations. *Molecular Simulation, 40*(1–3), 229–236. https://doi.org/10.1080/08927022.2013.843176

Schneider, T. R. (2000). Objective comparison of protein structures: Error-scaled difference distance matrices. *Acta Crystallographica. Section D, Biological Crystallography, 56*(Pt 6), 714–721. https://doi.org/10.1107/s0907444900003723

Schneider, T. R. (2002). A genetic algorithm for the identification of conformationally invariant regions in protein molecules. *Acta Crystallographica Section D: Biological Crystallography, 58*(2), 2. https://doi.org/10.1107/S0907444901019291

Shen, G., & Mao, H. K. (2016). High-pressure studies with x-rays using diamond anvil cells. *Reports on Progress in Physics, 80*(1), 016101. https://doi.org/10.1088/1361-6633/80/1/016101

Shen, G., Wang, Y., Dewaele, A., Wu, C., Fratanduono, D. E., Eggert, J., ... Wentzcovitch, R. M. M. (2020). Toward an international practical pressure scale: A proposal for an IPPS ruby gauge (IPPS-Ruby2020). *High Pressure Research, 40*(3), 299–314. https://doi.org/10.1080/08957959.2020.1791107

Smith, J. S., & Desgreniers, S. (2009). Selected techniques in diamond anvil cell crystallography: Centring samples using X-ray transmission and rocking powder samples to improve X-ray diffraction image quality. *Journal of Synchrotron Radiation, 16*(Pt 1), 83–96. https://doi.org/10.1107/S0909049508030859

Storm, S. L. S., Axford, D., & Owen, R. L. (2021). Experimental evidence for the benefits of higher X-ray energies for macromolecular crystallography. *IUCrJ, 8*(6), 6. https://doi.org/10.1107/S2052252521008423

Syassen, K., & Holzapfel, W. B. (1978). Isothermal compression of Al and Ag to 120 kbar. *Journal of Applied Physics, 49*(8), 4427–4430. https://doi.org/10.1063/1.325497

Takemura, K. (2021). Hydrostaticity in high pressure experiments: Some general observations and guidelines for high pressure experimenters. *High Pressure Research, 41*(2), 155–174. https://doi.org/10.1080/08957959.2021.1903457

Tian, W., Chen, C., Lei, X., Zhao, J., & Liang, J. (2018). CASTp 3.0: Computed atlas of surface topography of proteins. *Nucleic Acids Research, 46*(W1), W363–W367. https://doi.org/10.1093/nar/gky473

Urayama, P., Phillips, G. N., Jr, & Gruner, S. M. (2002). Probing substates in sperm whale myoglobin using high-pressure crystallography. *Structure (London, England: 1993), 10*(1), 51–60.

Vehlow, C., Stehr, H., Winkelmann, M., Duarte, J. M., Petzold, L., Dinse, J., & Lappe, M. (2011). CMView: Interactive contact map visualization and analysis. *Bioinformatics (Oxford, England), 27*(11), 1573–1574. https://doi.org/10.1093/bioinformatics/btr163

Williamson, M. P., & Kitahara, R. (2019). Characterization of low-lying excited states of proteins by high-pressure NMR. *Biochimica et Biophysica Acta. Proteins and Proteomics, 1867*(3), 350–358. https://doi.org/10.1016/j.bbapap.2018.10.014

Winn, M. D., Ballard, C. C., Cowtan, K. D., Dodson, E. J., Emsley, P., Evans, P. R., ... Wilson, K. S. (2011). Overview of the CCP4 suite and current developments. *Acta Crystallographica. Section D, Biological Crystallography, 67*(Pt 4), 235–242. https://doi.org/10.1107/S0907444910045749

Winter, R. (2019). Interrogating the structural dynamics and energetics of biomolecular systems with pressure modulation. *Annual Review of Biophysics, 48*, 441–463. https://doi.org/10.1146/annurev-biophys-052118-115601

Yamada, H., Nagae, T., & Watanabe, N. (2015). High-pressure protein crystallography of hen egg-white lysozyme. *Acta Crystallographica. Section D, Biological Crystallography, 71*(Pt 4), 742–753. https://doi.org/10.1107/S1399004715000292

Printed and bound by CPI Group (UK) Ltd, Croydon, CR0 4YY
28/08/2026
02205845-0002